TOTAL QUALITY MANAGEMENT FOR HOSPITAL NUTRITION SERVICES

M. Rosita Schiller, RSM, PhD, RD, LD
Professor and Director
Medical Dietetics Division
The Ohio State University
Columbus, Ohio

Karen Miller-Kovach, MBA, MS, RD
General Manager of Program Development
Weight Watchers International, Inc.
Jericho, New York

Mary Angela Miller, MS, RD, LD
Director
Nutrition and Dietetics
The Ohio State University Medical Center
Columbus, Ohio

AN ASPEN PUBLICATION®
Aspen Publishers, Inc.
Gaithersburg, Maryland
1994

Library of Congress Cataloging-in-Publication Data

Schiller, M. Rosita.
Total quality management for hospital nutrition services /
M. Rosita Schiller, Karen Miller-Kovach, Mary Angela Miller.
p. cm.
Includes bibliographical references and index.
ISBN 0-8342-0551-3
1. Hospitals—Food service. 2. Total quality management.
I. Miller-Kovach, Karen. II. Miller, Mary Angela. III. Title.
RA975.5.D5S37 1994
362.1′76′0685—dc20
93-39360
CIP

Copyright © 1994 by Aspen Publishers, Inc.
All rights reserved.

Aspen Publishers, Inc., grants permission for photocopying for limited personal or internal use. This consent does not extend to other kinds of copying, such as copying for general distribution, for advertising or promotional purposes, for creating new collective works, or for resale. For information, address Aspen Publishers, Inc., Permissions Department, 200 Orchard Ridge Drive, Gaithersburg, Maryland 20878.

Aspen Publishers, Inc., is not affiliated with the American Society of Parenteral and Enteral Nutrition

The authors have made every effort to ensure the accuracy of the information herein, particularly with regard to drug selection and dose. However, appropriate information sources should be consulted, especially for new or unfamiliar procedures. It is the responsibility of every practitioner to evaluate the appropriateness of a particular opinion in the context of actual clinical situations and with due consideration to new developments. Authors, editors, and the publisher cannot be held responsible for any typographical or other errors found in this book.

Editorial Resources: Ruth Bloom

Library of Congress Catalog Card Number: 93-39360
ISBN: 0-8342-0551-3

Printed in the United States of America

1 2 3 4 5

To My Family

—*MRS*

To My Husband, Kevin

—*KMK*

To My Sister, Jody

—*MAM*

Table of Contents

Panel of Editorial Advisors	ix
Preface	xi
Acknowledgments	xiii
PART I—BASICS OF QUALITY MANAGEMENT	1
Chapter 1—Overview of Quality Management	3
What Is Quality?	3
What Is Total Quality Management?	5
Quality Programs	5
The Quality Journey	7
Chapter 2—Quality Management in Health Care	9
Quality in Health Care Services	9
Structuring Quality Programs in Health Care	12
Implementing Quality Programs in Health Care	13
Educating for Quality Health Care	15
What Can One Reasonably Expect?	18
The Importance of Measurement	20
Chapter 3—Patient Satisfaction: A Mark of Quality	22
Work As a Process	22
Meeting Customer Wants and Needs	25
Tying the Work Process and the Customer Together	27
Managing Customer Expectations	29
Chapter 4—Continuous Quality Improvement	33
Joint Commission's *Agenda for Change*	33
Quality Assurance/Quality Improvement	34
Quality Improvement Using the Ten-Step Process	35

  Making the Transition 40
  Additional Action Steps 41
  The Institutional Approach 42
  Conclusion 43

PART II—DEPARTMENTAL SYSTEMS AND TOOLS FOR QUALITY MANAGEMENT **45**

Chapter 5—Preparing To Initiate a Quality Program **47**

  Leadership 47
  Work Group Readiness 50
  The Quality Readiness Survey 50

Chapter 6—The Nutrition Services Environment **55**

  Mission or Purpose 55
  Strategic Plan 56
  Standards of Practice 56
  Policies and Procedures 57
  Method of Screening for Nutritional Risk 57
  Method for Determining Patient Acuity Levels 58
  Documentation Procedures 58
  Method To Track Clinical Productivity 58
  Staffing Patterns 63
  Criteria-Based Performance Standards 64
  Monitoring and Evaluation System 65
  Supportive Environment 65

Chapter 7—Improving Quality through Group Processes **67**

  The Quality Mission Statement 67
  Problem-Solving Work Groups 67
  Organizing Ideas 72
  Cause-and-Effect Diagrams 75
  Flowcharts 77
  Force Field Analysis 81

Chapter 8—Quantitative Quality Management Tools **83**

  Check Sheets 83
  Pareto Charts 84
  Histograms 86
  Run Charts 89
  Control Charts 91
  A Case Study 92
  Benchmarking 92
  Scatter Diagrams 94

PART III—MONITORING AND EVALUATION **97**

Chapter 9—Departmental Planning for Continuous Quality Improvement **99**

  Assign Roles and Responsibilities 99
  Delineate the Scope of Services 100
  Identify Important Aspects of Care 101
  Cross-Functional Initiatives 104
  Quality Management Plan 105

Chapter 10—Quality Indicators, Criteria, and Monitors **107**

    Practice Guidelines Produce Quality Indicators 107
    What Is a Quality Indicator? 107
    Attributes of Indicators 107
    Types of Indicators 109
    How To Develop Quality Indicators 113
    Some Practical Pointers 121
    Appendix 10-A—Council on Practice Quality Management Task Force 123

Chapter 11—Collecting and Organizing Data **125**

    Data Collection Forms 125
    Selecting the Sample 127
    Data Collection 130
    Organizing Data 130
    Use of a Computer To Organize Data 133
    Conclusion 138

Chapter 12—Data Analysis and Evaluation **139**

    Purpose and Expected Outcomes 139
    Attributes of Effective Data Analysis 139
    Methods of Data Analysis 140
    Interpreting the Data 144
    Problem Clarification Methods 147

Chapter 13—Taking Action To Improve Care **151**

    Choosing a Game Plan 151
    Focus-PDCA: Find a Process To Improve 153
    Focus-PDCA: Organizing Your Team 161
    Focus-PDCA: Clarify Knowledge and Understand Reasons for Variation 166
    Focus-PDCA: Select and Plan 170
    Focus-PDCA: Do, Check, Act 173

Chapter 14—Communication: The Vital Link in Continuous Quality Improvement **179**

    Quality Steering Committee 179
    Department Interactions with the Quality Steering Committee 179
    Department Communication Responsibilities 181
    Documentation: Doing the Paperwork 181
    Paper Trail 187

PART IV—MANAGING A NUTRITION SERVICES QUALITY PROGRAM **189**

Chapter 15—Involving Clinical Staff in the Quest for Excellence **191**

    The Clinical Manager 191
    Role of the Dietitian As Trainer 201
    Dietitians and Dietetic Technicians As Star Performers 201

Chapter 16—Managing Change To Achieve Quality Improvement **206**

    Framework for Change 206
    General Nature of Change 207
    The Health Care Environment 208
    Barriers to Change in Health Care Institutions 209
    Resistance to Change 210

	Improving Performance and Outcomes	211
	Research Utilization As a Basis for Change	214

Chapter 17—Patient-Focused and Cross-Functional Teams ... 217

- Evaluating Patient-Centered Care ... 218
- Patient-Focused Care ... 218
- Case Management ... 219
- Definition and Roles of Cross-Functional Teams ... 220
- Starting a Cross-Functional Problem-Solving Team ... 220
- Use of a Facilitator ... 222
- Empowerment through Collaboration ... 222
- Examples of Cross-Functional Teams and Indicators ... 223
- Factors Contributing to Team Success ... 226
- Pitfalls and Problems ... 227
- Conclusion ... 227

Chapter 18—Evaluating Quality Management Programs ... 229

- Purpose of Evaluation ... 229
- Plan Early To Evaluate Later ... 229
- Evaluate Purpose, Objectives, and Outcomes ... 230
- Evaluate Efficiency ... 232
- Evaluation and Accreditation Surveys ... 239
- Evaluating Benefits of Continuous Quality Improvement ... 240
- Planning for the Future ... 240

Chapter 19—Quality Management in Hospital-Affiliated Services ... 244

- Guidelines and Standards ... 244
- Quality Management in Long-Term Care ... 245
- Child Care Services and Facilities ... 248
- Health Maintenance Organizations ... 248
- Quality Management of Home Care ... 249
- Quality Management for Hospice Programs ... 250
- Women, Infants, and Children Programs ... 250
- Organization for Quality ... 251
- Multidisciplinary Quality Programs ... 252
- Monitoring and Evaluating Quality Care ... 253

Appendix A—Sample Documents ... 257

Appendix B—American Dietetic Association Clinical Indicator Project Summary of Results ... 319

Appendix C—Indicators from University Consortium ... 327

Appendix D—Sample Monitoring Forms ... 341

Appendix E—Table of Random Numbers ... 357

Appendix F—Quality Management Resource Manuals ... 363

Appendix G—Glossary of Terms ... 367

Index ... 371

About the Authors ... 376

Panel of Editorial Advisors

Terry Bambrick-Mohorcic, MS, RD, LD
Clinical Dietitian
Department of Nutrition and Hospitality
Meridia Euclid Hospital
Euclid, Ohio

Donna M. Bashara, RD, LD
Quality Assurance Dietitian
Dietetic Services
Olin E. Teague Veterans' Center
Temple, Texas

Rhonda Billman, MS, RD, LD
Clinical Manager
Nutrition Services Department
Southwest General Hospital
Middleburg Heights, Ohio

Cheryl Bragg, MSA, RD, LD
Assistant Director, Clinical Services
Department of Food and Nutrition Services
Medical College of Georgia
Augusta, Georgia

Kathryn Burden, RD, LD
Director
Food and Nutrition Services
Medical Center Hospital
Chillicothe, Ohio

Susan DeHoog, RD, CD
Director, Clinical Nutrition
University of Washington Medical Center
Seattle, Washington

Marcia Durell, MEd, RD
Chief Clinical Dietitian
Holyoke Hospital, Inc.
Holyoke, Massachusetts

Deborah Ford Flanel, MS, RD
Assistant Director, Clinical Nutrition
Yale-New Haven Hospital
New Haven, Connecticut

Marilyn Lawler, PhD, RD
Director of Nutrition Services
University of Chicago Hospitals
Chicago, Illinois

Hildreth A. Macy, MA, RD
Associate Director
Food and Nutrition Services
Henry Ford Hospital
Detroit, Michigan

Joyce Price, RD
Chief of Clinical Nutrition
Duke University Medical Center
Durham, North Carolina

Laura Rackley, MPA, RD, CD
Director, Nutrition Care Services
University of Utah Health Science Center
Salt Lake City, Utah

Patricia Queen Samour, MMSc, RD
Director, Department of Dietetics
New England Deaconess Hospital
Boston, Massachusetts

Alice E. Smith, MS, RD
Director, Clinical Dietetics
The Children's Memorial Hospital
Chicago, Illinois

Preface

Quality management is the keystone of any health care organization today. Concern for quality transcends other significant movements in health care including reform at both the federal and state levels, pressures toward cost containment, and initiatives to restructure and downsize. The quality movement pervades the organization, engaging personnel at all levels from the board of directors to the lowest rank associate. Nutrition professionals are no exception.

Practitioners who devoted a great deal of energy to developing quality assurance programs must now learn the language and the ways of total quality management and quality improvement. Dietitians and dietetic technicians must begin to think of their work as a process and define appropriate outcome measures, learn new techniques for group participation in quality initiatives, use new tools and techniques for problem identification and analysis, and find ways to upgrade the quality of services provided. They must be actively involved in new approaches to quality management such as cross-functional teams, clinical pathways, and patient-focused care.

Patient satisfaction takes on new meaning in contemporary quality paradigms. Thus, dietitians need to identify the determinants of patient satisfaction for both food and nutrition services. They must establish indicators for patient satisfaction, set quality goals, and create ways to collect patient information that is pertinent, accurate, timely, and quantitative. Results need to be analyzed and continuously used to improve service quality.

This manual was designed to help nutrition professionals build and sustain an effective total quality management program for nutrition services in hospitals, skilled nursing facilities, nursing homes, and other organizations that may include nutrition services. In most cases, dietetic practitioners in these institutions have previously participated in a quality assurance program. Material in this book should assist such practitioners to make a smooth transition in the quality journey, moving away from the delimiting constructs of a quality assurance program toward the more global approach of total quality management and continuous quality improvement (CQI).

Few other publications provide the breadth and scope of content offered in this book. For example, earlier texts show how to develop a quality assurance program in patient care and education. Such books fall short of current Joint Commission expectations for demonstrated efforts toward quality improvement—a challenge we addressed in this book. A few books on quality management are self-published; they lack wide distribution within the dietetic community. Some references offer systematic approaches to high quality patient care. But these books are not designed as guides to quality improvement. A few new books on quality management offer a comprehensive approach to CQI but they lack examples and application to nutrition services.

This book provides a thorough discussion of quality assessment, monitoring, and evaluation gathered from business and health care literature and compiled into a single reference manual for the dietetic practitioner. It contains numerous references and examples from dietetic practice. Part I includes the Basics of Quality Management. This section includes background information on total quality management and its adaptation to health care settings. Because customer satisfaction is central to quality service, a chapter is devoted to customer wants, needs, and expectations in relation to nutrition services. There is an overview of the 10-step process for quality monitoring and evaluation and suggestions are given for making the transition from quality assurance to CQI or other contemporary quality management system.

Part II includes a discussion of departmental systems and tools for quality management. A quality readiness survey is included to help identify departmental or institutional limitations that may inhibit employees from embracing a quality improvement program. Departmental structures are described, such as screening procedures and a system for determining patient acuity. These structural components must be in place because they form the basis for quality nutrition services. Two chapters are devoted to quality tools and processes. They include suggestions for constructing and using problem analysis diagrams and charts, quality work groups, group problem-solving techniques, and benchmarking.

Part III deals with the quality monitoring and evaluation process. Although the 10-step process is not defined as such, chapters include all components of this approach to quality management. For example, contents include defining responsibility, delineating scope of services, identifying important aspects of care, developing quality indicators and setting thresholds of performance, collecting and organizing data, using various techniques for data analysis and interpretation, evaluating results, taking action to make and sustain improvements, and reporting quality-related information to appropriate groups and through designated channels. Each chapter contains several examples from real-life situations.

Part IV offers suggestions for managing the quality process and helps prepare readers for ongoing changes in health care delivery. Topics include involving dietetic staff in the quality management process, managing change, participating in cross-functional and patient-focused care teams, and evaluating the quality process. For many dietetic professionals, involvement in nutrition services extends far beyond the typical patient who is hospitalized or followed in the outpatient department. Hence, suggestions are given for quality management in long-term care, child care centers, health maintenance organizations, home care, hospice programs, and WIC clinics.

Numerous sample quality indicators, quality management plans, quality evaluation documents, and monitoring forms are given in the appendices. Also included is a table of random numbers, glossary of terms, and an annotated list of resource manuals.

Each of the authors has experience in nutrition services management and involvement in quality improvement processes. This book is rooted in their expertise, embellished by references from the current literature, and supported by input from an Editorial Advisory Panel, from institutions providing diverse approaches to quality care. The authors learned a great deal while preparing the manuscript. We invite others to use this book to support their efforts in the quality quest. Our goal is to enhance the quality of health care through superior nutrition services in both hospital and nonhospital settings.

Acknowledgments

A book such as this cannot be completed without the commitment and assistance of many individuals who contribute time, talents, insights, and expertise to the project. Sincere thanks to all who had a part in this work: contributors, reviewers, advisors, and mentors.

Special recognition goes to our Editorial Advisors and the institutions they represent. These professionals and their coworkers allowed their materials to be used as examples. They had the courage to share what they had, permitting others to scrutinize their work, to build upon it, to improve it where possible, and to adapt it where appropriate. Their leadership paves the way and serves as a guide for others as they embark on the quality journey.

We also wish to acknowledge the Clinical Nutrition Staff at The Ohio State University Medical Center for sharing their experiences and providing suggestions and examples that were used throughout this manuscript. We want to specifically thank Julie Jones, Michelle Yost, Martha Orabella, Terri Guanciale, Susan Uhl, and dietetic student Suzanne Tope, for their personal assistance.

Appreciation also goes to the Nutrition Services Staff at the Cleveland Clinic Foundation. These individuals are recognized especially for their embodiment of the spirit, belief, and values necessary to support the quality quest.

Finally, special thanks to Juanita Williams whose clerical expertise contributed to the successful completion of this book.

Part I

Basics of Quality Management

Chapter 1
Overview of Quality Management

The quality movement has taken hold in corporate America. Beginning in the manufacturing sector about two decades ago, the concepts and processes associated with quality management were quickly adapted to the service industries. Within the last decade, the health care sector has begun to examine whether or not quality management is appropriate for hospitals.

Whether the economic sector is manufacturing, leisure services, health care, or nutrition, the impetus for moving to quality management is the same—the need to be competitive in a market where competition is increasing and resources are becoming more scarce. Indeed, the advent of global markets for products and services has greatly increased competitive pressures on every American industry.

Quality management programs have been instrumental in improving the competitive position of many industries that have faced competitive challenges—from Walt Disney Company in the entertainment services sector to Ford Motor Company in the automotive market.

This chapter describes the basic concepts and tenets of quality management.

WHAT IS QUALITY?

Many people, when asked what quality means to them, will speak in abstract terms. Statements such as "I can't tell you what quality looks like, but I know it when I see it" and "You can't put a number on a quality product or service" are common. But the concept of quality must be clearly defined and completely understood if it is to become an integral part of the work produced by a department.

There are volumes of books and articles written about what quality is and how one can achieve it. Each of these publications, as well as each of the persons who have written them, has a definition of what quality means. Unfortunately, a clear, universally accepted definition of the term is still lacking.

Exhibit 1-1 provides some of the more commonly used definitions of quality. A review of this listing suggests that, while there is not a universal definition, there are themes about what quality means. Some authorities define quality in quantitative terms, and others focus on a narrative description that is based on the customer's point of view.

Quantitative Quality versus Qualitative Quality

Taking the quantitative approach, *Business Week* ("Quality Imperative" 1991) defines quality as "the absence of variation." By this definition, a quality product or service is one

Exhibit 1-1 Definitions of Quality

- Quality is the lack of variation.
- Quality is full compliance with the standards or requirements of the work.
- Quality is whatever the customer says it is.
- Quality is doing the right thing right.
- Quality is meeting or exceeding the customer's expectations.

that is always the same. This should not be confused with providing a product or service that is the best that it can be. Hamburgers provide a useful illustration of the quantitative definition of quality. A McDonald's hamburger is always the same. Whether it is purchased in Rome or Chicago, whether it is eaten at noon or midnight, one can be sure that the hamburger that is obtained at McDonald's will have a consistency that is not often seen. Therefore, using this definition, McDonald's makes a quality hamburger.

The other school of defining quality focuses on a narrative description from the customer's point of view. Using this definition, there could be a great deal of variation in products and services as the result of differing customer expectations. Again, however, this definition of quality should not be confused with making a product or service the best that it can be. Let us return to the hamburger illustration. If a customer defines a quality hamburger as one that is shaped by hand from fresh ground meat and broiled to medium rare, then, by definition, McDonald's does not produce a quality hamburger.

It should be noted that neither the quantitative approach nor the descriptive approach to defining quality includes the concept of perfection. Quality products and services should not be thought of as being all things to all people. Perfection is cost prohibitive. Rather, quality products and services should be thought of as being predictable in content and/or designed to meet or exceed the customer's expectations.

In conclusion, the meaning of quality is not universally agreed on. As one reads and learns about quality management, it is important to determine what definition is used by the source that is being studied. Moreover, one must have a clear departmental or, preferably, organizational definition and understanding of what is meant by quality before the initiation of any kind of improvement process.

Factual Quality versus Perceived Quality

The quality of any product or service has two components. They are quality in fact and quality in perception. These components are depicted in Figure 1–1.

Factual quality is achieved when the provider of the service has completed the task in accordance with the specifications of the job. For example, one could say that a house painter achieved factual quality in painting a living room if he or she met the employer's defined criteria for the task. Those specifications may include such things as

- arriving on time for the job
- using the amount and type of paint that had been specified on the previously completed quote
- patching the walls before painting
- returning any unused paint to the warehouse

Perceived quality is achieved when the receiver of the service has had his or her expectations met. For example, the expectations of a person having the living room painted could include

- washing the walls before patching
- returning wall hangings and furniture that were moved during the painting process
- having any unused paint left at the home
- having the painter's name and phone number available for questions

Factual quality and perceived quality are two distinct facets of any product or service. An important objective in any quality effort is to determine fully the perceived expectations on the part of customers and to evaluate those perceptions against the existing factual specifications. Discrepancies between factual and perceived quality require immediate attention.

Gaps between factual and perceived quality occur frequently. Indeed, discrepancies between what providers define as specifications for quality products and services and expectations that customers have for those same products and services are often significant. For example, take the scenario presented above. The painter's employer provided a quote that assumed that the room would be prepared for painting before the employee arrived, so washing the walls was not part of the job specifications. Likewise, the job specified that unused paint be returned to the warehouse for use on a future job. These factual specifications, while understandable, do little to alleviate the customer's disappointment when the painter's performance did not match the service that was expected.

Alleviating discrepancies between factual and perceived quality is a constant challenge for several reasons:

- There is an innate resistance to changing task specifications, particularly when the change is not scientifically valid.
- Every customer is unique and will have differing expectations about the service provided.
- Customer expectations change over time and are often situation dependent.
- Customers may not always want or be able to verbalize their expectations, especially for technical services.
- Customer expectations may not be realistic.

While difficult, differences in perceived and factual quality must be addressed. This can be done by changing the task's specifications to meet the perceived quality standards or by educating the customer about the factual quality in an effort to make the perceived level of quality more realistic. Although the latter may appear to be a preferable solution, it is much harder to accomplish.

Perceived quality has a firm foundation in an individual's value and belief systems. These systems are highly resistant to change. And although values and beliefs can be transformed over time as the result of educational efforts, values and be-

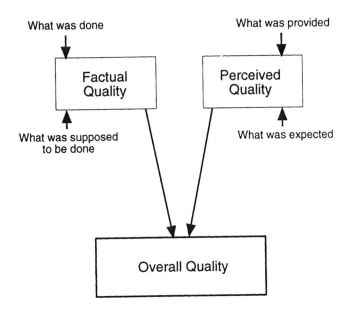

Figure 1-1 Perceived and Factual Quality

QUALITY PROGRAMS

To confuse the issue further, there are several "schools" of quality. Each has its followers and its critics. A summary of the most frequently cited initiatives is presented as Table 1-1.

A review of Table 1-1 helps to clarify some of the confusion about the definitions of quality, continuous quality improvement, and total quality management. Deming, with statistical process control, emphasizes the process aspects of quality (i.e., continuous quality improvement). The cultural transformation model advocated by Peters, Ouchi, and Naisbitt focuses on the adoption of the philosophical and conceptual aspects of quality (i.e., total quality management).

Three of the programs outlined in Table 1-1 have specific steps to implementation. The step plans for the programs fathered by Deming, Juran, and Crosby are defined as Table 1-2. Although each program has different steps, there is a remarkable similarity among them, as well.

Concepts that are common to virtually all of the theories of quality include the following:

- the need for the organization's top management team to lead the way
- customer-defined service levels
- the belief that most organizational problems are due to faulty systems, not problem employees
- the adoption of standardized methods for problem solving, including the use of data-derived decisions
- a management style that encourages those who perform a task to be involved in decision making about their work
- a commitment to education and training for everybody
- a belief that there is always potential for more improvement.

liefs can never be dictated. To paraphrase an old saying, perception is reality in the eyes of the beholder.

WHAT IS TOTAL QUALITY MANAGEMENT?

Just as there is no consensus about the definition of quality, neither is there universal agreement about the meaning of total quality management. It seems that one cannot read a group of articles on the subject without coming away with the impression that no one agrees on even the basic terms. Take, for example, the phrases *total quality management* and *continuous quality improvement*: to some, the phrases are synonymous; to others, they are very different. When a difference is perceived, it cannot be merely passed off as a matter of semantics.

Figure 1-2 illustrates the difference between total quality management and continuous quality improvement. Continuous quality improvement, represented as the oval within the rectangle, is used to describe the process part of quality. It means that the organization supports the use of methodological problem solving, usually in group or team settings. Continuous quality improvement also means that statistical analysis and other quality tools used in the collection and evaluation of data are in place.

Total quality management, the rectangle that includes the continuous quality improvement oval, describes the larger issue of how the organization conducts business. It includes the values, beliefs, and behaviors that uniquely define an institution's culture. Not only are the tools and methods of continuous quality improvement in use, but the organization's leadership and entire work force place the philosophies and concepts of quality at the center of all its operations.

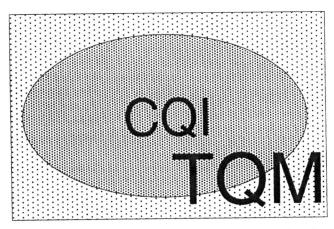

Figure 1-2 Continuous Quality Improvement (CQI) As a Function of Total Quality Management (TQM)

Table 1-1 Comparison of Often-Cited Quality Programs

Type of Program	Major Advocates	Underlying Beliefs	Implementation Methods	Critical Success Factors
Statistical process or quality control	Deming	• Focus on the process, not the product. • Everyone must incorporate process control methods into work. • Prevent problems, don't fix them.	• Learn and live by the 14-step plan. • Apply statistical control methods to all work processes.	• Equal involvement at all organization levels, top to bottom. • Total work force training in statistical control methods. • Abandonment of acceptable levels of error.
Cultural transformation	Peters, Ouchi, Naisbitt	• Customer satisfaction is the core of the organization's mission. • Complete work force commitment to core values unifies work processes and creates excellence at all levels.	• Learn from those organizations that have been successful. • Define organizational values by consensus. • Model the values in all actions and behaviors.	• Strong leadership commitment and skills at the top. • Evaluative methods that measure achievement of the values. • Communication and training of employees that enables them to meet the expectations.
Participative problem solving	Ishikawa	• The people who do the work are in the best position to make improvements. • Group decisions are better than those reached by an individual.	• Educate and train in group problem-solving and data-gathering techniques.	• Strong interpersonal communication and problem-solving skills throughout the organization. • Leadership commitment to the philosophy.
Quality work design	Juran	• Products and services must be designed based on the customer's wants and needs.	• Learn and live by the 10-step plan. • Focus efforts on the design of the end product or service. • Use teams to identify opportunities for improvement and problem solving.	• Knowledge of everything there is to know about your customers. • Commitment to cross-functional teamwork that is centered around the end products and services.
Quality cost management	Crosby	• It is less costly to do it right the first time. • Using costs as the basis of tracking leads to identification of opportunities for improvement. • Nothing less than zero defects is acceptable.	• Learn and live by the 14-step plan.	• Strong interpersonal communication, problem-solving, and analytical skills throughout the organization. • Leadership commitment to the philosophy.
Benchmarking	Garvin	• Define your customer's niche and focus on meeting those wants and needs (i.e., don't try to be all things to all people).	• Set goals based on customer's point of view. • Use teams to narrow the gap between the goals and current operations.	• Clear knowledge of who the customer is and what is desired. • A willingness to focus on strengths in defining the niche.

Table 1-2 Comparison of Quality Program Steps

Step	Statistical Process Control (Deming)	Quality Work Design (Juran)	Quality Cost Management (Crosby)
1	Create a constancy of purpose.	Build awareness of the need and opportunity for improvement.	Obtain management commitment.
2	Adopt a new philosophy.	Set goals for improvement.	Form quality improvement teams.
3	Cease dependency on inspection.	Organize to reach the goals (establish a quality council, identify problems, select projects, appoint teams, designate facilitators).	Determine by measurement where the current and potential quality problems are.
4	End the practice of awarding on price.	Provide training.	Evaluate the cost of quality.
5	Constantly improve systems.	Carry out projects to solve problems.	Raise quality awareness and personal concern for all.
6	Institute training on the job.	Report progress.	Take actions to correct problems.
7	Adopt a modern method of supervision.	Give recognition.	Establish a committee for a zero defects program.
8	Drive out fear.	Communicate results.	Train supervisors.
9	Break down barriers between departments.	Keep score.	Hold a "zero defects" day.
10	Eliminate slogans.	Maintain momentum by making annual improvement part of the regular systems.	Encourage goal setting.
11	Institute working standards and statistical process control.		Encourage employees to communicate obstacles to management.
12	Remove barriers that rob workers of pride.		Recognize and appreciate those who do the work.
13	Institute a vigorous program of education.		Establish quality councils.
14	Put everybody in the company to work on the plan.		Do it all over again.

THE QUALITY JOURNEY

Organizations that have chosen total quality management often speak of its implementation as being analogous to embarking on a long journey. These organizations stress that the decision to take the journey should not be made lightly because it requires traveling a long, winding, difficult road. There are roadblocks to traverse and inevitable detours; the trip takes years to complete.

It is worth noting that the success of quality programs varies a great deal. To date, it appears that those organizations that have had the most success in realizing and sustaining the benefits of quality are the same organizations that have opted for the "package" of total quality management. Indeed, leadership initiatives and actions that run counter to the core concepts of quality are likely to doom any program.

An example of an ill-fated quality program occurred at the Douglas Aircraft Company, a subsidiary of the McDonnell-Douglas Corporation. As reported in the *Plain Dealer* ("Training Stalls at Douglas Aircraft" 1992), Douglas chose to move forward with an aggressive implementation of its quality program and, at the same time, announced that almost 20 percent of the company's middle and executive managers would be cut. As the program was rolled out, its funding was cut and two more reductions in force took place. Morale sank and with it went the newly inspired commitment to quality; workers wondered what was going on in the executive suite. W. Edwards Deming, a renowned expert in the quality field, said that it was "too late" for quality to work at Douglas. According to Deming, once an organization has started a program and then sabotaged its success, it is impossible to try it again.

The above example illustrates that the decision to invest in either continuous quality improvement or total quality management should not be taken lightly. If it is, it will unlikely be anything more than a fad or the organization's management program of the month. Each enterprise, as well as each department that makes up the whole, must decide for itself how much it is willing (and able) to invest in any quality initiative.

REFERENCES

The quality imperative. 1991. *Business Week,* October 25, 8. Bonus issue.

Training stalls at Douglas Aircraft: Total quality management system a victim of reduction philosophy. 1992. *Plain Dealer,* March 29, 5-E.

SUGGESTED READING

Albrecht, K. 1988. *At America's service: How corporations can revolutionize the way they treat customers.* Homewood, Ill: Irwin, Professional Publishing.

Berry, T.H. 1991. *Managing the total quality transformation.* New York: McGraw-Hill Publishing Co.

Brocka, B. 1992. *Quality management: Implementing the best ideas of the masters.* Homewood, Ill: Richard D. Irwin, Inc.

Crosby, P.B. 1992. *Completeness: Quality for the 21st century.* New York: Dutton.

Hunt, D.V. 1992. *Quality in America: How to implement a competitive quality program.* Homewood, Ill: Richard D. Irwin Professional Publishers.

Juran, J.M. 1992. *Juran on quality by design: The new steps for planning quality into goods and services.* New York: Free Press.

Walton, M. 1990. Deming management at work. New York: Putnam Publishing Group.

Chapter 2
Quality Management in Health Care

Applying the concepts of quality management to health care is a relatively new phenomenon. As stated earlier, the quality movement originated in the manufacturing sector. It was there that Deming, Crosby, and Juran applied their theories and fine-tuned their programs.

With success in that area, the quality movement expanded into the service sector of the economy. Some adaptations to the manufacturing models were needed to accommodate the innate differences between the sectors. Altering mechanical conformance requirements into their human equivalents is one example of this adaptation process.

Now the quality movement is making its way into health care, and further adaptations are needed. For example, Deming's 14-step quality program outlined in Chapter 1 can be adapted for use in clinical nutrition services (Exhibit 2–1). This chapter explores aspects of quality management that are unique to health care.

QUALITY IN HEALTH CARE SERVICES

Dennis S. O'Leary, M.D. (1991), president of the Joint Commission on Accreditation of Healthcare Organizations (Joint Commission) has developed a definition of quality health care. He defines quality medical care as doing the right thing right—making the appropriate decision associated with the medical needs of the patient and then carrying out those decisions correctly regarding diagnosis or treatment. Dr. O'Leary goes on to define quality health care to include the following:

- access to the system
- safety throughout the health care environment
- continuity of care throughout the health care experience
- most importantly, quality in the relationships that are the result of interactions between patients and everyone with whom they come in contact within the health care system

A review of the quality literature for health care often separates initiatives that are associated with administrative or support services from those that are clinically oriented. This is sometimes referred to as those services that "make patients happy" and those services that "make patients healthy."

The majority of reported inroads for quality initiatives in the health care setting have involved administrative or support services. Examples include

- reducing the time it takes to get a patient through the admissions procedure
- simplifying the billing statement
- opening ombudsman offices

Support Services, Medical Services, and Clinical Products

Brent James (1991) breaks the provision of health care services into three categories. These categories are depicted in Figure 2–1.

James defines support services as the "hotel" services that every organization must provide. Cleaning rooms and delivering mail are two examples. Support services are usually the exclusive property of the hospital's administration. In most institutions, the provision of food is treated as a support service.

Exhibit 2–1 Deming's 14 Steps to Quality As Performed by Nutrition Services

1. **Create a constancy of purpose.** Clearly define your nutrition products and services. Establish current and future quality standards for each product and service that is based on the needs and expectations of the customers who receive them. In your quest to exceed customer expectations, find cost-effective, innovative ways to improve your products and services.
2. **Adopt a new philosophy.** An acceptable level of mistakes is no longer acceptable. Quality products and services are possible only when the people, materials, and the processes that combine them are working efficiently and effectively.
3. **Cease dependency on inspection.** Evaluating a nutrition product or service after it has been provided to the customer is too late. Proactively design your products and services to ensure the desired outcome. Develop systems to capture opportunities to improve as well as collect data about changes in your customers' needs and expectations.
4. **End the practice of awarding on price.** Your nutrition products and services can only be as good as the materials of which they are made. For external suppliers, deal only with those who can show quantitative evidence that they can and will give you—their customer—what you need and expect. With internal suppliers, provide ongoing data-driven feedback about providers' products as well as your needs and expectations as a customer. Work collaboratively to remove obstacles and barriers to the cross-functional flow of patient care.
5. **Constantly improve systems.** Work processes can always be made better. Believe it; do it.
6. **Institute training on the job.** The provision of nutrition services depends on people. Provide training that teaches all staff members how to go beyond the tasks of the job and into the identification and evaluation of supplier, process, and customer interactions. Education and on-the-job use of quality measurement tools are needed in addition to technical continuing education programs.
7. **Adopt a modern method of supervision.** Shift the fulcrum of supervisory efforts away from the low performers and toward improving the work processes for all. Emphasize facilitation of staff initiatives for process improvement, management by fact, and reliance on the quality tools of measurement. Walk the talk.
8. **Drive out fear.** Stop blaming people for system problems. By encouraging identification of potential improvements from those who perform the work, processes are improved and everyone benefits. Concentrating on the constancy of purpose assists in eliminating biases between employee groups (e.g., departmental, professional vs. technical).
9. **Break down barriers between departments.** Almost all nutrition products and services cross functional department lines. We are often called upon to function as supplier, provider, or customer, depending on the process. Recognizing and building on the interdependent relationships that are necessary to provide quality nutrition services is integral to success.
10. **Eliminate slogans.** Teamwork and communication regarding the accomplishments of process improvement are what makes meaningful results. As above, walk the talk.
11. **Institute working standards and statistical process control.** Evaluative measurements that are limited to the quantity of work performed are counterproductive. Focusing on how to do the work better—to reduce errors, abolish redundancy, and streamline the process—leads to qualitative gains. Performing work for the sake of work is meaningless unless it is accompanied by an understanding of its contribution to the hospital as a whole.
12. **Remove barriers that rob workers of pride.** By teaching and using the quality measurement tools, all people involved in the work process will speak a common language and work from a shared data set. This, combined with a consensual constancy of purpose, enables groups to get past the "who does what?" and get to improving the processes that create quality nutrition care.
13. **Institute a vigorous program of education.** Constantly work at improving the technical skill base of every person in the department. Greater knowledge has many benefits, including enhanced self-esteem, improved job performance, and greater work versatility.
14. **Put everyone in the company to work on the plan.** Quality cannot belong to a position or a function. Rather, it is the essence of the work that each of us does. Personal responsibility and accountability for the quality of work performed is essential.

Medical services, as defined by James, are the foundation of the hospital's infrastructure. This category includes those departments and services that provide medical support to physicians who, in turn, provide direct patient care. In the structure of most hospitals, the medical services departments are staffed by health care professionals and have a reporting relationship to the organization's administrative staff. Many of the functions performed by the clinical nutrition staff would be classified as medical services.

The third category is clinical products. James defines these as diagnostic and treatment processes that are provided directly to patients. Organizationally, clinical products are the result of physicians' actions and, as such, are placed under the auspices of the medical operations staff. It is important to note that, according to James, the support services and medical infrastructure form a base from which all clinical products are delivered to patients.

Application of this model to the provision of nutrition services is enlightening. Dietitians are directly involved in the provision of both support and medical services. Moreover, dietitians produce clinical products in their role as direct patient care providers. Examples of these nutrition products and services are presented as Table 2–1.

Dietetics spans the realm of health care services. This presents extraordinary challenges because it implies a responsibility to keep patients both happy and healthy. Indeed, providing patients with meals that meet their food quality expectations and ensuring that the integrity of the physician-prescribed diet is maintained can create conflicts. Dietitians are in a unique position to facilitate the accomplishment of quality initiatives in both administrative and clinical services.

Quality Patient Care Is Multidimensional

Another perspective about quality health care services is offered by E. C. Murphy, Ltd. (1991). This company's research found that quality patient care is multidimensional; there is no single factor that constitutes quality. Rather, this

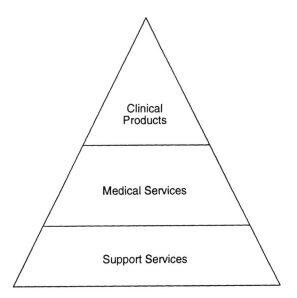

Figure 2–1 Categories of Health Care Products and Services. *Source:* Adapted from James, B.C., How Do You Involve Physicians in TQM?, *Journal for Quality and Participation*, January–February 1991, pp. 42–47, with permission of the Association for Quality and Participation, © 1991.

group's work identified seven behavioral indicators that are critical in determining quality in a clinical product or medical service. They reached this finding by evaluating sequential feedback from three groups—patients, physicians, and hospital employees. The seven critical factors, with a corresponding definition for nutrition care, are presented as Table 2–2.

When evaluating individual dimensions of quality, the research found that expertise was the highest rated behavior for physicians. For hospital employees, expertise was second to sensitivity. When individual indicators were evaluated through a multivariant analysis, however, the importance of single variables lost their relative standing. For example, when expert behavior was rated higher than sensitive behavior, the overall quality perception suffered. Conversely, when the ratings for sensitive and responsible behavior were equal to or greater than that assigned to expert behavior, overall quality was rated high. The research concluded that, when the patient care provided is seen as being performed in a sensitive and responsible manner, the overall feeling about the quality of that care is enhanced.

Table 2–1 Clinical Nutrition Products and Services

Support Services	Medical Services	Clinical Products
Selective menu	Calorie counts	Nutrition assessments
Meals	Therapeutic diets	Diet education
Snacks	Enteral formulas	Care plan recommendations

Source: Adapted from James, B.C., How Do You Involve Physicians in TQM?, *Journal for Quality and Participation*, January–February 1991, pp. 42–47, with permission of the Association for Quality and Participation, © 1991.

Table 2–2 Behaviors That Demonstrate Quality Nutrition Care

Behavior	Description
Accurate	The services provided are free of mistakes and errors. There is attention to detail throughout the process, including data collection, interviewing techniques, use of calculations, and documentation.
Coordinated	The services are delivered in an organized, predictable sequence that recognizes the importance of patient comfort as well as the efficient use of time and resources.
Expert	The provider of services is an educated, technically skilled professional who uses methods and equipment that optimize diagnostic and treatment procedures.
Responsible	The provider of services recognizes and supports the rights and wishes of the customer; patient care supersedes financial, personal, and procedural demands. Information is openly shared and the customer's involvement is solicited in all decisions that affect the services being provided.
Sensitive	The emotional and psychological ramifications inherent with the service provided are responded to with courtesy, empathy, and respect.
Thorough	The service has been provided comprehensively and using all relevant data. Attention to detail is evident throughout all phases of a patient's diagnosis, treatment, and follow-up.
Timely	The provider recognizes the value of the customer's time by being responsive to requests, by being punctual in meeting delivery expectations, and by keeping scheduled appointments.

Source: Adapted from *The Quality Manager Report,* by E.C. Murphy Ltd. with permission of E.C. Murphy Ltd., © 1991.

Content Quality/Delivery Quality

The concepts of perceived and factual quality were described earlier. These descriptors have been renamed for use in the health care environment in some of the quality literature.

Content quality can be roughly equated with factual quality. Content quality is concerned with the medical outcome that is achieved as the result of the clinical product or medical service that is provided to a patient. Content quality differs from factual quality in that it is recognized that there is more variability in health care services than in manufacturing. Patients cannot be equated with widgets. Each patient presents with different symptoms, medical histories, and risk profiles. As a result, the outcomes resultant from a clinical product cannot be evaluated by using the same kind of conformance requirements that are employed in a factory. Still, the basic concepts are the same and, by evaluating groups of similar patients, outcomes for clinical products can be assessed for content quality.

In traditional medicine, the clinical product and its corresponding quality indices have been the exclusive domain of physicians and their professional health care associates. An example of content quality for nutrition services would be the normalization of blood glucose levels as an outcome of education about the diet for diabetes.

Delivery quality is synonymous with perceived quality. Put simply, the delivery quality describes the patient's interactions with the health care system as defined by the patient.

STRUCTURING QUALITY PROGRAMS IN HEALTH CARE

As stated earlier, the advent of quality management as a discipline is relatively new in the United States. It is even newer to health care.

The "best" quality program, as well as the most successful implementation methods, are largely unknown. Some organizations have chosen to follow manufacturing models with only minor adaptations. The Hospital Corporation of America began its quality quest in the mid-1980s with a modified Deming method (Duncan et al. 1991). Others have chosen an amalgamation of several programs that have been uniquely combined to mesh with the hospital's needs.

The literature on the subject runs the gamut. There is no clear consensus. Exhibit 2–2 lists some of the areas where conflicting approaches are common.

Quality Councils—A Good Idea?

Organizational implementation of a quality program can occur in many ways. Every way requires top management's support and involvement. Quality is not something that a chief executive officer (CEO) can delegate to a subordinate. It takes hard work on everyone's part, especially the CEO, to make it work.

Some organizations choose to demonstrate their commitment to quality by establishing an "office of quality." These departments are generally headed by an individual who has formal training in quality programs and is staffed at the vice-president level. In hospitals, this unit may also be responsible for coordinating the Joint Commission accreditation process.

There are critics of the "office of quality" concept. When used inappropriately, these units can serve as little more than window dressing for the organization. If the CEO delegates the quality initiative to this area and resumes his or her duties using a traditional management approach, then a failure is imminent. Likewise, if the vice-president of quality does not have access to the CEO, as well as his or her complete support, then effectiveness is lost. Probably the biggest criticism to this approach lies in the key principle that quality is everybody's job. If quality is integral to the work that is performed by an organization, then the responsibility for it lies with the top management team.

Exhibit 2–2 Areas of Uncertainty for Health Care Quality Programs

1. Quality improvement councils
2. The spread of implementation
   —Top down, bottom up, or everybody at once
   —Transaction or transformation approaches
3. Demonstration projects vs. whole-hospital start-ups
4. The timing and depth of educational programs
   —Early classroom training vs. "just in time," tool-specific training when needed
5. The content of training
   —Conceptual understanding vs. specific skill development
6. Timelines for implementation and benefits realization
7. When to focus on clinical products vs. support services issues

Quality Steering Teams

Another approach is the quality steering team, also known as the quality improvement council. This committee is composed of approximately a dozen key individuals within the organization. The top leaders in administration, medical operations, and staff functions are represented. Most authorities agree that the CEO must be an active member to the team's work.

The primary role of a quality steering team is to be responsible for the successful implementation of a quality program. To be effective, the team must have the power to act. Duties performed by a typical team include the following:

- Establishing the organization's quality policies: This includes such things as creating a quality definition, mission, and core values; developing a strategic plan to implement the program; coordinating quality initiatives and progress throughout the complex; and monitoring goal achievement.
- Facilitating the implementation of the quality program at the unit level: This is done by mentoring departmental or cross-functional initiatives; removing barriers that are preventing progress; and advocating for change in management systems.
- Allocating sufficient resources to ensure success: Areas that require attention include funding in training and education programs; sponsoring celebrations for success; advocating that personnel time be dedicated to team efforts; and investing in trials of change.

Unlike the office of quality, the life span of the quality steering team is usually limited. Once the quality program is firmly in place within the organization, the team is disbanded or its charge is significantly altered. Industrial models suggest that the average life of a quality steering team is about five years.

The quality steering team also has its critics. If one believes that only the executive management team has the ability to establish organizational policy and allocate resources, then the concept of a steering team does not make sense. At best, the team becomes a separate meeting of the same top management people. At worst, parallel and inevitably conflicting committee structures are established. And, like the office of quality approach, a major objection to the steering team effort is that the responsibility for quality needs to be woven into the fabric of the organization. As such, the implementation of a quality program, including the details, must rest with the top management team.

The same decisions faced at the organizational level must be made by departments implementing a quality management program. The department leadership must decide who will be responsible, both literally and figuratively, for the program. Questions that need to be addressed are listed as Exhibit 2-3.

IMPLEMENTING QUALITY PROGRAMS IN HEALTH CARE

Another aspect of quality programs about which there is no consensus involves optimal methods of implementation. Although the experts agree that implementing a quality initiative is like a journey with a specific destination, they differ on the best route to get there. Specific areas of disagreement include the need to address cultural changes at the outset of the journey, the rollout through the hospital's hierarchy, and the relative benefits of conducting demonstration pilots.

Transformation or Transaction?

The first issue, addressing the need for early cultural change, goes back to the type of quality program that the hospital is considering. Many quality management advocates emphasize an all-inclusive transformation of the organization. The transformation includes changing the entire hospital's culture and values—in other words, the way it thinks about and provides patient care. This means directing initial quality efforts at making an internal, environmental change. Ensuring that managers' styles are uniformly participative and developing the leadership skills of facilitation, vision, and motivation are two examples in this model.

The quantitative tools of quality that make up the structure by which problems are solved and work methods are improved are secondary to these quality management experts. Indeed, transformation advocates often relegate this aspect of the quality program to simply providing a structural platform to assist in meeting the hospital's mission. Overemphasizing the use of the tools, some contend, can result in a delay of

Exhibit 2-3 Structuring Quality

- Will new, continuous, quality improvement initiatives be intermingled with existing quality assessment work?
- How is quality assurance (QA) handled now?
    By an individual?
    By a QA committee?
    By the department management team?
    By the department director?
- Who will do the staff training about quality?
    The department trainer?
    The QA person?
    A representative from human resources?
    The department management team?
- How will resources be allocated?
    Will the director decide and include it in the department budget?
    Will it be handled as a separate line item or together with other related areas (e.g., education)?
- If the department management team is not implementing the program, how will communications be handled?
    How will conflicts be resolved?

achieving the ultimate drive toward total quality within the organization.

An analogy can be drawn between this transformation model of quality implementation and the clinical nutrition care of a patient. In developing a patient's nutrition care plan, the transformation advocate would state that improving the patient's status is the goal and that all efforts should be directed at reaching that objective. The tools used to assess the patient's status, such as measurements of albumin, transferrin, and height and weight, are simply a means to reaching the desired outcome. Focusing too much on the nutrition assessment variables—which ones to use and how to interpret them—can detract from reaching the objective of improving nutritional status.

Other quality experts emphasize the use of quantitative quality tools at the initial stages of implementation. This school of thought is synonymous with the continuous quality improvement model defined earlier. The continuous quality improvement model focuses on the methods or tools of quality. The philosophy of this approach is that, by learning and using the tools, the commitment to quality will evolve. In time, the use of the quality tools will become ingrained in the workplace. Quality will become the center of the organization's mission and, as a result, the hospital's cultural transformation will occur. As one expert states, "The synthesis of statistical process control, teamwork, and process-driven activities is inevitably going to change the organization" (Kennedy 1992, 83).

One of the major benefits of this transactional model of quality implementation is that it is tangible. While transformation deals with philosophy, transaction is about facts. For the most part, there is a right way and a wrong way of doing things that can be readily taught and evaluated. Indeed, it is much easier to teach a manager how to construct a Pareto chart than how to value cultural differences between employees.

Returning to the nutrition care analogy, transaction advocates would not discourage a focus on assessment variables in the development of a patient's nutrition care plan. By becoming intimately familiar with the intricacies of using albumin, transferrin, and height and weight measurements in patient assessment, it is presumed that better decisions will be made and nutritional status will improve.

Top Down or Everybody All at Once?

The vast majority of quality experts agree that commitment by the hospital's top leadership is critical to a program's success. Not everyone agrees, however, whether or not the organization's top management team must have complete knowledge, familiarity, and experience with the specific quality program before it is implemented throughout the organization.

Advocates of the top-down implementation method base their belief on the need for ownership. Ownership for the program is critical for its success, and the only way to achieve true ownership, they say, is to have experienced it yourself. As already stated, leadership actions that run counter to the core concepts spell doom for the quality program's fate.

Using the top-down approach also tests top management's resolve. Michael Pugh, the CEO of a hospital that is a recognized leader in health care quality, touches on this issue in a round-table discussion about implementation issues. Health care management is notoriously faddish in its selection of models, he says, going jauntily from one program to the next (Kennedy 1992). Because a commitment to quality is a long-term effort that takes significant time and financial resources, having the hospital's top management fully aware of the ramifications makes sense before an organizational commitment is made. Indeed, managers must be willing to spend at least two hours per day educating themselves about quality management at the early stages of implementation (Kennedy 1992).

Another rationale behind the top-down approach has to do with the need for coaches. Because quality management is so different from the way in which hospitals have been traditionally managed, there is a great need for mentors and teachers. According to those who advocate this approach, a hospital's top managers are in the best position to teach middle managers. Middle management resistance is seen as a major barrier to successful implementation of quality programs. Using the executive team as mentors goes a long way, they believe, in minimizing this particular obstacle.

Criticisms of the Top-Down Approach

The top-down approach is not without critics, however. There are three major drawbacks to this implementation strategy. The first drawback is time. Implementation of a quality program is a long-term endeavor in the best of circumstances. Although estimates vary, many experts cite six years (Kennedy 1992) or more as a reasonable expectation. If the decision is made to delay implementation of the program throughout the organization until top management is fully operational, then the overall timeline is even longer. In the health care environment, this time delay could be detrimental to the organization's competitive standing.

Another drawback deals with the amount of time when different, and often conflicting, management practices would be in place. Committing to a quality program means using the tools and techniques all of the time. Top management interacts with middle management every day. It is unrealistic and counterproductive, the top-down critics contend, to expect the top management team to function under one set of principles when working on projects as a group and then switch to other methods when dealing with those who have a reporting relationship to them.

Finally, a core tenet of quality management is that quality is an ongoing learning process that permeates the organization. In some ways, isolated initial implementation with the top management team violates this principle. By learning the methods first, the hospital's executives are more likely to develop biases about what tools work best, what strategies are most effective, and so forth. This, in turn, can foster the "I know better than you know" culture that is a mainstay in traditional management and the antithesis of quality management. Advocates of the organizationwide implementation method contend that this process can be an important learning experience that assists in facilitating the transformation of the hospital's culture.

Do Demonstration Projects Facilitate Implementation?

A third area where experts disagree is the desirability of pilot or demonstration projects during the initial stages of implementing a quality program. The alternative would be implementation of the program simultaneously throughout the hospital. In many respects, the rationale used by demonstration project proponents are the same as those described for cascading implementation from the top management team down through the organizational levels.

Proponents of pilot projects cite several advantages to this implementation approach. First, demonstration projects can be carefully chosen to get the quality program off to a positive start. By selecting key, influential personnel to work on the pilot, the probability of success is enhanced. Early successes assist in creating an optimistic work environment for additional projects, which, in turn, builds momentum throughout the organization. The result is an upbeat program that cascades through the hospital at a faster rate than would otherwise be achieved.

Another advantage to demonstration projects is that they provide an early opportunity to discover problems and refine the methods of the quality program before widespread implementation. Moreover, pilot projects permit a final chance to ensure that the quality program and methods that were selected for the organization are going to be effective. If success is not achieved from a carefully chosen pilot project that is conducted by targeted staff, then the likelihood of organizational success is small.

The final advantage of demonstration projects is that they can be used to develop a sense of ownership for the process. Many experts believe that demonstration projects should be used to kick off the implementation of the quality program and that the top management team should be the project group. This is an effective strategy to ensure that the hospital's executive staff has a clear understanding of the quality process.

Disadvantages of Demonstration Projects

Conducting demonstration or pilot projects also has disadvantages, however. First, it extends the rollout of the quality program throughout the organization. Selecting the pilots and the project groups, completing the quality process, waiting for measurable results, and evaluating the projects take time. If changes in the planned program are made, further delays are likely.

Some hospitals have described unanticipated lags. Indeed, implementing a quality program takes a lot of time and results may not be seen quickly. If demonstration projects are done with a lot of fanfare, the outcome can be the creation of unrealistic expectations about results. Publicity about achievements leads to a heightened awareness by employees who may not be introduced to the quality program for some time. When their turn comes, the lack of publicity in the interim can contribute to an early cynical attitude.

While proponents argue that demonstration projects reduce the risk for failure, opponents argue that pilots carry significant risks of their own. If a pilot fails or the results are less than anticipated, the future of the entire program can be doomed. Furthermore, the time and resources dedicated to a pilot may not be realistic for a whole-house effort, thus setting up an environment for failure in the long run.

Finally, demonstration projects, like early executive training, can create biases that negatively affect the larger program. The methods and processes that worked best for a demonstration project team may not be optimal for others. Quality projects need to be a learning experience for each team. Rigid transference of the pilot team's experience to subsequent work groups should be avoided.

As with structuring a departmental quality program, the organizational dilemmas inherent to implementation issues must also be addressed on a departmental level. Questions about implementation that must be posed and decided by the department's leadership are listed as Exhibit 2–4.

Exhibit 2–4 Implementing Quality

- What is your primary objective in implementing the program?
  Using the structured quality process and tools to solve departmental problems?
  Transforming the culture of your work group?
- Does your objective fit with the larger organizations?
  If not, how do you plan to reconcile the differences?
- Will the departmental management team learn the quality program before others? If so,
  Will the department members be informed?
  How will managerial resistance, if encountered, be handled?
  How will you use the tools of quality in your everyday work if the department's membership is unfamiliar with them?
  What actions will you take to ensure that others are encouraged to experience the learning curve?
- How will you make the time for department members and yourself to learn?
- Will demonstration projects be an initial step? If so,
  How will the projects be determined?
  How will you choose the participants?
  What will you do if a project fails, especially if the quality program was organizationally selected?
  How will you convey the success of the project without building unrealistic expectations?

EDUCATING FOR QUALITY HEALTH CARE

Implementing any quality program means doing things differently. This, in turn, requires educating those who are expected to change how they do their work.

The importance of education and training is integral to all quality programs. Deming, Juran, and Crosby all emphasize the importance of training in their quality program steps (Table 1–2). In health care, the Joint Commission has made education of top management the initial standard in its transition to a continuous quality improvement mandate (Koska 1991).

Not everyone agrees, however, on the best way to implement education into the workplace. Differences of opinion exist in the areas of the education program's content, the number and organizational depth of participants, and the timing of training in the quality program's rollout.

Concepts or Skills Focus?

The first area of difference concerns the focus of the training program's content. Some experts emphasize the need to teach concepts; others devote training time to learning specific skills. The choice between conceptual and skills-based educational efforts is closely linked to the organization's strategic decision about which quality path to follow. Hospitals that choose to follow a transformation implementation model will most likely focus educational efforts on conceptual learning. Conversely, hospitals that buy into the transaction quality model will emphasize mastery of specific tools and problem-solving methods in their training initiatives.

For organizations that are working toward changing internal cultural values and norms, training should be focused on teaching concepts that will result in behavioral changes, especially on the part of managers. Educational programs in the areas of leadership, visioning, conflict resolution, and coaching are needed. Curricula that enhance self-understanding, appreciating diversity, and assessing values are also important.

Skills-based training about how to use the tools of quality are secondary in the transformation model. The teaching of these tools may be limited to their application within a larger lesson. For example, a hospital may develop a program on the importance of managing by fact. As an exercise to stress the importance of this concept, the program may include presenting a case history that would lead the manager-participant to believe that an employee is responsible for a recurrent problem. By collecting information and then plotting it on a run chart, however, the manager may discover that the work process, not the employee, was problematical and that a change in the system could stop the problem. As shown in this example, the training effort is focused on the concept of data management and not on learning the quality tool.

The underlying philosophy in concept-based quality training is that education leads to behavioral changes. By focusing educational efforts on such topics as the value of employee work groups, the importance of serving the customer, and the opportunities for improved systems, it is believed that the hospital's employees will adopt these concepts into their work life. In time, the organization's culture will be transformed into one that has a commitment to quality as a core value. As a corollary to this, the teaching of tools and methods are secondary because it is believed that the quality commitment must be present for the tools to be effective. In other words, the tools are merely a means to the end; the end is the quality concept.

Skills-Based Training

Skills-based training differs significantly from concept-based programs. Hospitals that spend their education dollars on teaching skills are investing in a transaction quality model. Program offerings in these organizations are focused on statistical process control, methodological problem solving, and data representation. Whereas a concept-based program has a lot of "touch and feel" components, a skills program does not.

Standardizing work and managerial efforts is often the prime objective of skills-based training. Educational offerings are often directed at those who do the work, as well as their managerial counterparts.

Skills training is the bread and butter of transaction quality programs. Learning the uses and fine points of quality tools is the core of the curriculum. For example, a hospital may develop a program on management by fact. As an exercise in this program, the participants may look at data in a variety of formats to pinpoint opportunities for change. Putting data into run charts, Pareto charts, and cause-and-effect diagrams is an example of the classroom training in this model.

The foundation of skills training is to teach a specific way of solving problems and managing a process. Because the tools work, success follows. Systematically building quality into the hospital's products and services instills pride in the work force. In time, quality is an innate part of every aspect of the organization. In the transaction model, the tools and methods, when used by individuals skilled in their application, make quality.

Who and How Many Are Trained?

There is not a consensus among quality experts about the depth and breadth of training programs that an organization needs. Although it is generally agreed that training of the top management team is important, some experts strongly believe that every employee needs comprehensive training. Others contend that this across-the-board approach is not cost effective.

Making the decisions about who and how many employees need to participate in formal quality training requires consideration of several variables. One variable that is frequently cited is the presence of limited financial resources. Education proponents state that the resources must be made available if the program is to succeed. Critics say that effectiveness can be achieved with a limited investment that is targeted to specific employee groups. Other variables that need to be considered are

- whether the training will be done by internal people or by an outside consultant
- the geographic distribution of the work force
- the timetable for the quality program rollout
- the kinds of employee work groups, if any, to be used
- the planned content of the training program (i.e., concepts or skills).

Advantages of Comprehensive Education

There are several advantages to taking the approach of training everyone. Perhaps the greatest advantage is that whole-house training provides consistency; there is greater control over the training process. Because education is usually the first major step in implementing a quality program, standardized training provides the best opportunity for getting everyone at the starting block at the same time with the same message.

Comprehensive education of the organization also may lessen the likelihood of pockets of resistance. All employees receive the same message. Resistance on the part of a single worker is not likely to hamper the efforts of others. Whole-house training ensures that no one feels left out. If training efforts are limited to the management group, there may not be a feeling of ownership or involvement by those who do the work.

Another significant advantage is that organizational training demonstrates commitment to the process and the program on the part of the top management group. Education is expensive in both time and money. Training all employees is positive proof of the investment being made in the quality initiative.

There are also disadvantages to training everyone. Because of logistical difficulties, whole-house education programs are usually done en masse. This means that the curriculum may be diluted in order to make the challenge manageable. Rather than dealing with specific functional issues, generalized examples are used in the classroom.

In addition, mass education increases the probability that there will be a significant lag time between when the education takes place and when it is actually applied on the job. This is particularly true if the quality program is phased in, with education being a distinct initial phase in the rollout. Mass education programs also increase the likelihood that employees will be taught more than is needed for them to do their jobs effectively.

The decision to provide organizational education often has a big impact on how the program cascades through the organization. This approach is counterproductive if the hospital plans to implement its quality program from the top down through the organizational hierarchy.

Targeted Training

Targeting training to managers and those employees who will function as facilitators in project teams also has several advantages. Taking this approach provides the hospital with an opportunity to do more effective training with smaller groups. The materials and course content can be made more specific and meaningful to the individual participant groups.

Targeted training also can provide more flexibility with timing of the quality program's implementation. Groups can be trained just prior to beginning a project; initiating projects will not be delayed because whole-house education is not yet complete.

There may be additional advantages when those who are target trained are used to teach others on their project team. Team members made up of the larger work group may learn more readily from a person with whom they are familiar and while working on a work process that they know well. Likewise, the manager or facilitator may be better able to develop a more cohesive work team if he or she leads the training efforts. Finally, targeted training is usually less costly to the organization.

There are also several disadvantages to the targeted training approach. It is exclusionary for a program that is, by definition, inclusionary. By targeting training programs to selected individuals within the organization, the wrong message can be sent to the larger work group. Unless the intent of the education program is communicated clearly, it is likely that the hospital's employees will be introduced to the quality process with the belief that it belongs to only a few. This is not the way to start.

Another disadvantage is that standardization is less likely to be maintained if training is done by many people over a protracted period of time. Moreover, people who are not successful in training are likely to pass misinformation or their biases on to others. This can set an entire project team or department up for failure.

Where and When To Train

The final educational area in which there is a lack of consensus is where education should occur. Some experts believe that traditional classroom training works well. Others advocate a "just in time" approach. This strategy implies that training is done on a project team basis either immediately before the time begins its work or as the work proceeds.

There are several reasons that traditional classroom training makes sense in a quality program. First, the training is dedicated to the topic at hand and can concentrate a specific curriculum in a defined period of time. This can be a big help in the hospital setting, where people are often pulled in several directions at once. By scheduling time away from the work area, classroom training assists in ensuring that the staff is made available for concentrated learning.

Another advantage to classroom training is that it is conducive to educating large groups of people at one time. Thus this form of training tends to be less costly than the "just in time" alternative. Perhaps the greatest reason for traditional classroom training is that it maximizes control over the program content. Because the environment is controlled, standardization of the curriculum, content, and teaching methods is optimized.

At the same time, however, there are disadvantages to classroom training. Some employees may have difficulty in translating the concepts or methods that were part of their classroom learning into their jobs. This is especially true if the course material is generalized. Similarly, some workers are intimidated by the traditional, structured, classroom environment. They may be less likely to ask questions, clarify confusions, or debate issues that do not correspond with their experience or values.

Another classroom disadvantage deals with the evaluative aspect of learning. In a classroom training situation, particularly in large groups, it is difficult to ensure that all of the participants are successful in learning, much less mastering, the material. This can be problematical, particularly if the participants will be expected to lead others through the quality process.

Classroom training is often tied to education of the entire work force early in the implementation of the quality program. Some hospitals have not been successful with this approach. Indeed, one hospital executive described his experience with whole-house classroom training as having dressed up a thousand people to go to a dance and having no dance to send them to because there were not enough projects identified for those trained to work on them (Kennedy 1992).

"Just in Time" Training

In "just in time" training, the project team learns what they need to know when they need to know it. This approach is gaining support throughout the health care quality community (Kennedy 1992; American Hospital Association 1992).

Supporters of "just in time" training in health care believe that it works for two reasons. First, health care professionals are specialized, and learning the quality process is not the focus of their work. Therefore, interest in and motivation to use a quality process are most effective when the process is tied directly to their specialization. In addition, hospital clinicians, on average, are well educated, accustomed to continuous learning, and used to change. "Just in time" training works well because the group tends to learn the quality tool quickly. This is particularly true when the project team is homogeneous and made up of professional staff.

The overriding advantage of "just in time" training is that learning and doing are connected in a timely fashion. This focus on timeliness leads to the disadvantages of this educational strategy. Indeed, the problems of "just in time" training are similar to some of those encountered with targeted training. Specifically, the "just in time" approach is likely to sacrifice some consistency, standardization, and control in order to maximize timing.

Educational issues associated with training department members must be addressed. Some of the decisions may be made in conjunction with the organization's quality strategy. Others, however, will be made at the departmental level. For those who are implementing a quality initiative within their area of responsibility, it must be recognized that the decisions that are made about training will make a big difference in how the overall program unfolds. Exhibit 2–5 asks some questions that must be answered by the department's leadership early on in the planning of a quality program.

WHAT CAN ONE REASONABLY EXPECT?

Implementing a major organizational change is never done lightly. Before embarking on the quality journey, it is reasonable to define expected outcomes as the result of the change,

Exhibit 2-5 Educating for Quality

- What resources will you have available?
    - An outside consultant?
    - An organizational consultant?
    - Whatever you develop on a departmental basis?
- Who will do the teaching?
- What will be the focus of educational efforts?
    - The tools of quality?
    - The importance of quality in health care?
- At whom will your training efforts be directed?
    - Yourself?
    - The facilitators who will work within project teams?
    - The management team?
    - The professional staff?
    - Everyone in the department?
- Will quality training be mandated or offered on a voluntary basis?
- How will training groups be determined?
    - On a project team basis?
    - By level of reporting relationships?
- How will the timing of educational efforts be managed?
    - Everyone all at once?
    - In phases, with each level of employees completing the training before the next begins?
    - In segmented phases, with a new level of employees starting as soon as another level is completing a segment?
    - On a "just in time" basis as a project team is working on a problem?
- Will education take place in a classroom or on the job?
- What evaluative methods will be used to assess whether learning is being integrated into the completion of work?
- How important is consistency in your training efforts?

as well as when the results should be realized. Unfortunately, although there is a consensus among experts about the benefits of implementing a quality program, there is not agreement about the type, timing, and degree of payoffs that can be reasonably expected.

Monetary Gains

Quality gains are instrumental in improving the hospital's bottom line. An enhanced bottom line, however, can be the result of increased services, reduced expenses, or a combination of the two.

Reduced expenses are thought to be an inherent outcome of continuous quality improvement efforts. By evaluating work processes in a systematic, in-depth way, tasks can be restructured to increase efficiency. If work is structured properly, the need to fix mistakes and replicate previous steps is eliminated. This, in turn, leads to reduced costs and improved productivity. This concept is embodied in the Deming chain reaction, which states that you will decrease costs when you increase quality (Kennedy 1992).

Estimates of the potential cost savings in health care are impressive. In a round-table discussion on the subject, Michael Pugh estimated that 25 percent to 50 percent of hospital expenses are due to system inefficiencies that "create waste, rework, needless complexity, malpractice medicine, and wrong clinical decisions" (Kennedy 1992, 95). Anderson and Daigh estimate that quality costs as much as 40 percent of health care revenues (Orme and Parsons 1992).

Reducing Costs

There are also opportunity costs for potential savings that are never discovered. Research has shown that 60 percent of employees feel that they are significantly underutilized (Esty 1992). If one could capture the savings available from employee ideas for process improvements, as well as for new products and services, the potential for gains could be much greater.

Expense reduction, especially when done on a project basis, is probably the easiest way to achieve financial gains. Most people who work in the health care industry believe that there are many opportunities for significant improvements. Rapid identification and implementation of process improvements can lead to early payoffs. The University of Michigan Hospitals cite a nearly $14 million savings in their initial two years with a quality program (Coffey 1992).

Early savings can have the potential of undermining the long-term success of a quality program, however. To succeed, quality programs must be conducted in an environment of trust. There is little incentive for a project team to streamline operations at the cost of their jobs or those of their coworkers. According to Deming, driving out fear and blame is at the core of total quality management (Kennedy 1992). Therefore, the preferred method is to enhance efficiency and increase productivity without outplacing staff. Reallocating personnel into more productive positions and/or to handle an increased patient volume is better than taking the quick savings made possible with staffing cuts.

Another way that fast-tracking a quality program's implementation can backfire lies in forcing payoffs without making a sufficient investment first. This happened at Wright-Patterson in Dayton, Ohio, where projects that were expected to lead to big financial gains were aggressively pursued without appropriate training. The result was disappointing monetary gains and discouraged employees (Kennedy 1992).

While harder to quantify directly, it is believed that enhanced revenues are another outcome of quality. By assessing each work process as a function of meeting customers' wants and needs, it is believed that greater patient satisfaction will be achieved. Improved satisfaction enhances the hospital's image, which, in turn, leads to increased patient utilization. This results in increased revenue for the organization.

Changing the Way Business Is Done

Transforming the culture of the organization takes three to six years. Juran believes that tangible results can begin to be realized in about three years, with greater gains as more time elapses (Kennedy 1992). To expect results before three years is, according to Juran, not reasonable. Others disagree, citing the University of Michigan Hospitals' experience, as well as others, as proof that tangible results can be achieved quickly. Whether the early successes found in some organizations are due to the uniqueness of the hospital, the extraordinary opportunities available in health care, or pushing too hard too soon (thus undermining the long-term potential) cannot yet be determined.

When do you know that a transformation has occurred? One clue is when the automatic response to a situation is compliant with the quality mission. An example would be when an error in patient care elicits a "why did this happen?" as opposed to a "who did it?" investigation. Another clue is when organization employees at all levels speak in terms of data instead of feelings. A third clue is an unsolicited concern for serving the customer. In summary, the transformation has occurred when employees believe that they can (and should) change their work to better the organization. This does not mean that the work is complete, however, because the quest for quality is a continuous process.

Working on Clinical Products

Opinions about when to direct efforts toward improving clinical products are diverse. Some experts advocate using the quality tools and principles on clinical products from the onset. Others believe that initial efforts are best directed at support and medical services.

Early clinical product advocates base their position on the fact that clinical products are the business of health care and that quality efforts should be directed at the core of the organization (Kennedy 1992). Another argument for this position lies in the importance of physician involvement. Many experts believe that physician involvement is critical to the ultimate success of any hospital quality program. Physicians are unlikely to take the time and effort required to learn and use the process unless it will make a difference in the work that they do. In other words, involvement in areas other than clinical products is unlikely.

This concept can be readily transferred to the practice of clinical nutrition services. Clinical dietitians are more likely to involve themselves in improving processes that will make a difference in patient care and outcomes. Although improving the way in which food is purchased or trays are assembled is important, it is unlikely to garner the interest or involvement of the clinical staff.

Others believe that clinical products should not be included in quality efforts until experience and familiarity are gained in areas that are easier to manage. Those who advocate this position believe that there is much to be learned from industry when it comes to quality programs. Because there is nothing comparable to clinical products in manufacturing, other opportunities to learn should be explored first. Likely initial projects include those products and services that are common to all businesses, including billing, purchasing, and other administrative functions.

Another reason cited for beginning quality projects with support and medical services has to do with the position that these items play in the provision of health care. If one believes that support and medical services are the platform on which clinical products are provided, then it makes sense to devote initial efforts to that foundation (Kennedy 1992). To use an analogy, it does not make any sense to build a faster race car if the racetrack is full of potholes.

Clinical Products in Nutrition

This concept is easily transferred to the work performed in a dietetics department because the full range of support and medical services, as well as clinical products, is produced at this level. Using a clinical nutrition example, it may be preferable to improve the accuracy and reliability of calorie count information (i.e., a medical service) as a prelude to evaluating nutrition assessments (i.e., a clinical product). If the quality of nutrition assessments is dependent on data concerning the patient's intake, it will be difficult to improve the latter unless the former is already well done.

Finally, there are less experience with and fewer validated reports of the tangible benefits of using quality tools with clinical products. Some organizations have chosen to use medical and support services to demonstrate that quality programs are based in management science, are logical, and can be done within the context of the hospital. This done, the barriers to using the techniques on clinical products are thought to be reduced (Kennedy 1992).

Regardless of when it is introduced in the implementation period, many believe that the greatest gains in quality management will be derived from improving clinical products. There is a great deal of variability in the way practitioners approach patient care. Indeed, the way in which one dietitian approaches a patient can be very different from the way another treats the same patient. The amount of time spent per patient and the points stressed during a counseling session are just two examples of practitioner variability. Few data are available, however, to determine what is most effective. This is where the gains from quality have the most potential.

Developing a yardstick by which to measure the effects of a quality program is important. Expectations, however, must be realistically defined in terms of both time and money. How one approaches the implementation of the program can make a difference with respect to expectations. Exhibit 2–6 asks some questions that can assist in defining the criteria by which the department's quality program can be assessed.

Exhibit 2–6 Setting Expectations for Quality

- How do you anticipate improving the bottom line?
    By reducing expenses?
    By increasing revenues?
- Do you have a numerical value that you expect to achieve in a specific period of time?
    Have you allotted sufficient time and resources for learning?
    Have you taken into account the failures that will occur?
- How will you handle any labor savings that are realized as a result of your quality efforts?
- How will you know when you have "made it"?
- On what products and services will you focus?
    Clinical products?
    Support services?
    Medical services?
    A combination of the three?
- Are you being realistic in your expectations?

THE IMPORTANCE OF MEASUREMENT

However an organization defines quality, it needs to be measured. Measurement yields data that, in turn, yield information on which to make objective decisions. The quality process cannot be successful without information on which to achieve and evaluate progress. Ideally, each work process should have multiple measurement variables. For medical services and clinical products, there are three major measurement sources. They are

1. clinical indicators
2. patient/customer satisfaction
3. cost

Clinical indicators measure patient outcomes. Although they are not usually direct measures of the care provided, clinical indicators give valuable information about what is happening in a patient population. When measured before and after the implementation of a work process change (particularly when no other identifiable factor can account for a difference), any alteration in a clinical indicator can be indirectly ascribed to the change.

The average length of hospital stay and the nosocomial infection rate are two clinical indicators that are often used in the medical setting. Examples of clinical indicators that indirectly measure nutrition care are weight and serum transferrin values in patients receiving enteral support.

Patient and/or customer perception about the quality of care provided is another measurement source. Patients may not be trained medical specialists, but they are amazingly perceptive at recognizing quality care (Press et al. 1992). Moreover, there is a great deal of truth in the saying, "Perception is reality in the eyes of the beholder." The perceived quality of a product and the delivery quality of a service are critical aspects for identifying opportunities for process improvement.

Cost, because of its finite nature and importance in the health care system, is a natural measurement source. An underlying premise in the quality initiative is that improving quality can reduce costs. Measuring cost savings that result from work process improvements is vital to the success of any organizational quality program.

In many ways, measuring quality is like performing a nutritional evaluation. There are objective data (clinical indicators and cost) and subjective data (patient satisfaction) that are collected. The evaluation is the culmination of assessing these data. The more data and more data types that are available, the better the nutrition evaluation that is produced. The same holds true in determining quality.

REFERENCES

American Hospital Association and the Joint Commission on Accreditation of Healthcare Organizations. 1992. *Strengthening medical staff involvement in quality improvement* (teleconference workbook). Chicago: American Hospital Association. April 16.

Coffey, R.J. 1992. Costs and savings of CQI: A real-life example from National Demonstration Project and Joint Commission forums celebrate successes and address future needs in quality improvement, II: Joint Commission's fourth annual national forum on health care quality improvement. *Quality Review Bulletin* 18, no. 3:111.

Duncan, R.P., et al. 1991. Implementing a continuous quality improvement program in a community hospital. *Quality Review Bulletin* 17, no. 4:106–112.

Esty, K. 1992. Productive work environments: How healthcare executives can stimulate productivity—an eight-step model. *Healthcare Executive* 7, no. 3:22–23.

James, B.C. 1991. How do you involve physicians in TQM? *Journal for Quality and Participation* 14, no. 1:42–47.

Kennedy, M. 1992. A roundtable discussion: Hospital leaders discuss QI implementation issues. *Quality Review Bulletin* 18, no. 3:78–96.

Koska, M.T. 1991. New JCAHO standards emphasize continuous quality improvement. *Hospitals* 65, no. 15:41–44.

Murphy, E.C., Ltd. 1991. *The quality manager report*. Amherst, NY.

O'Leary, D.S. 1991. Accreditation in the quality improvement mold—a vision for tomorrow. *Quality Review Bulletin* 17, no. 3:72–77.

Orme, C.N., and Parsons, R.J. 1992. Customer information and the quality improvement process: Developing a customer information system. *Hospital & Health Services Administration* 37, no. 2: 197–212.

Press, I., et al., 1992. Patient satisfaction: Where does it fit in the quality picture? *Trustee* 45, no. 4:8–10, 21.

Chapter 3

Patient Satisfaction: A Mark of Quality

Successful implementation of quality management requires a thorough understanding of one's area of responsibility. Although many clinical nutrition managers believe that they understand their department's roles, functions, and responsibilities, this understanding is often based on the paradigm of traditional medical management.

Quality management requires a different paradigm, one that sees the provision of nutrition services in a work-flow model. According to Fritz (1992), there are four steps that must be taken to make this transformation:

1. Define the products and services that are produced within your department.
2. Delineate what you need and expect from others in order to produce your products and services, as well as what you need or expect from those who receive them.
3. Come to a mutual understanding about needs and expectations with those whom you provide and from whom you receive services.
4. Monitor whether or not the mutual understanding is being met.

This chapter discusses the elements of quality nutrition products and services. It also provides tools to assist you in making the transformation from traditional quality assurance to quality management.

WORK AS A PROCESS

One of the fundamental tenets of any quality program is that all work can be defined as a process that, in turn, can be broken down into parts. A corollary of this tenet is that, if every part of the process meets or exceeds its well-defined standards, then the outcome is a quality product or service.

Well-defined standards are a key to quality. It is imperative that all of the quality standards by which a product or service is evaluated have been agreed upon by both the customer and the supplier prior to delivery.

Figure 3–1 depicts the chain that develops in the flow of work processes as they occur in a hospital. One can view these processes as a series of handoffs between people and departments that eventually leads to the provision of a clinical product or medical service. For most hospital services, the flow of work is cross-functional. In other words, the process flows between departments.

Customers and Suppliers

A key concept in this flow is the customer and supplier relationship. A customer is anyone who receives a product or service from a supplier; a supplier is anyone who provides a product or service to a customer. The work process is, in essence, the sum of these transactions. The service or product that is delivered to the final customer is the culmination of these inter- and intradepartmental processes. For example, one can examine the customer and supplier transactions necessary to complete the work process of providing a patient with one day's supply of tube feeding. By working backward, it can be surmised that

- The patient (external, final customer) is administered the tube feeding (clinical product) from his or her nurse (internal supplier).

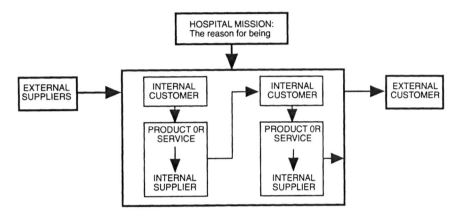

Figure 3-1 Work As a Process

- The nurse (internal customer) gets the patient's tube-feeding product from the unit's supply room, which has been stocked by food service worker A (internal supplier).
- Food service worker A (internal customer) has delivered the patient's tube-feeding product based on instructions from his or her co-worker, food service worker B, who works in the formula room (internal supplier).
- Food service worker B (internal customer) has filled the day's tube-feeding order based on directions provided by the dietitian (internal supplier).
- The dietitian (internal customer) has directed the delivery based on the patient's diet order, which was provided by the unit's secretary (internal supplier).
- The unit secretary (internal customer) transcribed the diet order from the medical record. The order was written by the attending physician (external supplier).

This simple example demonstrates the multiple customer and supplier relationships that exist for this relatively simple task to be performed. It also illustrates the cross-functionality inherent to most hospital procedures. Indeed, getting a tube-feeding product into a patient requires coordination and execution of specific steps by the medical staff and nursing departments, as well as the food and nutrition services functions. And to ensure that the service level is of a high quality, there must be clear specifications established and adhered to at every point of the chain.

Defining and understanding your products and services, as well as your customer and supplier roles, are critical to any quality management program. Exhibit 3-1 is a tool that you can use to assist you in clarifying these key roles.

Value Added

If work is a process that entails a series of transactions between suppliers and customers, then why are so many steps required to complete the process? The answer to this question lies in the premise that every transaction in the chain has a distinct purpose; as the work is transferred from one area to the next, there is some tangible value that is added to it. This concept is depicted in Figure 3-2. The final medical service or clinical product is, in fact, the sum of a series of steps in which sequential value has been added. Most medical products and services have several steps because they require a variety of specialized skills and knowledge.

To illustrate this concept, let us examine the value that was added during each step of the work process outlined earlier. The specialized skills and knowledge that are involved in providing a patient with his or her tube feeding include the following:

- medical knowledge and authorization to provide the feeding, which is provided by the physician when the order is written
- accurate and timely transcription of the physician's order by the nursing unit secretary
- translation of the physician's order into its product equivalent as determined by the dietitian
- provision of the product to the nursing unit by the food service workers
- safe administration of the product to the patient by the nurse

Each step in the chain has a defined purpose, and each person involved with processing the work is uniquely qualified to complete his or her part of the transaction. Put simply, the value added is the difference between the product or service that is supplied and the product or service that is received.

Just as it is important to define your customer and supplier roles, it is critical that you delineate the value that is added as the result of your department's involvement in the provision of nutrition services and products. Exhibit 3-2 is a tool that can assist in clarifying the value added from the work that you supply to your customers.

Exhibit 3–1 Defining Your Role As a Customer and Supplier

What Are Your Products and Services?	Who and or What Do You Need To Do This Work? **You Are Their Customer.**	Who Is the Recipient of This Product or Service? **This Is Your Customer.**

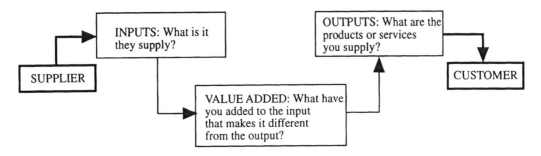

Figure 3-2 The Value Added in a Work Process

MEETING CUSTOMER WANTS AND NEEDS

Customer-driven products and services are an integral aspect of any quality program. Indeed, some definitions of quality deal solely with this issue, stating that the quality of any product or service is "whatever the customer says it is." Most experts agree, however, that organizations consistently fall short in this aspect of managing their business. Some reasons for this organizational failure include the following:

- a belief that the customers' wants, needs, and expectations are already known
- a lack of respect for market research methods (Often top-level managers view the data derived from these efforts as being a waste of resources, yielding little in the way of tangible, precise information for a very high price.)
- a perception that the work is not worth the effort because many of the factors that influence customer decisions are outside the direct control of the organization.

What transpires within the functional units of the larger organization is even more dismal. Although some hospitals have market research specialists who collect general information about what patients want, need, and expect from their organizations, rarely are these individuals able to devote time and resources to customer expectations about specific, smaller, product and service lines. To find organizational market research data that extend beyond customers' satisfaction with the hospital's food service and into the realm of clinical nutrition services is rare indeed. This means that defining customers' expectations for nutrition products and services must occur at the departmental level. But the same obstacles that get in the way of larger organizational initiatives also serve as strong inhibitors for departmental managers. In addition, most clinical managers have little background and training in market research techniques.

Market Research Is about Listening

Just as many people believe that all nutritionists are dietitians, so do many people believe that high levels of patient satisfaction equate to customer-driven products and services. This is not the case, however. Being satisfied with what is provided can be very different from getting what is wanted.

Asking customers about their level of satisfaction with a product or service that is provided can tell you a great deal about what you are already doing. It gives you no feedback, however, about unmet needs that customers may have. If survey initiatives are limited to the current products and services provided, opportunities for improvement may never be uncovered.

For example, a nutrition department routinely surveys patients who received diet instruction as part of their hospitalization. One of the questions asks the degree of satisfaction with the written materials that were provided. On a four-point scale, with four being excellent, the average rating was consistently three. One could conclude that, on average, patients were satisfied with the materials. By changing the question to ask what information was needed that was not provided in the written materials, the department discovered that patients wanted recipes. This was an unmet need. By developing a cookbook that complemented the diet instruction materials, the nutrition department was able to fill a previously unmet customer need and, at the same time, add a revenue source for the department.

To summarize, market research is not about telling your customer what you do and then asking for feedback. Market research is about listening. It is only by asking the right questions and then encouraging the customer to identify his or her unmet needs that customer-driven products and services can be developed.

How To Hear

There are several market research tools that can be used to assist your efforts in clearly hearing your customers' wants, needs, and expectations. Before choosing a specific tool, however, it is important that you organize your efforts. Figure 3-3 details this process.

The first step is to target the customer group whose feedback is desired. Key customer groups may include past,

Exhibit 3-2 Defining the Value Added to Your Products and Services

What Is the Product or Service?	What Is the Input That You Get from Your Supplier?	What Is the Output That You Provide to Your Customer?	What Differentiates the Input and Output? **This Is the Value Added.**

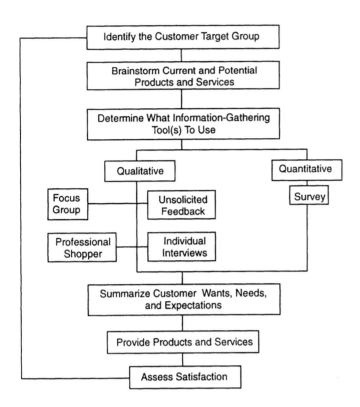

Figure 3-3 Hearing What Your Customers Have To Say

present, and/or future users of the nutrition products and services that are produced in your department. The customers that were identified in Exhibit 3–1 may be a good starting point.

Once the customer group has been identified, the next step is to anticipate loosely what wants, needs, and expectations may exist for your products and services. Extreme caution must be used here. The goal of this step is to identify current or potential products and services that can then be used to develop probing questions to ask the customer. Brainstorming, without evaluation of the ideas generated, is an appropriate method of completing this step.

Collecting Information

Deciding what method or methods to use to collect the information comes next. A summary of the most common market research tools is presented as Table 3–1. Tips for developing appropriate survey questions are included as Exhibit 3–3.

The research methods provide either qualitative or quantitative results. Both are useful in identifying opportunities for improvement. Tool selection is usually determined by a multifactorial analysis that includes available time and financial resources, the degree of business risk involved, and the kinds of information desired.

Based on the results obtained, the next step is to summarize the wants, needs, and expectations of the customer group. It is important to limit the summation to the kinds of results that were obtained; qualitative research data should not be summarized in a quantitative format. For example, stating that 67 percent of customers believe that written diet materials are difficult to understand is appropriate only if a survey was conducted. Reaching this same conclusion from a focus group is wrong. If the data do not yield clearly defined needs, sorting them into segment groups for reanalysis may be necessary.

Providing Customer-Driven Products and Services

At this point, you have heard what your customer wants. The next step is to provide the services and products that will meet that need. This may mean using existing resources, adapting current products, and/or developing new approaches. Whatever it takes, however, is what needs to be done.

The final step in this ongoing process is to assess how well you have done. This means returning to the targeted customer group and assessing its perceptions about your products and services. If you have been successful, the wants and needs that were identified previously will have been met and replaced with new ones. And the cycle begins again.

Keep in mind that the customers who receive nutrition products and services are numerous and varied. There are external customers and internal customers; there are multiple customers for a single product or service. Every customer has wants, needs, and expectations that need to be met or exceeded. The process of hearing what your customers have to say is not limited to patients. Listening to all customer groups and providing products and services that meet their needs is the essence of a customer-driven organization.

TYING THE WORK PROCESS AND THE CUSTOMER TOGETHER

Two critical facets of any quality program have been elucidated—viewing work as a process and providing customer-driven products and services. These two concepts should not be treated as separate entities, however. Rather, using customer-defined wants and needs as an evaluative tool for assessing existing work processes is critical if the goal is to maximize quality.

Figure 3–4 illustrates this marriage of customer-driven products and services to the value that is added as the result of the department's work. Every product or service that is supplied by the department should be tied directly with optimally filling a known customer want or need. In some cases, the work is directed at meeting the next internal customer's expectation. For other products, the value that is added is required to satisfy the external customer's need.

For example, let us look at how one hospital produces a common clinical nutrition product, the calorie count. By breaking down this product, it is determined that the input is the patient's menus that have been marked by nursing (i.e., the

Table 3-1 Frequently Used Market Research Tools

Tool	Appropriate When	Critical Success Factors
Unsolicited feedback	It is used as an initial source to develop a hypothesis. Few resources are available for research. There is a low-risk factor in the process being investigated.	All feedback sources should be considered. The comments, suggestions, and complaints have been collected over a sufficient period of time to represent the process being examined. Feedback should be categorized into themes to assist in analysis. Prospective, ongoing collection is preferable to retrospective, selective reviews. Quantitative outcomes are not derived from the data.
Professional shopper	Information about the feelings or psychological impact of the product or service is desired.	The shopper must have exceptional communication skills to be able to report the experience fully. The shopper must have a clear understanding of the kinds of information wanted prior to the experience. Quantitative outcomes are not derived from the data.
Individual interviews	It is desirable to get reactions without a group influence. The subject matter is personal. A list of already defined features needs to be rank ordered.	The selected individuals must represent a good cross-section of the target customer group. The interviewer is informed but unbiased about the topic being discussed. The interviewer is skilled at the technique. The outline has been carefully prepared to ensure that interview data are collected consistently, yet provide opportunities for more information. Audiotaping and/or videotaping the interview enhances data analysis. Quantitative outcomes are not derived from the data.
Focus group	You are in the initial stages of developing a new product or service. Discussion between a group of customers would be helpful. There are funds available for in-depth research. Identifying a list of key features is part of the desired outcome.	The selected group must represent a good cross-section of the target customer group. The discussion must follow a carefully scripted outline to ensure that all topics are adequately covered. The facilitator is informed but unbiased about the topic being discussed. The facilitator is skilled at running this type of group. Audiotaping and/or videotaping the group enhances data analysis. Quantitative outcomes are not derived from the data.
Survey	Quantitative data are needed to make accurate projections about the customer group as a whole. There is a high risk and/or cost to the process being investigated.	Sample size must be large enough to represent statistically the target customer group. Questions must accurately represent the desired information.

supplier) with the approximate amounts of food consumed. The output, in the existing process, is daily documentation in the medical record of the protein, fat, carbohydrate, and calorie intake of the patient for the previous 24-hour period. The customer is the physician. The value added is the translation of the food consumed into its nutritional equivalents. The work is performed by a clinical nutrition staff member because of the specialized knowledge of foods that is required to complete the task. When combined, these components define the work as a process.

Turning to customer expectations, let us presume that individual interviews were conducted with key members of the physician staff about their wants and needs from calorie counts. These interviews found that the physicians use the nutrition-provided documentation to make decisions about the need to initiate tube feeding. The physicians are not interested in the daily intake of the patient. Rather, those interviewed indicated that they took the information and then evaluated it to determine a general trend about the patient's intake of calories and protein.

Exhibit 3–3 Tips for Developing Good Survey Questions

- Limit each question to the measurement of one factor.
    Bad question: How competent and friendly was the dietitian?
    Good question: How competent was the dietitian?
- State questions in clear, easy, simple language; avoid any jargon.
    Bad question: How well did the carbohydrate distribution in the meal pattern accommodate your usual dietary patterns?
    Good question: How similar was the eating plan that was developed for you to your previous eating habits?
- Offer a range of responses; do not ask questions that can be answered yes or no.
    Bad question: Did you have all your nutrition questions answered?
    Good question: How satisfied were you with the dietitian's answers to your nutrition questions? Very satisfied, somewhat satisfied, somewhat dissatisfied, or very dissatisfied?
- Provide options that are balanced and unbiased.
    Bad question: How would you rate the cost of the nutrition counseling that you received? Extremely affordable, very affordable, reasonable, or unreasonable?
    Good question: How would you rate the cost of the nutritional care that you received? Very reasonable, reasonable, unreasonable, or very unreasonable?

Assessing the current calorie count work process in relation to the defined physician-customer needs identifies opportunities for improvement. The process of providing calorie counts in this hospital may be enhanced by forgoing the daily documentation of specific data and including a periodic summary of average protein and caloric intake with pertinent trending information. Documenting current intake as a function of estimated patient needs may be another enhancement. By changing the process of producing calorie counts to better meet the physicians' needs, the nutrition department will have significantly improved the quality of its product.

Evaluating the work that is done within your department as a function of meeting customer needs can be an enlightening experience. It helps to focus the priorities of the tasks that make up the department's work and, in some cases, question whether the current process is efficient and effective. Exhibit 3–4 can assist in this evaluative effort.

MANAGING CUSTOMER EXPECTATIONS

Whenever a customer receives a product or service, he or she has expectations of what it will be. These expectations have a strong bearing on the overall degree of satisfaction that the customer will perceive as the result of having received the product or service. For example, if patients come into a hospital believing that the food will be inedible, they may be highly satisfied with mediocre food. If, on the other hand, patients enter the hospital with an expectation that meals can be ordered as room service, their satisfaction with good food served at regular times will be low.

Customer expectations are formed from a variety of sources, including the following:

- the previous experience of the customer in similar situations
- the reputation of the product or service supplier.
- promotional efforts, such as advertising and media stories, that have created an awareness on the part of the customer about the supplier.

Identifying and meeting customer expectations is important. It is also helpful, however, to take steps that will assist in setting customer expectations at a realistic level. This is particularly important in the provision of clinical nutrition products and services because customers often do not have much experience on which to base expectations.

General actions that assist in minimizing customer dissatisfaction due to unrealistic expectations include the following:

- communicating effectively to the customer about what the nutrition product or service is and is not before it is provided

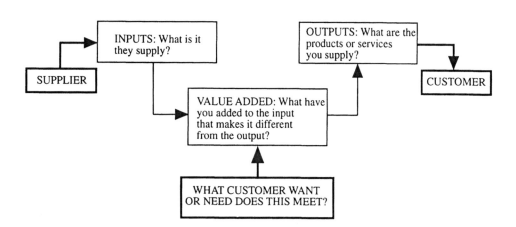

Figure 3–4 Meeting Customer Needs through the Work Process

Exhibit 3-4 Defining the Customer Wants and Needs That Your Products and Services Meet

What Is the Product or Service?	What Value Have You Added?	What Customer Want or Need Has Been Satisfied As the Result of the Value Added?

- making good on any and all commitments that have been made to the customer, regardless of the cost
- designing messages about your products and services that complement and reinforce the overall organization's strategy, mission, and values
- controlling the environment in which the products and services are provided, including staff appearance and attitude, as well as the physical surroundings.

The methods used to manage expectations can vary depending on the source of the customer. Specifically, techniques for communicating realistic expectations to external customers versus internal customers are different.

External Customer (e.g., Patient) Expectations

Managing the expectations of patients who receive clinical services is a vital component of a nutrition department's quality program. The services that are provided by the dietetic department reflect directly on the patients' overall satisfaction with their hospital experience.

Patients are adept at differentiating support services from the medical care that they receive during the hospital stay. Perceptions of medical care, particularly interpersonal skills demonstrated by health care providers, correlate highly with overall patient satisfaction. Dissatisfaction with support services does not show the same correlation, however.

Ganley Associates Inc. is a patient satisfaction–monitoring firm. This company's survey instrument assesses 48 variables and has been administered to more than 134,000 patients. Its results have demonstrated a high correlation (0.80 to 0.89) between the attitude, sensitivity, and friendliness of allied health professionals and overall satisfaction. In addition, four direct food and diet questions are included on the survey. The variables (rank order, correlation) are the likelihood of getting the food selected from a menu (36, 0.64), receiving information about the diet (44, 0.55), the quality of the food (46, 0.50), and the temperature of the food (48, 0.42). Clearly, the management of the patient's experience, especially by the professional staff who provide clinical products and medical services, affects overall organizational satisfaction scores (Press et al. 1992).

Exhibit 3–5 Sample Service Agreement

Customer: H50 nursing unit
Supplier: Nutrition services
Service covered: Diet changes
Effective period: June 1, 1994 to May 31, 1996

General description of services

This service agreement covers the handling of diet change forms between the H50 nursing unit and the nutrition services department. Diet change forms reflect changes in physician-prescribed diet orders that have occurred since the last pickup time. Diet change forms are processed three times a day, 365 days per year. Information conveyed on the forms becomes effective at the meal following notification and continues until another change is processed.

Service standards

1. The dietetic technician will pick up the diet change forms from the nursing unit desk between 6:00 and 6:15 A.M., 9:45 and 10:00 A.M., and 2:00 and 2:15 P.M.
2. The dietetic technician will notify the unit secretary that the change forms are being taken.
3. The unit secretary will have the forms completed by 6:00 A.M., 9:45 A.M., and 2:00 P.M.
4. The unit secretary will legibly write the name, room number, and history number on all admissions, transfers, and diet changes.

Signatures

Nutrition Services Director: _____

Dietetic Technician: _____

H50 Head Nurse: _____

H50 Unit Secretary: _____

For example, a referral for nutrition counseling is often a patient's first interaction with a dietitian. Many—if not most—patients come to the appointment with a lack of similar experiences on which to draw. Being unfamiliar with dietitians, patients may have a lack of knowledge or even misconceptions about the service that is to be provided.

For the counseling dietitian, guiding the patient's expectations at the beginning of the session is an effective means of managing outcome satisfaction. This can be done by

- asking whether the patient has any specific objectives that he or she wants to accomplish as the result of the appointment (if the stated objectives are unrealistic, they can be addressed up front)
- outlining the amount of time, any fees, and the steps that will be included in the session (e.g., determining previous eating patterns; setting goals for diet modification; providing educational materials and suggestions to achieve the goals; and making a follow-up appointment)
- reinforcing the fact that nutrition counseling is an important part of the patient's overall medical care.

Confusion, misunderstandings, and dissatisfaction are minimized if the patient is informed about what to expect from the service, the service is provided as described, and, finally, the link between the two is made at the end of the service provision.

Managing Internal Customer Expectations

Internal customers generally are more educated than patients about the clinical nutrition services that are provided in the hospital. Knowledge about the general scope of services, however, is not always synonymous with having reasonable expectations. For this reason, it is a good idea for internal suppliers and internal customers to develop clearly documented, measurable service levels that meet their mutual needs and constraints.

Written service agreements are an effective tool for managing internal customer expectations (Iacobelli and Lawrence 1991). These agreements detail the responsibilities of each party in the provision of a specific product or service. In essence, written service agreements are the culmination of the shared process described earlier in Figure 3–4 and Exhibit 3–4.

The steps needed to develop a written service agreement are these:

1. Identify the internal customer(s) who receive the product or service.
2. Meet with the customer(s) to identify the expectations, wants, and needs associated with the product or service.
3. Develop a consensus about the product or service that is reasonable and measurable.
4. Document the agreement, including a beginning and a review date.
5. Obtain signatures of both the supplier and the customer(s) on the agreement.
6. Renegotiate the agreement on the review date.

Written service agreements assist in clarifying internal customer expectations and can serve as a foundation for monitoring service provision. Exhibit 3–5 is an example of an internal service agreement.

REFERENCES

Fritz, L.R. 1992. Changing the structure of daily operations: The work-flow model of the future. *Healthcare Executive* 7, no. 4:24–26.

Iacobelli, L.P., and W.P. Lawrence. 1991. Service agreements: An integral tool for ensuring customer satisfaction. *Journal of Healthcare Material Management* 9, no. 9:26–34. October.

Press, I., et al. 1992. Patient satisfaction: Where does it fit in the quality picture? *Trustee* 45, no. 4:8–10, 21.

Chapter 4
Continuous Quality Improvement

Quality can be neither assured nor legislated. However, quality can be improved steadily when an organization's leadership is dedicated to the concept and when employees are trained and motivated to perform their work according to established standards.

This chapter provides an overview of the Joint Commission's role and expectations regarding the improvement of quality patient care and an outline of the ten-step monitoring and evaluation process.

JOINT COMMISSION'S *AGENDA FOR CHANGE*

Methods of quality assurance put in place during the 1970s facilitated development of structures for monitoring quality, but they emphasized the wrong things. For example, the very word *assurance* gave the false impression that an organization could guarantee the delivery of quality patient care. When expectations for quality were not met, both regulators and the public felt shortchanged. Quality assurance procedures required managers to rely on inspection, rather than support and guidance, of employees to ensure that high standards were met (Berwick 1989). The typical employee reaction was that of conformance tinged with a good deal of frustration and fear, reactions that were often carried over to encounters with patients. Thus employees centered their attention more on the sullen avoidance of overt mistakes than on cooperative efforts to deliver first-rate services.

Recognizing the folly of attempting to achieve high-quality care by focusing primarily on deviant behavior, the Joint Commission (1988a, 1988b) drafted its *Agenda for Change*. The "change" is intended to shift the emphasis from "assurance" to "improvement"—a conversion from compliance with standard procedures to attentiveness toward activities known to affect patient health and satisfaction. For those working in health care, this "change" demands a transformation in thinking.

What the *Agenda for Change* Means

The objective of the *Agenda for Change* is to modernize the accreditation process (O'Leary 1991). In its newsletter, *Perspectives*, the Joint Commission stated three major goals for the *Agenda for Change* (1992). These goals are as follows:

1. to reformulate accreditation standards
2. to improve the process by which accreditation decisions are made
3. to develop valid, reliable, data bases that can serve as indicators of effective organizational performance.

As hospitals modify their quality management programs to comply with the *Agenda for Change*, many processes will be carried over from the previous accreditation process (Koska 1991b). For example, survey decisions will continue to be based on an organization's ability to meet or exceed stated standards. An integral part of the traditional accreditation process has been the evaluation of the hospital's services for compliance with published standards. There has been and will continue to be an evolution in the stated standards, however.

Hospital functions that have the greatest impact on patient outcomes are now the focus of standard selection. It is no longer enough to show that one is doing what is expected. There is an emphasis on the effective use of quality tools to measure and continuously improve these important functions. A demonstration that one is doing what needs to be done, doing it well, and doing it in a manner that considers patient feedback is the expected norm now.

Focus on Meaningful Data

Surveyors will continue to emphasize monitoring and evaluation as an integral part of the accreditation decision. If a hospital is to demonstrate that it is doing what needs to be done and is doing it well, then there must be evidence that attention to the details of effective health care delivery is deliberate and ongoing. Again, the effective use of quality tools, not random samples and the investigation of single events, is needed to show evidence of outstanding patient care.

The accreditation process also requires a demonstration that the collected data are actually used to monitor and improve work performance. Raw data are meaningless unless they are evaluated, understood, and, most importantly, translated into actions that are meaningful to those health care professionals that perform patient services.

Evolving Expectations

The *Agenda for Change* embraces the concept of cross-functional processes. Health care delivery is the culmination of clinical products and support services that result in the provision of medical care. In other words, patient care can (and should) be viewed as a series of "handoffs" between functional units such as departments. It is, after all, the sum of the parts that determines the quality of care that a patient receives. Evaluation of key patient care processes, not departmental units, is at the heart of the *Agenda for Change*. Therefore, cross-functional (i.e., interdepartmental) work groups that evaluate and monitor these processes are needed. See Chapter 17 for further discussion of the cross-functional approach to quality care.

It is not the intention of the *Agenda for Change* to require that hospitals embrace the "religion" of total quality management (O'Leary 1992). Neither does the Joint Commission require hospitals to implement a prescribed management style, a specific quality program, or individual quality tools (e.g., flowcharts, control charts).

The Joint Commission recognizes that an organizational commitment to quality can never be mandated, only encouraged. It is the intention of the Joint Commission, however, to facilitate and stimulate this transition. Furthermore, although the Joint Commission believes that quality programs are one (and possibly the best) way of achieving improved performance, it acknowledges that as an external reviewer it is not in a position to mandate a hospital's values (O'Leary 1992).

Specific quality concepts are increasingly seen in the Joint Commission's standards, however. For example, the surveyors expect to see the use of quality tools as an integral part of improvement efforts. The need for each health care organization to commit to some kind of quality initiative is unequivocal.

Education for Quality

The *Agenda for Change* mandates that a quality improvement effort be in place at a hospital for accreditation to occur. How the organization chooses to implement this mandate is unspecified, but the directive to each organization's top leadership is clear.

Effective on January 1, 1993, the Joint Commission required evidence of education about total quality management/continuous quality improvement. This requirement grew out of the recognition that, to be successful, a quality initiative needs strong leadership and top management commitment. While focusing on the organization's leadership, the Joint Commission is assessing the provision of education throughout the organization (Koska 1991a). It is likely that quality improvement education and opportunities for involvement for all staff will be required at some time in the future.

QUALITY ASSURANCE/QUALITY IMPROVEMENT

The Joint Commission believes that quality improvement is conducive to good medicine. The *Agenda for Change* embodies this belief. The transformation from quality assurance to quality assessment and improvement is a significant transition in the Joint Commission's support of quality management.

Quality assurance and quality improvement are not synonymous. As adroitly stated by Townsend and Gebhardt (1991), "Healthcare systems that take the new Joint Commission guidance as merely a call for a larger, tougher police force with stricter quality control/assurance rules will be missing the point."

Quality assurance lives in the past because it evaluates and compares work processes after the fact. Using the quality assurance method means designing the flow of work and then measuring the outcomes. Results are reported in terms of either meeting or exceeding predetermined thresholds. Those results that are outside the threshold are classified as outliers. Almost without exception, the outlier results are ascribed to a person who, by virtue of the method, is labeled as "the problem." Corrective measures aimed at the person follow. The measures can be punitive or educational, but they are limited to the outlying individual. The system is considered back in control when the identified person has achieved results that are within the threshold. Perhaps the single biggest flaw with quality assurance efforts is that they completely miss the identification of opportunities for improvement of the work process as a whole. As summarized by O'Leary (1991), quality assurance is punitive, outlier oriented, hard work, inefficient, and frustrating.

Continuous quality improvement is not about rules; it is about making things better. Developing and filling out forms is not the ticket to quality improvement. Rather, improved work processes are the outcome of communication and teamwork that has, in turn, come about as the result of data evaluation. As stated by O'Leary (1991), the idea of continuous quality improvement is for us to become masters of the data, not the data's slaves.

Exhibit 4–1 describes differences between traditional quality assurance and newer quality improvement models. The contrasts are stark. A review of Exhibit 4–1 reveals, however, that the tenets of quality improvement parallel those cited for quality management elsewhere in this text. Specifically, the Joint Commission's expectations for quality improvement include the following concepts:

- A commitment to quality exists at all levels of the health care organization.
- Customer needs are central to the provision of quality medical care.
- Every work process provides opportunities for improvement.
- Quality is a responsibility that is shared by all who perform the work.
- Process improvements should be based on data.
- Group processes that include those who do the work create improvements.

Defining Quality

Many institutions have developed definitions of quality, such as the ones shown in Exhibit 4–2, as a foundation for their quality management framework. These definitions are used throughout the institution to guide departmental policies, procedures, and plans for quality improvement. Other conceptual frameworks evolve from quality definitions. For instance, Henry Ford Hospital, Detroit, Michigan, drafted the institutional values and guiding principles shown in Exhibit 4–3 to undergird departmental quality improvement programs. Such documents demonstrate institutional commitment to continuous quality improvement, and they offer a framework for departmental planning.

QUALITY IMPROVEMENT USING THE TEN-STEP PROCESS

Hospitals are not required to use the Joint Commission's ten-step process for monitoring and evaluation (Joint Commission 1993), but it can serve as an anchor for a quality management program. Although segments of the ten-step process remain similar to those used in quality assurance, there is a clear evolution in the meaning of the distinct parts (Puckett 1991). As shown in Exhibit 4–4, there is movement toward continuous quality improvement in each of the steps. Indeed, the use of quality tools, group problem-solving methods, and identifying opportunities for improvement in clinical care are evident.

Other key differences are as follows:

- Responsibility is assigned as part of the process.
- Cross-functional teams are encouraged.
- Departmental scope of care is defined and all aspects of care are included in the quality process.

Exhibit 4–1 The Contrasts between Quality Assurance and Quality Improvement

QUALITY ASSURANCE	QUALITY IMPROVEMENT
Focuses on outliers.	Focuses on work processes.
Is retrospective.	Is prospective and continuous.
Negative outcomes are punished.	Negative results imply a suboptimal process.
A person determines the outcome.	A person is only one input variable in the process that results in an outcome.
Spotlights system failures.	Strives to improve all systems.
Assumes that the process is correctly designed.	Assumes that all processes can be improved.
Thrives on rules.	Thrives on making things better for everyone.
Emphasizes forms and data collection.	Emphasizes data-driven action.
Focuses on conformity to predetermined standards.	Focuses on customer needs.
Is the responsibility of one (or a few) quality assurance specialists.	Is everyone's responsibility.
Relies on inspection.	Relies on identifying and fixing problems before they occur.
Infers that the quality assurance specialists know what is wrong.	Infers that those who do the work know best how to improve it.

Exhibit 4–2 Institutional Definitions of Quality

> Quality is continuous improvement and innovation of our work processes to meet the needs and exceed the expectations of patients, families, employees, volunteers, the community and related organizations.
> —*Henry Ford Hospital, Detroit, Michigan*
>
> Quality is team achievement of excellence in customer service and satisfaction.
> —*Courtesy of Mercy Medical Center, Springfield, Ohio*

- Important aspects of care are defined, focusing on high-volume, high-risk, high-cost, or problem-prone activities.
- Indicators of quality are determined and methods are established to monitor the important aspects of care.
- Reasonable standards (thresholds or triggers) are set and used as a baseline for assessing performance, upgrading standards, and noting trends in activities.

Like the quality assurance cycle, part of the ten-step process is cyclical in nature, as shown in Figure 4–1. The first five steps of the process are a prelude to the repetitive sequence of collecting, organizing, and evaluating data to take action as necessary to improve services continuously. At times it may be necessary to modify indicators or to establish new thresholds to challenge personnel to work toward higher standards of excellence. Open communication is, of course, integral to the quality process, and regular reports to designated groups are an essential part of the cycle.

Exhibit 4–5 gives an outline of the monitoring and evaluation process, and applications for nutritional services are briefly described below.

Step 1: *Assign responsibility for monitoring and evaluating activities related to the quality process.* Using institutional guidelines, departments must first develop a departmental plan and quality assurance policies and procedures. A plan, such as those displayed in Appendixes A–1 to A–3, includes assignments for committee membership, scope of care, data collection and analysis, action plans, reporting, and program evaluation. Interdepartmental relationships may be included in the plan or in a flowchart, as shown in Figure 4–2.

Step 2: *Delineate the scope of care provided by the organization or department.* The scope of care will be defined generally in the department's purpose or mission statement, often included in the quality plan (see Appendixes A–1 to A–3 and Exhibit 6–1). The purpose or mission statement is followed by a more detailed delineation of key functions or procedures, such as specific treatments and procedures, units where services are delivered, types of patients served, responsibility for educating students, and team involvement. Chapter 9 contains a further description of this step and some practical examples.

Exhibit 4–3 Henry Ford Hospital "Values and Guiding Principles" for Quality Management

> The Values and Guiding Principles of Henry Ford Hospital serve as the foundation for conduct throughout the hospital. These guidelines actively promote the Quality Management Process. Continuously improving quality health services is our primary focus. We are a team and recognize that successful achievement of these principles will require our long-term commitment.
>
> **Customer Service**
> - Quality patient care is our first priority.
> - We strive to continuously improve the quality and value of services to all our internal and external customers, dedicating ourselves to excellence in all that we do.
> - We seek and support new and innovative ways to add value to our services.
>
> **Employees and Volunteers**
> - We respect and value each individual.
> - We appreciate the importance of each individual's unique contribution as a member of the team.
> - We encourage and support individual development.
>
> **Leadership**
> - We promote trust and understanding through honesty, consistency, open communication and compassion.
> - We promote practices which empower people to improve their work processes.
> - We view our shortcomings not as an occasion to find fault, but rather as an opportunity for improvement.
> - We will recognize efforts toward quality improvement and innovation and take time to celebrate our successes.
>
> **Methods**
> - We will use customer feedback, data and statistical methods to guide our decision making and to improve our processes.
> - We seek to continuously improve our processes through reduction in variation.
> - We will develop mutually beneficial relationships with external suppliers who demonstrate commitment to continuous quality improvement.
>
> Courtesy of Henry Ford Hospital, Detroit, Michigan.

Step 3: *Identify the most important aspects of care provided by the organization or department.* The purpose of this step is to compel departments to select from among their multiple and diverse activities those services that have greatest impact on patient outcomes and to give priority to these activities. Generally, departments of food and nutrition services will include some of the following as important aspects of care:

- assessment and evaluation of patients' nutritional needs
- development of nutritional care goals and plans
- dietary interventions such as modified diets and enteral feedings

Exhibit 4–4 The Ten-Step Process: Old and New

TEN STEPS	PAST EMPHASIS...	...IS GOING TO
Responsibility	Designated quality assurance staff and department directors	Everybody with special attention to top management's support of team endeavors
Scope	Clinical products and medical service providers	All systems and processes that lead to patient care
Aspects of Care	Set priorities based on importance	Continued attention to prime areas
Indicators	Measure clinical outcomes and events	Focus on systems and processes that lead to patient outcomes through trend measurement and evaluation
Thresholds	If met, review is triggered	Continuous efforts to improve
Data Collection	Organized	Organized by using established quality tools
Data Analysis	Case focus with some trend analysis	Statistical trend analysis using process control tools
Action	Department-initiated and intrafunctionally implemented	Cross-functional action based on the process being improved and using specific quality methods
Evaluation	Measured as the solution to a defined problem	Measured as improvement to systemic processes and trends
Communication	Structured to correspond with functional reporting relationships	Cross-functional as a part of interdisciplinary teams

- periodic assessment of the effects of nutritional therapy
- patient and family education and counseling
- food preparation, storage, and service following procedures to ensure safety and sanitation.

These aspects, as well as interdepartmental activities in which dietetic team members participate, are further described in subsequent chapters.

Step 4: *Identify indicators and appropriate criteria for monitoring the important aspects of care.* Having identified important aspects of care in Step 3, the next step is to define measurable factors by which each aspect of care can be evaluated. Such indicators may relate to

- outcomes of care providing evidence of how services resulted in improved patient health, e.g., lowered serum cholesterol levels following intensive nutrition counseling or improved albumin levels resulting from specialized nutrition support
- processes indicating the use of appropriate methods or timeliness of care, e.g., screening and assessments conducted within 72 hours of admission
- structures relating to departmental organization or personnel, e.g., services provided by a qualified dietitian holding a valid license to practice.

Chapter 10 is devoted to a discussion and examples of both departmental and interdepartmental indicators and criteria for evaluating important aspects of care.

Step 5: *Establish thresholds (levels, patterns, and trends) for the indicators that trigger an evaluation of care.* Business leaders use many strategies in their quest for world-class quality goods and services. Three such strategies ("Achieving World-Class Quality" 1991) include the following:

1. Set stretch goals that specify dramatic changes expected over the next one to five years. For example, "exhibit a tenfold improvement in four years," or "reduce the number of errors by 50 percent in the next 12 months."
2. Use benchmarking to counteract the feeling that stretch goals can never be met. This is done by using quantitative criteria already achieved by competitors as a basis for company goals.

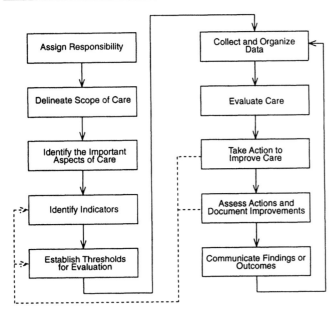

Figure 4–1 Overview of the Ten-Step Process Showing Its Cyclical Nature

Exhibit 4–5 Outline of the Monitoring and Evaluation Process

Step 1: **Assign responsibility**
  a. Identify organization leaders
  b. Design and foster approach to continuous improvement of quality
  c. Set priorities for assessment and improvement
Step 2: **Delineate scope of care and service**
  a. Identify key functions and/or identify the procedures, treatments, and other activities performed in the organization
Step 3: **Identify important aspects of care and service**
  a. Determine the key functions, treatments, processes, and other aspects of care and service that warrant ongoing monitoring
  b. Establish priorities among the important aspects of care and services chosen
Step 4: **Identify indicators**
  a. Identify teams to develop indicators for the important aspects of care and service
  b. Select indicators
Step 5: **Establish means to trigger evaluation**
  a. For each indicator, the team identifies how evaluation may be triggered
  b. Select the means to trigger evaluation
Step 6: **Collect and organize data**
  a. Each team identifies data sources and data-collection methods for the recommended indicators
  b. Design the final data-collection methodology, including those responsible for collection, organization, and determining whether evaluation is triggered
  c. Collect data
  d. Organize data to determine whether evaluation is required
  e. Collect data from other sources, including patient and staff surveys, comments, suggestions, and complaints
Step 7: **Initiate evaluation**
  a. Determine whether evaluation should be initiated
  b. Assess other feedback (e.g., staff suggestions, patient-satisfaction survey results) that may contribute to priority setting for evaluation
  c. Set priorities for evaluation
  d. Teams undertake intensive evaluation
Step 8: **Take actions to improve care and service**
  a. Teams recommend and/or take actions
Step 9: **Assess the effectiveness of actions and assure improvement is maintained**
  a. Assess to determine whether care and service have improved
  b. If not, determine further action
  c. Repeat a) and b) until improvement is obtained and maintained
  d. Maintain monitoring
  e. Periodically reassess priorities for monitoring
Step 10: **Communicate results to relevant individuals and groups**
  a. Teams forward conclusions, actions, and results to leaders and to affected individuals, committees, department, and services
  b. Disseminate information as necessary
  c. Leaders and others receive and disseminate comments, reactions, and information from involved individuals and groups

Source: Copyright 1991 by the Joint Commission on Accreditation of Healthcare Organizations, Oakbrook Terrace, Illinois. Reprinted with permission from *The Transition from QA to CQI: An Introduction to Quality Improvement in Health Care.*

3. Visualize quality improvement at a revolutionary rate, making thousands of quality improvements all at once. Such a strategy requires an institutionwide infrastructure for quality management and a process for choosing which projects will be tackled first.

In hospitals, similar strategies may be used to achieve high-quality services. Thresholds specify the point at which intensive evaluation of performance for a certain criterion MUST occur. Using the strategies outlined above, the long-range goal may be to have thresholds of 95 percent to 100 percent in most areas of monitoring. Immediate thresholds, requiring intensive effort for compliance, would be viewed as attainable if benchmarks show that "the best" hospitals in the country have achieved the desired levels of performance.

Thresholds are given in percentages, indicating the number of cases expected to meet the stated criteria. Thresholds may be high or low, such as "the need for enteral feedings will be documented 100 percent of the time," or "2 percent complications associated with tube feeding administration." When performance falls below or above the respective threshold, a process of intensive review is put into motion to determine why deficiencies or problems occurred and what can be done to achieve desired standards at the next monitoring period. Thresholds can also be used to track overall levels of performance, trends, and patterns of activity. Chapter 10 includes further discussion and examples related to this procedure.

Because of their importance, sentinel events (i.e., food poisoning) are reviewed at each occurrence. Triggers for evaluation are set at 0 percent (food poisoning) or 100 percent (detection of nutritional risk) for such events, indicating the need for intensive review for every deviation.

Step 6: *Monitor the important aspects of care by collecting and organizing the data for each indicator.* Periodically, quality assurance data are collected from appropriate sources and organized into charts, graphs, or tables to compare findings with thresholds or to identify patterns and trends. Inherent in these activities is a prior determination of what data are needed, sources of data, sampling procedures, data collection methodologies, timetables, audit schedules, and parties responsible for data collection and organization. Chapter 11 contains further discussion of this topic.

Step 7: *Evaluate performance in relation to thresholds to identify opportunities to improve care or to correct problems.*

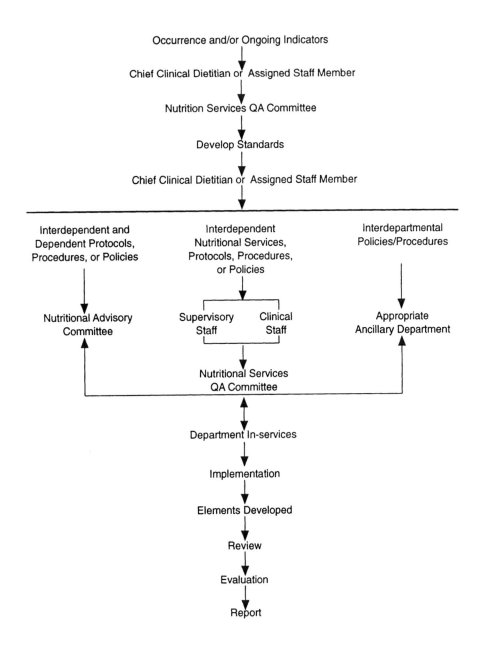

Figure 4-2 Flowchart of Quality Assurance (QA) Process in Nutrition Services Department. Courtesy of Southwest General Hospital, Middleburg Heights, Ohio.

Those accountable for evaluating performance use data-collection forms and summary reports to evaluate performance in relation to established thresholds. In the words of the president of the Joint Commission, "You must become good analysts. You must be able to recognize patterns and trends, and you must determine at what point data raise a red flag—a threshold for concern about what the data show" (O'Leary 1991). If performance is in line with expected levels, no action is necessary. Thresholds that are consistently met need to be adjusted up (for positive aspects of care such as patient satisfaction scores) or down (for negative indicators such as readmission of diabetic patients to control blood glucose levels) to challenge personnel to greater limits of quality performance.

When data show a deviation from thresholds, further analysis is necessary. Usually a small team or committee, representing diverse job classifications and/or departments, is charged with evaluating the aspect of care under scrutiny to determine reasons for departures from the norms. This evaluation team may review trends in periodic reports and may also return to original sources of data, consult other documents, perform further observations, or conduct interviews to identify reasons for deviations from expected performance. Finally, the committee develops an action plan to address the problem area.

Step 8: *Take actions to improve care or to correct identified problems.* High-quality performance results when qualified personnel perform their assigned responsibilities within quality systems, as shown in the following equation (Schermerhorn 1987):

Excellent Performance = Quality Systems × Abilities of Personnel × Individual Effort

Problems may arise in any part of this equation.

Action plans are designed to improve quality where there is a consistent gap between expected and actual outcomes. Depending on results of problem analysis, plans are often related to one or more of the three areas that comprise the equation:

1. Systems. Most problems can be traced to weaknesses in communication, organizational structures, work assignments, scheduling, facilities, equipment, or similar systems in the department. This a good example of why there must be organizationwide commitment to quality; long-standing and deep-seated problems can be addressed effectively only when everyone in the department works together to overcome systemic obstacles to quality.
2. Knowledge or abilities. Because employees are the first contacts with patients and other customers, quality service demands that each has a thorough knowledge of how to perform all tasks inherent in their scope of responsibilities (Harris 1989). Educational action plans may require improved orientation, focused in-service education offerings, or participation in continuing education programs.
3. Behaviors or individual effort. On occasion, deviations in performance are due to neglect on the part of a particular employee. Edelstein (1991) used the example of a department falling short of its 90 percent threshold for diet counseling because one dietitian consistently failed to give diet counseling to patients before discharge from the hospital. Overt behavioral problems are only rarely the basis for quality deficits in the department. However, when individual performance regularly lags behind established standards, the action plan may include recommendations for informal or formal counseling, changes in work assignments, or disciplinary action.

In addition to detailed information about what needs to be changed, action plans should specify who takes responsibility for implementing the change and the timelines for completion.

Step 9: *Assess the effectiveness of the actions and document the improvement in care.* Generally the effectiveness of corrective plans are integrated with the usual monitoring and evaluation system, with particular attention paid to improvements in the "flagged" area. In other words, consistent audit topics and procedures are used but the frequency of data collection may be increased, at least in the short term, to ascertain the benefits of implemented changes. Thus, if the usual plan for checking trayline accuracy is once a month, the action plan may call for weekly checks of a smaller sample size to monitor improvement until the next regularly scheduled evaluation. Although tray error rates of 25 percent to 36.6 percent have been reported in the literature, one hospital watched its tray error rate of 12.9 percent drop to 9.6 percent when performance data were supplied to trayline employees and quality circles were set up to address problems associated with tray accuracy (Dowling and Cotner 1988). This is a good example of how careful analysis of data and effective action plans can be used to effect remarkable improvements in quality.

Step 10: *Communicate results of the monitoring and evaluation process to relevant individuals, departments, or services and to the organization's quality assurance program.* Reports may be generated weekly, monthly, quarterly, or annually for submission to designated groups. For example, the quality assurance action plan of Holyoke Hospital, Holyoke, Massachusetts, includes reporting stipulations shown in Exhibit 4–6. It should be noted in the example that reporting is both verbal and written, both departmental and interdepartmental. The format and content of reporting are addressed more fully elsewhere in this manual.

MAKING THE TRANSITION

Although quality assurance and quality improvement are distinctly different concepts, the years spent developing quality assurance programs have not been without value. Indeed, the Joint Commission (1991, 8) reinforces "that the concept of continuously improving quality incorporates the strengths of quality assurance, while broadening its scope, refining its ap-

Exhibit 4-6 Quality Assurance Plan—Section on Reporting

> The written findings, conclusions, recommendations and actions taken by the department are reported verbally to the Continuous Quality Improvement Committee on a quarterly basis. The Department's QA activity is then reported to the Patient Care Assessment Committee by the Quality Assurance Manager. Necessary information is communicated among the departments and services when problems or opportunities to improve patient care involve more than one department.

Courtesy of the Continuous Quality Improvement Committee, Holyoke Hospital, Inc., Holyoke, Massachusetts.

proach to assessing care, and dispensing with the negative connotations sometimes associated with quality assurance."

How can dietitians adapt their quality assurance programs that are already in place? Sandy Metzler, of the American Hospital Association (American Hospital Association 1992), cautions against both abandoning traditional quality assurance methods and placing these methods at odds with quality improvement tools. Indeed, it is not surprising that professionals who have worked long and hard for many years to improve the quality assurance process may feel frustrated or threatened by the focus and excitement surrounding quality improvement (Koska 1990).

Traditional quality assurance programs concentrate on technical excellence, process, and structure (Marshik-Gustafson et al. 1981). Quality improvement differs in that it expands the demand for change to include customer needs (McLaughlin and Kaluzny 1990). This concept of patient input into accreditation standards can be quite disagreeable to health care providers who resist the patient's ability to question professional practice methods or make suggestions for improvement (Schmadl 1979). Still, patient satisfaction is a major determinant of successful implementation of a quality improvement program.

Incorporating the Patient's Point of View

To illustrate this point, a dietitian may think that clinical competence is what matters most in the delivery of nutrition care. A patient or family may think that the delivery of care in a compassionate, understanding manner is the priority. This dietitian must be reassured that quality improvement does not negate the need for clinical skills. Rather, it requires that the hospital staff demonstrate compassion and understanding in addition to clinical competence for the criterion of quality care to be realized (Quality Quest 1991). In this case, adding the patient-defined components to the previous quality assurance program results in quality improvement.

Clinical practitioners encounter many situations that demonstrate that a shift is taking place in the hospital workplace. For example, it is common practice in many hospitals to make every effort to obtain the patient's selections from the daily menu. This practice is based on the assumption that, if the patient selects the food, then satisfaction with the meals received will be greater. From a professional and functional point of view, this makes sense.

However, the customer's viewpoint can be very different. In one hospital, clinicians were asked to participate in a customer contact program. The customers were former hospital patients who had rated their stay exceedingly low on the patient satisfaction survey. During one session, the dietitian was surprised to learn that the efforts to obtain a menu selection were viewed as intrusive. The case involved a patient who was exhausted after a very tiring procedure. She repeatedly asked that her door be closed and that she be left alone to rest and recover her strength. Despite her wishes, the patient reported that she was "visited" by dietetics staff no fewer than three times, each time in an effort to obtain her menu selections. As expressed by the patient, "What difference did it make if I ate, if I couldn't even sleep?"

A similar example deals with menu accuracy, a common indicator in quality assurance programs. Dietitians may define menu accuracy as compliance with the physician's order for a modified diet. Changes made to the menu, such as adjusting beverages to restrict fluids or substituting fresh fruit for a brownie to reduce fat content, may make the meal more accurate from a clinical perspective.

The patient may have a very different definition of menu accuracy, however. The patient may feel that the meals are inaccurate if they do not match the selections made from the menu, are different from the meals normally eaten at home, or do not agree with what the physician has communicated to him or her. Indeed, comments made to the patient, such as "your arteries are fine," can create menu accuracy confusion when coupled with a cholesterol-lowering diet order.

Consider how either of these two issues, obtaining patient menu selections and providing accurate meals, would be evaluated in a traditional quality assurance program. The more menus obtained, the greater the chance of staying within the established threshold. Auditing menus for compliance with therapeutic diet guidelines is another valid quality assurance measure. Yet neither of these measures addresses the opportunity to improve the process from the patient's point of view. When quality assurance protocols and customer needs are mutually exclusive, then quality patient care does not occur—even when audits are done in accordance with quality program schedules and thresholds are met.

The void between technical expertise and customer satisfaction can be filled. Indeed, inclusion of customer expectations in the existing quality program is a big first step. Expanding an existing ten-step quality assurance program to address customer needs can be accomplished by

- defining customer groups who receive nutrition products and services in the scope of care
- including customer satisfaction as an important aspect of care
- developing indicators, criteria, and thresholds that include measurable data about customers' perceptions.

ADDITIONAL ACTION STEPS

While focusing on customer needs is a critical element in quality improvement, it is not the only one. Making the transition from quality assurance to quality improvement requires more. Four additional strategies that can be readily incorporated into an existing ten-step process are described below.

1. Assign responsibility. In traditional quality assurance programs, responsibility is in the hands of the department's management team. Although these individuals are ultimately responsible for the quality program, there is no reason why the assignment of responsibility cannot be extended to those who do the work. If quality is everyone's job, then the responsibility of quality within the written plan should reflect this tenet.
2. Incorporate group processes into the quality improvement program. Teams made up of those who do the work can be involved in identifying indicators for the important aspects of care, establishing thresholds, monitoring performance, evaluating performance, improving systems, assessing effectiveness, and communicating results. Including the roles, functions, and responsibilities of work group efforts in the written quality improvement plan ensures recognition of their valuable contributions.
3. Use structured process improvement methods and measurement tools. Techniques such as benchmarking, flowcharting, and brainstorming fit readily into the monitoring and evaluation model. Quantitative measurement tools, including Pareto charts, control charts, and histograms, assist in comprehensive data analysis, interpretation, and communication of results. All of these can be used to broaden the scope and depth of an existing quality assurance program.
4. Expand monitoring beyond departmental walls. Functional areas can, in fact, streamline the quality improvement process by collaboratively developing indicators that are cross-functional. Working as an interdepartmental team spreads the work of data collection, analysis, and process improvement among more people. The end results are patient care improvements that are greater than those that could be achieved on a departmental basis, with less individual cost and effort. Also, if nutrition-related criteria are not reflected in nursing unit quality improvement plans, nutritional dimensions of care may be overlooked completely, even on hemodialysis units (Patton and Stanley 1993).

THE INSTITUTIONAL APPROACH

A fundamental change is required in one's overall approach to nutrition care services. The department can no longer be viewed as a box with heavy walls separating it from other departments in the organization, responsible for providing self-regulated, self-monitored, unique services. Albeit some procedures, such as conducting nutrition assessments and evaluations, are within the purview of registered dietitians, other responsibilities, such as attaining high levels of patient satisfaction, are shared by all departments. Thus "cooperation, coordination, and most difficult of all, communication are the keys to success" in quality improvement systems (O'Leary 1991). In its efforts to facilitate the accreditation process, the Joint Commission guidelines tend toward an institutional, rather than a departmental, approach and format. Departments of food and nutrition services need to work with other departments to address four main focuses of the new accreditation process (O'Leary 1991).

1. Defining organizational tasks. Every hospital has a responsibility to meet designated health care needs of the community it serves. Those needs are often quite diverse, ranging from highly specialized diagnostic procedures and high-tech treatment options to outpatient surgery or other procedures and home health care. Once community needs are identified, the institution has a responsibility to provide up-to-date facilities, recruit sufficient qualified staff members, and provide support services so that the public can obtain the designated services in a timely manner.

    Organizational tasks will influence services and procedures offered by food and nutrition departments. For example, 24-hour food service availability may be needed if families are encouraged to stay with acutely ill patients through the night. Sophisticated nutrition assessment procedures, such as metabolic cart measurements, may be required if the facility has a large, specialized, nutrition support service. Comprehensive screening for nutrition risk may be warranted when the served population includes many chronically ill, elderly, or poor patients. The quest for quality should include continual assessment of the need for expanded services to complement and support services offered by the hospital.
2. Determining key functions. In addition to defining the departmental scope of care and identifying the most important aspects of care, the hospital as a whole is expected to take a broad view of its important functions such as credentialing, diagnosing disease, providing food services, or maintaining a safe and sanitary environment. Each of these functions typically falls under the jurisdiction of more than one department, especially when related subprocesses (such as purchasing, storage, sanitation, equipment maintenance, food service for employees, tray delivery, feeding selected patients, specialized nutrition support, and the like) are taken into consideration. Again, food and nutrition departments make important contributions to cross-functional teams and should be represented on multidisciplinary groups, especially since nutrition services often play a critical role in good patient care.

3. Creating a data-based system. The new accreditation process requires upgrading of organizational performance norms, not punishment of outliers or deviations from standards. Performance norms, measured in quantitative terms, give a basis for both comparisons and tracking to reveal trends and patterns over time. Such quantitative measures allow both consistent evaluation of organizational performance and identification of behaviors inconsistent with established norms.

   In dietetics it is a challenge to define in quantitative terms measurement criteria for many critical facets of departmental operations (e.g., food quality, use of sanitary procedures, effects of nutritional therapy, patient understanding of diet instructions, appropriateness of enteral/parenteral nutrition, and employee understanding of universal precautions policy and procedure). The *Agenda for Change* requires that dietitians, with the help of others, create quantitative systems for measuring the important aspects of nutritional care.

4. Using top-down leadership to establish a constructive environment. Unlike the old system of quality assurance, the *Agenda for Change* is uplifting. With continuous quality assessment and improvement, all members of an organization are empowered to do what they can to enhance quality, to please the patient or other customer, and to feel a sense of personal satisfaction from contributing to goal achievement. For the *Agenda for Change* to work, however, all management levels must be committed to quality improvement.

   How committed are dietitians to quality improvement? One industry study report showed that a whopping 76.9 percent of employees felt that middle management was most resistant to quality improvement ("Quality Movement" 1991). On the other hand, 41.7 percent and 37.5 percent of these employees said senior management and service workers, respectively, were most receptive to quality improvement. It goes without saying that dietetic managers and department heads not only must cooperate with organizational efforts to improve quality but must examine their stand and make an attitude change, if necessary, to favor total commitment to quality improvement.

CONCLUSION

Highlights of the *Agenda for Change* and its inherent ten-step process summarily contain a few brief concepts of importance for nutrition managers:

- Commitment to the process of quality improvement is essential, and this commitment must be clearly evident by demonstrating strong departmental leadership for the process.
- The process is founded on principles advocated by such leaders as Deming, Juran, and Crosby, who were able to achieve high levels of quality in industrial settings.
- Cooperation, collaboration, and communication are fundamental to success in this regard; interrelationships both within and outside the department are key to improving the overall quality of care in the institution.
- Emphasis should be placed on continuous quality improvement, measured in quantitative terms, not on mere problem identification, as was often the practice under old quality assurance programs.
- Nutrition services personnel participate as active members of quality teams to define, assess, measure, and improve key functions of both the department and the institution, thereby upgrading performance norms within the hospital.
- Once achieved, new levels of performance should be maintained and continually improved over time; once standards of performance are met, the standards should be elevated to further maximize quality of services rendered.

REFERENCES

Achieving world-class quality. 1991. *Fortune* 124:203. September 23.

Agenda for change Q & A. 1992. *Joint Commission Perspectives* 12, no. 1:1, 5.

American Hospital Association and the Joint Commission on Accreditation of Healthcare Organizations. 1992. Strengthening medical staff involvement in quality improvement. Teleconference, April 16. Chicago: American Hospital Association.

Berwick, D.M. 1989. Continuous improvement as an ideal in health care. *New England Journal of Medicine* 320, no. 1:53–56.

Dowling, R.A., and C.G. Cotner. 1988. Monitor of tray error rates for quality control. *Journal of the American Dietetic Association* 89:450–453.

Edelstein, S.F. 1991. Using thresholds to monitor dietetic services: The JCAHO 10-step process for quality assurance. *Journal of the American Dietetic Association* 91:1261–1265.

Harris, R.D. 1989. Returning to the basics: The bottom line to quality improvement. *Hospital Topics* 67, no. 2:19–23.

Joint Commission on Accreditation of Healthcare Organizations. 1988a. *Agenda for Change*. Oakbrook Terrace, Ill.

Joint Commission on Accreditation of Healthcare Organizations. 1988b. *Agenda for Change*. Oakbrook Terrace, Ill.

Joint Commission on Accreditation of Healthcare Organizations. 1991. *The transition from QA to CQI: An introduction to quality improvement in health care*. Oakbrook Terrace, Ill.

Joint Commission on Accreditation of Healthcare Organizations. 1993. *Accreditation manual for hospitals, 1993.* Oakbrook Terrace, Ill.

Koska, M.T. 1990. Adopting Deming's quality improvement ideas: A case study. *Hospitals* 64, no. 13:58–64.

Koska, M.T. 1991a. Is traditional quality assurance on its way out? *Trustee* 44, no. 9:3–23.

Koska, M.T. 1991b. New JCAHO standards emphasize continuous quality improvement. *Hospitals* 65, no. 15:41–44.

Marshik-Gustafson, J. et al. 1981. Planning is the key to successful QA programs. *Hospitals* 55, no. 11:67–73.

McLaughlin, C.P., and A.D. Kaluzny. 1990. Total quality management in health: Making it work. *Health Care Management Review* 15, no. 3:7–14.

O'Leary, D.S. 1991. Accreditation in the quality improvement mold—a vision for tomorrow. *Quality Review Bulletin* 17, no. 3:72–76.

O'Leary, D.S. 1992. Agenda for change fosters CQI concepts. *Joint Commission Perspectives* 12, no. 1:2–3.

Patton, S., and J. Stanley. 1993. Bridging quality assurance and continuous quality improvement. *Journal of Nursing Care Quality* 7, no. 2: 15–23.

Puckett, R.P. 1991. JCAHO's agenda for change. *Journal of the American Dietetic Association* 91:1225–1226.

Quality movement growing rapidly in Europe. 1991. *Fortune*, 124:173, September 23.

Quality quest. 1991. A briefing of health care professionals. Teleconference. Chicago: American Hospital Publishing, Inc.

Schermerhorn, J.R. 1987. Improving health care productivity through high-performance managerial development. *Health Care Management Review* 12, no. 4:49–55.

Schmadl, J.C. 1979. Quality assurance: examination of the concept. *Nursing Outlook* 27, no. 7:462–465.

Townsend, P., and J. Gebhardt. 1991. Will continuous improvement work here? *Journal for Quality and Participation* 14, no. 1:6–9.

Part II

Departmental Systems and Tools for Quality Management

Chapter 5

Preparing To Initiate a Quality Program

The road that leads to quality management is not short. Rather, the road is a long and winding path that requires dedication and skill to master. And just as one must prepare to embark on a long journey, so too must preparations be made to initiate a quality management program. Prior to starting the journey, it must be assured that competent leadership is at the helm and that the work group is ready to be led.

This chapter examines the issues surrounding a department's preparedness for a quality program. It defines the characteristics of quality leaders and explores the unique challenges faced by clinical nutrition managers. A survey tool that assesses work group readiness, as well as the results obtained from its administration in clinical nutrition departments across the country, is also included.

LEADERSHIP

Quality management requires strong, unwavering leadership. Strength is needed to get a program off the ground; commitment is required to see it through.

In many ways, the successful implementation of any quality program is analogous to the "wave" that is performed by spectators at sporting events. If you have ever seen a "wave," or been a part of one, you know the power that it can hold.

The "wave's" motion is put into place by an individual or a small group. He, she, or they stand up and raise their arms to the sky. Responding to this, the people adjacent to the instigators follow. And on it goes. Before long, an entire stadium of thousands of people are standing and stretching their arms upward. They act in synchrony. Everyone joins in. There is laughter, and smiles are shared between strangers. When it is done, people return to their seats with a sense of accomplishment that they have participated in a group effort of monumental proportion and have succeeded.

So it is with a quality program. The leader has to stand up to start the motion that culminates in the quality "wave."

Are You Ready To Lead?

Assuming a leadership position in implementing a quality program is not easy. In fact, it is very hard. Quality management requires that the leader never be satisfied with the status quo. Overcoming significant obstacles, believing in the skills and capabilities of one's work group, and never accepting excuses for subpar behavior are just a few of the personal attributes that nutrition managers must possess if they hope to achieve departmental quality.

Exhibit 5-1 is a short self-assessment that can be taken to ascertain one's current orientation to maintaining high standards. The tool was developed by Leebov and Scott (1990). According to the authors, negative responses to the first five questions and affirmative answers to the remaining questions indicate a manager who has high standards. Managers who are satisfied with maintaining the status quo are likely to provide opposite responses. Clearly, leaders who have high standards are needed to implement a quality program successfully.

Does It Matter Who Leads?

An examination of organizations with successful quality management programs usually reveals a chief executive of-

Exhibit 5–1 Self-Assessment of Managerial Standards

1. Do you believe that excellence is impossible today given staffing shortages and scarce resources?	YES	NO
2. Do you believe that customer demands are becoming more and more unrealistic?	YES	NO
3. Do you avoid confronting employees even when you know you should?	YES	NO
4. Do you think that most people are doing a decent job?	YES	NO
5. Do you believe that when it comes to excellence, people have it or do not?	YES	NO
6. Do you believe that negative people should not be permitted to bring down the entire group?	YES	NO
7. Do you often communicate a desire that employees stretch toward excellence, your vision of excellence?	YES	NO
8. Do you exemplify excellence in your own interactions with customers and employees?	YES	NO
9. Do you set and communicate ambitious performance expectations to all employees?	YES	NO
10. Do you monitor employee performance on a regular basis and intervene when you see substandard behavior?	YES	NO

Source: Reprinted from *Health Care Managers in Transition: Shifting Roles and Organizations* by W. Leebov and G. Scott, pp. 40–41, with permission of Jossey-Bass, Inc., ©1990.

ficer (CEO) who has led the way. These individuals share the characteristic of being quality believers. From their perch at the top of the organization, these CEOs have been the catalysts of transformation. Based on this case-study approach, it is often stated that the CEO must initiate the quality program. Indeed, some go so far as to say that if the CEO does not initiate the action, there is no way that quality can make its way into an organization.

At the same time, however, there are many quality believers who exercise leadership within smaller organizational units. They are sometimes referred to as departmental champions. By proactively implementing quality tools and, more importantly, the tenets of quality management, these champions have led their work groups to positive results.

It is not unusual for departmental champions to go unrecognized for their quality initiatives by the organization's top management. One cannot, after all, give credit to something that is not understood. The champions are simply seen as effective leaders, the managers that can be counted on.

Indeed, it is often not until the organization as a whole decides to travel the quality journey that these departmental champions' achievements are seen for what they are. Departmental champions then become the CEO's adjacent supporters as the quality "wave" makes its way through the hospital.

How Departmental Champions Work

Departmental champions sometimes play a vital role in introducing the larger organization to quality management. A quality team that produces results in a process that has been troubling the organization for years can do much to convince skeptics, including the hospital's top administrative group, that quality management produces improvements that are not attainable with traditional management methods.

A case study that illustrates the ability of departmental champions to convince upper management that quality teams produce results is provided by Sullivan and Frentzel (1992). They assembled a group of employees at Massachusetts General Hospital to examine the process of transporting patients from the nursing units to and from tests and procedures. Using quality management tools and methods, this group produced measurable improvements in the process. The success of their efforts provided the impetus for senior management to dedicate the time and resources needed to spread quality processes throughout the organization.

Most of us are not in the position to decide that our institution is going to choose quality as its core value. We are, however, in an excellent position to become departmental champions.

The steps to becoming a departmental champion include the following:

- never being content with the way things are
- learning as much as possible about the philosophy of quality management and the use of quality tools
- changing your personal behavior to reflect these values
- changing to quality process methods within your area of responsibility inasmuch as your organization's infrastructure allows

It is important to keep in mind that different quality programs have somewhat different methods. Moreover, when an organization chooses a specific quality program, it is important that all of the functional and cross-functional units act in unison. This means that you, as a departmental champion, will probably need to change to synchronize your department's quality efforts with those of the larger organization. But, by having learned the basics and seen the potential of quality by experiencing success, you will be further ahead than if you had waited until the "big wave" was in motion.

Unique Challenges for Nutrition Managers

Nutrition managers face unique challenges as they struggle with the quality issue. For many, the challenges are personal. Conflicting directives seem to come from everywhere—the need to cut back staffing levels is stressed by your administrator; requirements for better, more thorough, documentation come from regulators; and your staff cries out for relief from the need to care for sicker patients who are in the hospital for fewer days. Doing more with less, the health care slogan on the 1990s, can take its toll. Add to this the implementation of a quality program that will require more of your time and a great deal of effort, and it is difficult to control feelings of pessimism.

Everything can be done successfully, however. E. C. Murphy, Ltd. (1991) conducted an eight-year study that evaluated the behavior of more than 1,000 health care managers. The company's research concluded that effective managers, seen as those who thrive in these chaotic times, shared common behaviors. More specifically, these managers worked from a set of four core principles:

1. maintaining an unwavering personal commitment to serving the customer
2. valuing associates as individual people, not as a means to getting the work done
3. appreciating the individual contributions that are made to achieve a group effort
4. following predetermined responses to common situations

Integrating these principles simplifies the manager's life. Conflicts around satisfying diverse demands are minimized when the customer is consistently given top priority. Likewise, interactions with departmental members on an individual basis fosters trust, cooperation, information sharing, and delegation. By not reacting spontaneously to specific circumstances surrounding an event, the manager is able to get control of the situation more quickly and resolve the problem. In summary, time management becomes easier when there is a clear focus, a programmed response, and shared respect for work.

Even greater than the personal challenges faced by the nutrition manager are the professional questions raised by a quality management program. As nutrition specialists, dietitians are trained in the traditional medical management model.

The Traditional Medical Management Model

The characteristics and ramifications of the medical management model have been explored by McLaughlin and Kaluzny (1990). By applying this work to the field of dietetics, one can conclude the following:

- Dietitians are specialists in nutrition. They are hired based on the achievement of recognized technical skills that have been validated through an outside certification process (e.g., registration and/or licensing).
- There is a "gold standard" of nutrition care that is determined and understood by all of the specialized practitioners. It exists within the profession and is innately adopted by the hospital's administration (i.e., they do not know what good nutrition care is).
- As part of this "gold standard," major professional functions have set protocols. The protocols are not changed until or unless there is a consensus of the scientific literature that adjustments are indicated. Technological advancements are the impetus for most protocol changes.
- The dietitian has a great deal of autonomy in providing patient care. Because he or she is an expert who knows the professional standards, there is a trust that the work will be performed conscientiously with minimal administrative supervision from outside the department.
- Deviation from professional standards lies with the individual practitioner. Failure to comply with protocols is treated as an individual offense, and corrective action is taken in response to complaints.

Conflicts in Management Models

There are several areas of potential conflict between the traditional medical management and quality management models.

First, and perhaps most important, is the customer focus that is at the foundation of all quality programs. The tenets of quality management mandate that efforts should be directed at meeting the wants and needs of customers. For dietetic services, the customer is rarely a person with specialized knowledge of nutrition. Therefore, the standards of good nutrition care are not limited to what is defined exclusively by the hospital's dietitians or even within the larger nutrition community. Patients, nurses, physicians, administrators, and others become customers whose input with respect to the quality of nutrition products and services provided must be addressed.

Second, there are no "gold standards" in quality management. The quest for improvement is continuous and relentless. One can never be satisfied that a protocol is being followed. Rather, opportunities for streamlining the process or improving the outcome must be constantly explored. Although some of this can and will be done within the context of professional organizations, it is more likely that most of the work will be organization specific, because of the unique systems and cross-functional aspects of the work that is performed.

The clinician's autonomy is also affected. Process improvements are generated by teams. As such, the individual's contribution to patient care is viewed as a part of the process that is open for review and suggestion by anyone else who is part of the patient care delivery system. One need not deviate from a protocol to be asked to do something differently; the notion of departmental or specialized territory is erased.

Dietitians, like all health care specialists, can find these areas of potential conflict threatening. In many respects, it is even more disconcerting for nutrition managers. Like most technical professionals, nutrition managers are often dietitians who were promoted because of their clinical expertise. Shifting from the role of lone managerial expert to that of a contributor of specialized knowledge within a team made up of members who may not share one's value of nutrition can prove frustrating.

Making the Transition

One cannot move from the traditional medical management model to the quality management model overnight. The process requires a transition. The impetus for taking the necessary steps to make the transition can occur at the organizational, departmental, or individual level. Specific action steps that help ease the transition include the following:

- Focus on decision-making and group process skills as well as technical skills when hiring and evaluating the professional staff. Self- and departmental educational efforts should concentrate on preparing to be a professional for the future. This means knowing how to use statistical process techniques, facilitate team meetings and cultivate leadership and negotiation skills.
- Find an upper management believer in quality who is willing to advocate on your behalf, support your efforts, and assist in securing needed resources.
- Actively work at thinking about the tasks that you perform in a new way. By expanding your analysis beyond what you are doing to how and why you are doing it, the importance of process becomes ingrained.
- Learn to lead and facilitate group interactions more, while directing and controlling less.
- Modeling the way is critical; mentoring the way is expeditious. Identify others who are likely to become intradepartmental champions. Openly discuss the threats and opportunities inherent to quality management. By providing these potential champions with information, support, and the chance to experiment, the ease of transition is shared.
- Be sure that personal and/or departmental goals attend to both performance and process objectives.
- Look beyond the potential threats that may come with redefining how nutrition products are used and services are performed to the opportunities that lie ahead when your input is considered in the delivery of other patient services.
- Become more comfortable with data collection. Try using the quality tools on tasks that you perform at home or in the office. Analyze data and look for ways to improve your own work processes. Encourage others to do the same.

By recognizing the personal and professional challenges that a quality program presents to nutrition managers, proactive steps can be taken to circumvent normative conflicts and point out areas of progress throughout the transition period.

WORK GROUP READINESS

Implementing quality management in an organization is always a challenge. Quality management should not be introduced to a department without a comprehensive strategic plan.

An important aspect of strategic planning is conduction of a thorough examination of the existing organization. By understanding the thoughts, feelings, and behaviors of members of the current work environment, it is possible to make better decisions about the implementation timetable, potential areas of concern, and the probability for success of a quality management initiative. The introduction of change into a favorable environment increases the probability for success.

THE QUALITY READINESS SURVEY

The Quality Readiness Survey (Exhibit 5–2) was developed to assist in performing an environmental analysis for implementing quality management in a clinical nutrition department. It is based on systems theory that includes organizational development concepts (Melcher 1976).

The premise for the Quality Readiness Survey is illustrated as Figure 5–1. Several variables, including personality, communication channels, and the work environment, influence how a person feels about his or her job. The way a person feels, in turn, influences how he or she performs. In other words, the way that a person feels determines behavior—as an individual performing a task, as a member of a work group, and in reporting relationships. Quality management initiatives are presumed to be more successful when positive work behaviors are present.

A department manager cannot control the feelings of the people who work in the department. The manager does, however, have a great deal of influence over several environmental aspects in which the work is performed. This influence includes how the department's work is designed and how the manager chooses to lead. These two environmental factors af-

Exhibit 5-2 The Quality Readiness Survey

I am a member of the:
1. Dietitian staff
2. Technical staff
3. Clerical staff
4. Managerial staff
5. Other (specify): _____

Please circle the number that most closely expresses your perception (1 = Never; 5 = Always):

1. Hiring decisions are made by the manager without input from the people who will be working with the new employee.   1 2 3 4 5
2. I call everyone in the hospital by his or her first name.   1 2 3 4 5
3. I get current, accurate information about what's going on from my supervisor.   1 2 3 4 5
4. I only hear about the need for doing a quality job when we are due for a Joint Commission visit.   1 2 3 4 5
5. I am committed to my job.   1 2 3 4 5
6. I trust the people in my peer group.   1 2 3 4 5
7. I trust my supervisor.   1 2 3 4 5
8. Performance expectations are set by the people who are doing the work.   1 2 3 4 5
9. When an organizational achievement is made, the credit goes to a person (not a work group or department).   1 2 3 4 5
10. I am encouraged by my supervisor to apply ideas that I've learned outside of work into the work that I do.   1 2 3 4 5
11. I would recognize, by name, the top executives of the hospital if I saw them in the hall.   1 2 3 4 5
12. I get a sense of accomplishment from the work that I do.   1 2 3 4 5
13. I discuss how our work could be done better with my peers.   1 2 3 4 5
14. I prefer a supervisor who tells me what to do and how to do it.   1 2 3 4 5
15. The responsibility for our quality assessment and improvement is assigned to a person.   1 2 3 4 5
16. I know the financial condition of this hospital.   1 2 3 4 5
17. We spend a lot of time tracking useless information.   1 2 3 4 5
18. If the hospital's top executives saw me in the hall, they would recognize me by name.   1 2 3 4 5
19. I like to see changes in the way things are done.   1 2 3 4 5
20. I keep information about my personal life away from my co-workers.   1 2 3 4 5
21. I make suggestions to my supervisor whenever I see a way that my job can be done better.   1 2 3 4 5
22. Peer review is an important part of our evaluation process.   1 2 3 4 5
23. There is stability in the top management team of this organization.   1 2 3 4 5
24. In this department, what is said and what is done are the same.   1 2 3 4 5
25. I hear about new developments at this hospital from the local paper or the evening news.   1 2 3 4 5
26. I attend continuing education programs to ensure that I can maintain my license or registration.   1 2 3 4 5
27. I count on my peers to help me when I'm having trouble getting my work done.   1 2 3 4 5
28. My supervisor trusts me.   1 2 3 4 5
29. Contributions to team efforts are considered when pay increases occur.   1 2 3 4 5
30. Organizational policies are applied in the same way to everybody.   1 2 3 4 5
31. Departmental goals are clearly stated.   1 2 3 4 5
32. Hospital policies interfere with my ability to get my job done.   1 2 3 4 5
33. I think about leaving this position.   1 2 3 4 5
34. I like the people in my peer group.   1 2 3 4 5
35. I candidly respond to questions asked by my supervisor.   1 2 3 4 5
36. When what I am supposed to do and what I think I should do are different, I do what I think I should do.   1 2 3 4 5
37. We celebrate our successes.   1 2 3 4 5
38. The department's management makes decisions based on feelings, not facts.   1 2 3 4 5
39. After I've participated in an organizational survey or program, I find out what was done with the results.   1 2 3 4 5
40. I feel secure about my job.   1 2 3 4 5
41. Any of my co-workers could do my job as well as I do.   1 2 3 4 5
42. I feel part of a team that includes my supervisor.   1 2 3 4 5
43. My clinical judgment is the most important factor in providing patient care.   1 2 3 4 5
44. The grapevine is the best way to find things out.   1 2 3 4 5
45. I feel free to do whatever is best for the patient.   1 2 3 4 5
46. In this hospital, what is said and what is done are the same.   1 2 3 4 5
47. Monitoring the quality of the work that I do is someone else's job.   1 2 3 4 5
48. In my work group, we respect each other.   1 2 3 4 5
49. My supervisor doesn't know very much about my job.   1 2 3 4 5
50. Overcoming obstacles from other departments (e.g., nursing) takes a lot of my time.   1 2 3 4 5
51. Social gatherings are done on a departmental basis.   1 2 3 4 5
52. In my department, people are self-motivated, doing their best because they want to.   1 2 3 4 5
53. Major organizational decisions are made for factual (not political) reasons.   1 2 3 4 5
54. I am comfortable with measuring the results of the work that I do.   1 2 3 4 5
55. My co-workers and I are proud of the work that we do.   1 2 3 4 5
56. My supervisor respects me.   1 2 3 4 5

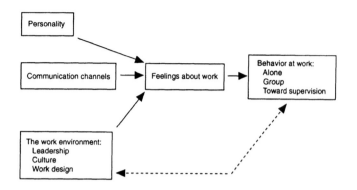

Figure 5-1 Variables That Influence Individual Feelings and Work Behavior

fect the way that people feel about work. This, in turn, affects behavior.

A department manager's influence at the organizational level is limited. The manager cannot determine the leadership style used by the hospital's top executives. In addition, a department manager's ability to affect the overall organization's culture is small. These two work environment factors, executive leadership and organizational culture, also influence how employees feel and behave.

The Quality Readiness Survey (Exhibit 5-2) asks employees to respond to questions that reflect both the four characteristics of the work environment and the three characteristics of work behavior that are described above. These can be summarized as the following areas:

- departmental work design
- organizational culture
- departmental leadership
- top management leadership
- individual behavior
- behavior in the work group
- behavior toward supervision.

By scoring the results on a numerical scale (Exhibit 5-3), the department manager can assess employee perceptions about their current behaviors as well as the work environment factors that influence those behaviors.

Exhibit 5-3 Scoring the Quality Readiness Survey

1. For those questions marked with an asterisk (*), take the reported score and subtract it from 6. Enter this number. For example, if "5" was circled for question 1, then enter "1" in the scoring sheet below.
2. For those questions not marked with an asterisk, enter the reported score below.
3. Sum the columns for each questionnaire and enter in the totals sections.
4. Determine average scores for each variable by summing the totals and then dividing by the number of surveys. Average scores may be calculated for the department and/or for work groups.

Organizational:		Leadership:		Behaviors:		
Work Design	Cultural	Departmental	Top Management	Individual	Peer	Toward Supervision
1* ___	2 ___	3 ___	4 ___	5 ___	6 ___	7 ___
8 ___	9* ___	10 ___	11 ___	12 ___	13 ___	14* ___
15 ___	16 ___	17* ___	18 ___	19 ___	20* ___	21 ___
22 ___	23 ___	24 ___	25* ___	26* ___	27 ___	28 ___
29 ___	30 ___	31 ___	32* ___	33* ___	34 ___	35 ___
36 ___	37 ___	38* ___	39 ___	40 ___	41* ___	42 ___
43 ___	44* ___	45 ___	46 ___	47 ___	48 ___	49* ___
50* ___	51 ___	52 ___	53 ___	54 ___	55 ___	56 ___

Totals (40 maximal points for each column):

Organizational Work Design: _____

Organizational Culture: _____

Departmental Leadership: _____

Top Management Leadership: _____

Individual Behaviors: _____

Peer Group Behaviors: _____

Behavior Toward Supervision: _____

Interpreting the Results

Each of the Quality Readiness Survey's seven characteristics has a maximal score of 40 points. Each characteristic should be viewed as a separate entity for evaluative purposes. In other words, the method of calculating work group means is to average the sums of each characteristic for all of the surveys collected.

The Quality Readiness Survey was taken by several of the clinical nutrition departments represented on this book's panel of advisors. A total of 201 clinical nutrition personnel, representing 13 hospitals, participated in the survey. Eighty percent of the respondents were either clinical dietitians or technicians; the remaining 20 percent included management, clerical, and student staff members. Mean departmental scores, including standard deviations, are presented as Table 5–1.

A review of Table 5–1 indicates a high level of readiness for quality management initiatives on the part of the clinical nutrition departments that were surveyed. Perceived levels of positive behavior are high. In all but a single case (i.e., dietitian results for department leadership and individual behavior), behavioral scores exceeded work environment attributes. This finding suggests that other variables that affect feelings about work, such as personality and communication channels, are acting favorably on performance behavior.

A quality management program should not be initiated unless a minimum score of 20 is achieved for each behavioral characteristic. By administering the survey and comparing the results with minimal scores and those presented in Table 5–1, the nutrition manager can assess the readiness of his or her work group.

Using the Survey Results

The Quality Readiness Survey is a tool that can assist the manager in answering the following questions:

- Are the current levels of behavior likely to be conducive or detrimental in the implementation of a quality management program?
- Do behavioral scores differ significantly between work groups (e.g., dietitians, technicians)?
- Do the three areas of behavior differ significantly?
- Are opportunities to change the work environment (identified by a low score in one of the four characteristics) more likely to be found at the departmental or organizational level?
- As a department manager, are there specific changes in work design and/or leadership that can be made to enhance behaviors that are favorable toward quality management?
- As a department manager, are there opportunities to influence organizational culture (e.g., increasing participation in influential committees)?

Responses to these questions assist in developing a strategic plan for quality. The strategic plan may involve a rapid timetable for implementing quality management or action steps geared to changing behavior in preparation for introducing a quality program. The survey results are valuable in planning actions that are targeted to specific work groups, behavioral characteristics, and/or environmental influences. Once the strategic plan is implemented, the survey can be repeated to ascertain whether the desired behavioral changes have occurred and/or to monitor effectiveness of the plan over a longer period.

Assessing readiness for a quality program need not be limited to using surveys. Indeed, Esty (1992) strongly recommends that focus groups be utilized to ascertain employee readiness. She contends that staff issues can be readily identified by using this technique with a relatively small number of employees.

The Quality Work Environment

As delineated in the Quality Readiness Survey, work design plays a role in preparedness for quality management. The elements of work design include an infrastructure, or departmental structure, that complements the tenets of quality. According to McCabe (1992), one of the major elements of structure that assists in the success of quality management is vertical alignment.

Vertical alignment means linking goals, plans, and responsibilities together throughout the organization. A comprehensive evaluation of the department's current structure should be conducted by a quality-oriented leader. Specifically, the successful manager of a vertically aligned department has

- communicated the organization's and department's mission to every employee in terms that are understood and embraced

Table 5–1 Mean (Standard Deviation) Scores of Clinical Nutrition Personnel Taking the Quality Readiness Survey

Characteristic	All Personnel (n = 201)	Dietitians (n = 108)	Technicians (n = 52)
Work design	24.23 (2.26)	24.22 (2.60)	23.83 (3.39)
Culture	25.25 (2.04)	24.94 (2.77)	24.36 (2.85)
Department leadership	27.64 (2.80)	27.94 (3.41)	27.36 (4.28)
Top management leadership	23.58 (2.24)	23.65 (2.68)	23.27 (2.47)
Individual behavior	28.35 (1.50)	27.69 (1.36)	28.32 (1.90)
Behavior in work group	29.18 (1.63)	29.40 (2.04)	28.11 (2.25)
Behavior toward supervision	30.47 (2.82)	30.51 (4.35)	28.53 (3.28)

- stated how the mission will be accomplished through the development of policies and procedures that emphasize the importance of meeting customer needs, taking individual responsibility, and striving for continuous improvement
- developed strategic objectives that describe in specific terms what needs to occur for the mission to be achieved
- converted the objectives into measurable goals that include the elements of customer satisfaction, clinical outcomes, and departmental operations.

Vertical alignment is not complete, however, until the above linkage is deployed in such a way that it touches departmental members in a meaningful and ongoing way. Indeed, specific responsibilities and expectations for quality need to be written into the position descriptions and performance management tools for every departmental job classification.

REFERENCES

Esty, K. 1992. Productive work environments: How healthcare executives can stimulate productivity—an eight-step model. *Healthcare Executive* 7, no. 3:22–23.

Leebov, W., and G. Scott. 1990. *Health care managers in transition: Shifting roles and changing organizations.* San Francisco: Jossey-Bass, Inc. Publishers.

McCabe W.J. 1992. Total quality management in a hospital. *Quality Review Bulletin* 18, no. 4:134–140.

McLaughlin, C.P., and A.D. Kaluzny. 1990. Total quality management in health: Making it work. *Health Care Management Review* 15, no. 3:7–14.

Melcher, A. 1976. *Structure and process of organizations: A systems approach.* Englewood Cliffs, NJ: Prentice Hall.

Murphy, E.C., Ltd. 1991. *The quality manager report.* Amherst, NY.

Sullivan, N., and K.U. Frentzel. 1992. A patient transport pilot quality improvement team. *Quality Review Bulletin* 18, no. 6:215–221.

SUGGESTED READING

Puckett, R.P. 1992. Continuous quality improvement: Where are we going in health care? *Topics in Clinical Nutrition* 7, no. 4:60–68.

Wilson, C.K. 1991. A climate for excellence—an impossible dream? In P. Schroeder, ed. *Issues and strategies for nursing care quality*, 7–19. Gaithersburg, Md: Aspen Publishers, Inc.

Chapter 6

The Nutrition Services Environment

A successful system of quality management requires a framework in which all dietitians take responsibility for the delivery and management of high-quality nutrition care services. Internal structures must be in place as a foundation on which to build a quality management system. Quality does not occur in a vacuum. However, quality services can both flourish and continually improve within an environment where daily operations are consistent with sound principles of management and accountability.

This chapter provides an overview of the basic elements within such a framework. Some of these elements were described by Ford and Fairchild (1990) and include

- a departmental statement of mission or purpose and a strategic plan for delivery of services congruent with the mission, a plan that can be changed as needed in response to adjustments in either the internal environment or the external environment, or both
- nutritional standards of practice customized to meet the needs, resources, and milieu of the department
- written, up-to-date policies and procedures to guide major responsibilities such as screening and nutritional status assessment, care planning, documentation, nutritional counseling, consultations, responsiveness to nutritional needs of patients, team involvement, and the like
- a method and tool for screening patients to identify efficiently those at nutritional risk and to set priorities for nutritional care services

- a system of documentation that facilitates both communication and data retrieval
- a method for determining patient acuity levels as a basis for setting clinical priorities and managing both time and resources
- an observable and measurable system for tracking productivity, especially of professional staff members
- appropriate and effective staffing patterns, maximizing the potential of each dietetic team member
- criteria-based performance standards to serve as the basis for competent practice, performance appraisal, and professional development
- an evaluation system to assess and monitor compliance with mandated regulations of government, health care agencies, and commissions
- an environment in which personnel are committed to the mission statement and strategic plan, kept informed, supported with recognition and reward systems, and empowered to take responsibility for the quality of services provided.

This chapter offers a brief exploration of these elements.

MISSION OR PURPOSE

A statement of mission or purpose is the foundation of any system for the delivery of high-quality nutrition services. This truism applies whether the focus of concern is effectively managing nutrition services or managing the quality of those services.

The mission statement should be short, 15 to 20 words, encompassing the general functions of the unit (Schiller et al. 1991). Exhibit 6–1 gives two mission statements that can be used as a guide in reviewing, revising, or developing such a statement.

STRATEGIC PLAN

A strategic plan is needed to ensure that structures and services keep pace with a changing environment. For example, a system for screening to identify nutritional risks was unheard of prior to 1974, when iatrogenic malnutrition was first identified among hospitalized patients (Butterworth 1974). The prevalence of hospital malnutrition (Kamath et al. 1986), benefits of nutrition services (Smith et al. 1991), and the pressures imposed by cost containment (Smith and Smith 1988b) influenced development of procedures to identify quickly and expeditiously those most in need of professional dietetic services.

These internal and external forces created immense pressures for nutrition care services. Nonetheless, only those departments cognizant of and responsive to these forces developed strategies and changed their modes of operation in response to those pressures.

Strategic planning is a continuing process (Brylinski 1993). Through it managers stay attuned to environmental factors, identifying shifts and trends that affect their operations, set goals to ensure viability both now and in the future, and develop strategies and objectives to guide accomplishment of the stated goals. A strategic plan is essential in today's fast-paced, rapidly changing health care milieu. For example, the American Dietetic Association (1992) is positioned to ensure that any new health care reform in the United States contains provisions for health maintenance and disease prevention. Such legislation will have a tremendous impact on dietetic services. How will your department adjust its operations to take advantage of new opportunities for reimbursement, nutrition education, nutrition counseling, and wellness?

Although it is a valuable tool, the strategic plan needs to be flexible and futuristic (Klopfenstein 1993). All professionals within the department should participate in developing the plan and promote their involvement in activities associated with it. A few techniques for preparing a strategic plan can be found in the article by Brylinski (1993) and in *Handbook for Clinical Nutrition Services Management* by Schiller et al. (1991).

STANDARDS OF PRACTICE

Quality practice is impossible without standards outlining the essential components of individualized quality care to patients. These standards can be constructed as standards of practice, standards of care, or practice guidelines. The American Dietetic Association's *Standards of Practice* (1986b) provides a basis for designing nutrition care systems that meet expected standards of quality. These standards require exten-

---

Exhibit 6–1 Sample Mission Statements

CLEVELAND CLINIC FOUNDATION

The Department of Nutrition Services meets its clients' clinical nutrition needs in an efficient, effective manner. This is achieved by: evaluating individuals' needs, recommending interventions, facilitating and providing nourishment (e.g., food, parenteral solutions), and monitoring the nutrition care of patients; educating patients and others about the role of nutrition in the prevention and treatment of disease; and collaborating with others (e.g., Food Services, Nursing) to provide appealing, nutritious food choices in patient meal service operations. Cost leadership is achieved by productivity improvement strategies and tight financial controls.

The Department of Nutrition Services is dedicated to being an active, collaborative partner in support of specialized medical care. Services are provided applying current scientific principles supported by reliable research to optimize patient/client care and service.

Courtesy of The Cleveland Clinic Foundation, Cleveland, Ohio.

**THE OHIO STATE UNIVERSITY MEDICAL CENTER
DEPARTMENT OF NUTRITION & DIETETICS MISSION STATEMENT**

The Department of Nutrition & Dietetics provides optimum nutrition care and quality food service and intensive caring for our patients, staff and visitors throughout the medical complex, university and community.

The Department's Goals mirror the hospital's mission by:
- Striving to exceed patient and customer expectations by providing the best products and services
- Promoting the best working environment for our staff
- Being committed to student education and research through liaisons with the academic community
- Being leaders and innovators in our professions

Courtesy of The Department of Nutrition & Dietetics, The Ohio State University Medical Center, Columbus, Ohio.

sive work on the part of practitioners to individualize the standards and customize them for use in any particular setting. For example, consultant dietitians in health care facilities used these Standards as a framework for developing 27 outcome-oriented guidelines for quality assurance in long-term care facilities (Gilmore et al. 1993). Practice guidelines were developed recently for outpatients with non–insulin-dependent diabetes mellitus (Franz 1992). Work continues on many fronts to develop practice standards and guidelines (Finn et al. 1993).

Other published standards of practice may also be adapted for institutional use. For example, the American Society for Parenteral and Enteral Nutrition has developed nutritional support standards for hospitalized adults (1988), hospitalized pediatric patients (1989), and home therapy (1992). Dietitians at Fresno Community Hospital and Medical Center, Fresno, California have written standards of practice for clinical dietetics (Menashian 1992), which may be purchased and/or modified to meet the needs of a specific hospital.

To be useful, standards of practice need to reach the level of protocols or standards of care (Huyck and Rowe 1990) before they can truly serve as a guide for individualized quality care to patients. Such guidelines are used to

- standardize procedures and reduce unnecessary variation in patient services such as screening, assessment, care planning, documentation, and monitoring
- ensure uniform, equitable, quality care for all patients with similar characteristics, such as those with similar methods of feeding, level of care, diagnosis, or treatment (Huyck and Rowe 1990).

If no standards exist in the institution, they should be developed before launching a quality management program. It is impossible to assess and monitor quality care without first knowing what constitutes quality care in the minds of the caregivers.

POLICIES AND PROCEDURES

In the past, policy and procedure manuals were routinely scrutinized during accreditation survey visits. Thus most hospitals will have such a manual in place. The policy manual provides the basis for many quality management activities. As such, its contents should be reviewed to make sure that the following are done:

- Policies and procedures are delineated for how major aspects of care are to be carried out.
- Procedures are sufficiently detailed to provide guidance to employees who carry out the policies.
- Policies include a quality management plan.

Both Huyck and Rowe (1990) and Schiller et al. (1991) offer suggestions for developing or revising a policy and procedure manual. Exhibit 6–2 gives a sample policy and procedure for quality improvement.

METHOD OF SCREENING FOR NUTRITIONAL RISK

Many hospitals do not regularly screen patients for nutritional risk, and in other institutions the process, although in place, could be improved (Foltz et al. 1993). An effective screening program should be quick, reliable, and cost effective (Smith and Smith 1988a). These characteristics mean that screening

- occurs at least within 24 to 48 hours of hospital admission (DeHoog 1985). The screening process should take no more than five to ten minutes (Hunt et al. 1985).
- identifies those likely to benefit from nutritional intervention, as well as those who do not need the services of a registered dietitian. Depending on circumstances of the particular institution, screening criteria will be sufficiently sensitive to identify nutritional problems without being so specific that costs are exorbitant. These measures of sensitivity and specificity are usually met by including in the screening process a limited diet history, quick review of the medical record, and diagnosis (Smith and Smith 1988a).
- saves money spent on unnecessary screening and comprehensive nutritional assessments. Patients hospitalized fewer than three days are not screened, and only those with known nutritional risk factors undergo a complete nutritional assessment.
- identifies early those who need nutritional care, thereby allowing early intervention for best results
- identifies patients who might qualify for extra payments for co-morbid or complicating conditions
- pinpoints diagnoses for which nutritional intervention improves patient outcomes, including fewer infections and complications and shortened hospital stay
- effectively utilizes time of both dietitians and support personnel.

Nutritional screening is facilitated by new procedures recently disseminated. One-page forms were developed for rapid screening; any health professional can complete the process and refer potential at-risk individuals to a dietitian for assessment (*Nutrition Interventions Manual* 1992; White et al. 1992). Computerized programs are also available for nutrition screening of hospitalized patients (Grossbauer 1993).

Dietetic literature includes numerous recommendations for items to be included in screening procedures (Christensen and Gstundtner 1985; Campbell and Behm 1988; DeHoog 1985, 1988; Smith and Smith 1988a; Ford and Fairchild 1990). At Yale-New Haven Hospital (Ford and Fairchild 1990), screening includes seven risk factors, defined in Exhibit 6–3. Others include recent changes in appetite and food intake, weight history, and changes in gastrointestinal function such as diarrhea (Campbell and Behm 1988). Depending on personnel re-

Exhibit 6–2 Sample Policy and Procedure for Quality Improvement

---

POLICY

There is a quality improvement program which monitors and evaluates key nutrition care functions in an effort to continuously improve the level of quality provided to patients and others receiving this service.

PROCEDURES

1. Standards are based on the facility-wide quality improvement program and the Ten Step process for monitoring and evaluation is used to implement this program.
2. All key nutrition care functions are monitored with special emphasis on important and critical aspects of care.
3. Program ratification, evaluation of results, and corrective action are determined by a multidisciplinary team of individuals who can best contribute to the program.
4. The multidisciplinary team consists of individuals from the facility's Nutrition Committee which consists of representatives from medicine, nursing, nutrition, pharmacy, administration, occupational therapy, speech therapy, and patient representative or ombudsman. Quality improvement issues are addressed by this team on at least a quarterly basis and other team members are elicited as determined appropriate such as experts on safety and infection control as needed.
5. An annual schedule is followed for monitoring and evaluation activities.
6. The director of nutrition care is responsible for participating in program development and integrating quality improvement activities in the routine management functions of the department to include training, scheduling, performance appraisal, financial allocations, procurement of goods and services, decision making, and other processes.

Source: Reprinted with permission of American Nutri-Tech, Inc. from *Continuous Quality Improvement for Nutrition Care.* Copyright © 1992 by Rita Jackson.

---

sources and patient loads, each institution needs to determine which screening criteria meet its individualized needs for sensitivity and specificity. The chosen criteria should be included in the policy and procedures for screening.

METHOD FOR DETERMINING PATIENT ACUITY LEVELS

Patient acuity levels as a basis for allocating dietetic resources were first described by Lutton et al. (1985). Since that time, others have developed their own systems based on characteristics and needs of specific institutions. At The Ohio State University Medical Center, four nutrition care levels are used (Campbell and Behm 1988), whereas Yale-New Haven Hospital has developed a seven-level classification system (Ford et al. 1990).

Patient acuity levels have been used in nursing services to determine staffing levels and to describe conditions that may limit practitioners' ability to achieve desired outcomes of professional intervention. A project was initiated under the auspices of the American Dietetic Association to develop similar standardized acuity levels for use in dietetic practice (Lutton 1990). Results of this project, unfinished at the time of publication, may be used to develop or revise patient acuity levels and improve patient outcomes.

DOCUMENTATION PROCEDURES

Medical record entries provide legal documentation of care provided. Charting should conform to institutional guidelines; entries should be brief, clear, and precise. Grant and DeHoog (1991) delineated some practical guidelines for charting, shown in Exhibit 6–4. The problem-oriented medical record and SOAP (subjective-objective-assessment-plan) format provide a framework for organizing and recording pertinent information. Exhibit 6–5 highlights content for each section of the SOAP format as well as specific suggestions for writing objective, assessment, evaluative, and intervention information. A checklist for evaluating medical record entries can be found in Appendix D-1.

Not all institutions use SOAP formats because they tend to foster repetitiveness. As an alternative, the problem, intervention, and evaluation (PIE) method of documentation has been recommended (Schiller et al. 1991; Scollard 1993). In this system, shown in Exhibit 6–6, assessment information is consolidated and a problem statement formulated as the basis for care. Interventions (such as patient education or specialized nutrition support) are documented, and results are both evaluated and documented. This approach encourages practitioners to remain alert to the effects of nutritional therapy and to modify plans when the intervention fails to solve the underlying nutritional problem adequately.

METHOD TO TRACK CLINICAL PRODUCTIVITY

Productivity is measured as a ratio of inputs over outputs. In food service management, productivity is measured by such things as meals per labor hours worked or total labor hours per patient day (Brown and Hoover 1990). Tracking productivity requires both a measurement system and a monitoring system.

Exhibit 6–3 Screening Criteria: Risk Factors, Definitions, and Examples

Risk Factors	Definitions	Examples
1. Age	< 18 or > 50 years of age.	Children, the elderly.
2. Diagnosis/Treatment	Highly probable effect on nutritional status or nutritional requirements.	Cancer of the tongue, pregnancy, radiation therapy to the abdomen, intestinal resection.
3. Diet	Medically advised/ordered restrictions or modifications, or self-imposed dietary restrictions that are not inherently nutritionally balanced or appropriate, or diets that require additional monitoring to assure Cal. and pro. needs are met.	2 gm. sodium diet; macrobiotic diet; regular diet for a diabetic; NPO or clear liquids > 3 days; high protein, high Calorie diet.
4. Metabolic or Mechanical Problems	Highly probable effect on food ingestion, digestion, absorption, and/or utilization.	Diabetes, coronary artery disease, ulcerative colitis, obesity, alcoholism, persistent vomiting, blindness.
5. Significant Lab Data	Abnormal indices reflecting conditions that can require dietary modification or may reflect abnormal nutritional status.	High glucose, low albumin, high amylase/lipase, high BUN/creatinine, low Hct/Hgb.
6. Pertinent Medications	Those that: a. require additional monitoring of lab indices to determine efficacy of, or need for, dietary intervention. b. include or preclude the ingestion of certain foods. c. may or do seriously affect the ability to ingest, digest, absorb or utilize nutrients.	Insulin; potassium-losing diuretics; certain antidepressants, antibiotics, or chemotherapeutic agents.

7. Weight for Height	a. Adults > 10% underweight.
	            > 20% overweight.
	            > 10% weight loss over past 6 months.
	
	b. Pediatrics < 25th % wt/ht
	                < 25th % wt/age or ht/age
	                > 95th % of wt/ht
	                > 95th % wt/age
	Any weight loss or growth deceleration
		5'4" female weighing 90 lb.;
		5'9" male weighing 215 lb.;
		6' male, pre-illness weight of 190 lb., now weighs 160 lb.
		
		5 y.o. male weighing 14.5 kg at 101 cm height
		3 y.o. male weighing 12 kg at 90 cm height
		8 y.o. male weighing 53 kg at 115 cm height
		6 y.o. male weighing 30 kg

Source: Reprinted from Ford, D.A., and Fairchild, M.M., Managing inpatient clinical nutrition services: A comprehensive program assures accountability and success, *Journal of the American Dietetic Association*, Vol. 90, pp. 695–702, with permission of the American Dietetic Association, © 1990.

Measuring Clinical Productivity

Productivity can be measured in terms of labor, materials, and energy. Most clinical dietetics productivity systems focus on labor because labor costs constitute a large majority of clinical nutrition costs in hospitals, and because labor can be measured in units that do not vary with time (American Dietetic Association 1986a). Thus clinical productivity programs are based on such ratios as

- ratio of labor hours or minutes per patient day
- patient visits or counseling sessions per dietitian hours
- dollars billed per dietitian hours

Methods for establishing productivity standards can be found in articles by DeHoog (1985) and McManners and Barina (1984), and in *Productivity Management for Nutrition Care Systems* (American Dietetic Association 1986a). Because productivity is greatly influenced by both the percentage of seriously ill patients and the time needed to provide appropriate services for patients at different levels of acuity, many institutions base productivity standards on acuity levels (Chafin 1990; DeHoog 1985; Ford and Fairchild 1990; McManners and Barina 1984). Productivity standards of 95 percent to 100 percent have been established (Chafin 1990).

Productivity levels ought not be based on what is currently done. Rather, clinical managers with input from staff dietitians should ask the following questions:

- What work **should be** done?
- Who is the least expensive but appropriately skilled individual to complete this work?

Exhibit 6–4 General Guidelines for Charting

1. Medical records are permanent legal documents therefore all entries should be written in black pen or typewritten. No soft felt pens, multicolored pens or pencils should be used.
2. Documentation should be complete, clear and concise.
3. Documentation must be legible. Printing is preferred. If any entry cannot be read, it can be argued the services did not occur.
4. Charting must be accurate.
5. Entries should be documented by service, date and time. Each medical record page should be identified by the patient's stamp or written name and hospital number.
6. Entries should be in chronological order and on consecutive lines.
7. No entry should be made in advance of the "procedure" performed.
8. The first word of every statement should be capitalized. Periods should be used at the completion of each thought. Complete sentences are not necessary. Grammar and spelling must be correct.
9. All entries should be consistent and non-contradictory.
10. Only institution-approved and authorized abbreviations should be used. Abbreviations with duplicate meanings should be avoided. Each institution should have a document for abbreviations.
11. All entries must be signed at the end of the chart note. The signature should include the first name initial, complete surname and status (e.g., R. Brown, MS.RD)
12. No one should ever chart or sign the medical record for another individual.
13. All student notes must be co-signed either by the clinical staff or the clinical instructor. Never countersign an entry without reading or confirming the entry for accuracy. A counter-signature attests to the authenticity and shares equal responsibility.
14. Charting must be objective and void of conclusions. Words should be carefully chosen so there is not any room for conjecture, doubt, or misunderstanding.
15. Charting must be specific. Many words can have different meanings.
16. Personal positions or points should never be argued in the chart. Neither should criticizing or casting doubt on the professionalism of others be in the documentation.
17. Time gaps must be avoided. Documentation must occur as soon as the actual procedure/service is rendered. The frequency is dependent upon the patient's degree of illness, therapies administered and standards of nutritional care.
18. Late entries should be identified as such, reflecting the date and time of the actual entry and then state in the date and time the entry should have been recorded.
19. Medical record entries should never be made non-legible.

Source: Reprinted from *Nutritional Assessment and Support*, 4th ed., by A. Grant and S. De Hoog, p. 221–222, with permission of Susan De Hoog, © 1991.

Enhanced efficiencies may be realized by aligning task assignments with the best-suited personnel. Job descriptions may need revisions to reflect roles consistent with delivery of high-quality nutrition services in a tough economic environment.

To establish productivity levels, three steps are necessary.

1. Identify duties and services consistent with the departmental mission and strategic plan. This is best done by delineating tasks enumerated in updated job descriptions of clinical dietitians, dietetic technicians, and diet aides/clerks. Fischer and Olmstead (1981) conducted a review of direct patient care activities and identified the following areas of responsibility and the estimated time needed to complete each task at their institution:

   - Menu production—collecting, marking, checking, and changing menus; forecasting food orders; and calculating meal patterns and tube feedings. Estimated time was 11 minutes per menu.
   - Basic care requirements—screening, basic chart review, initial patient interview, and care plan. Estimated time was 25 minutes per patient.
   - Diet therapy—in addition to "basic care," patients who require professional intervention need an assessment and care plan (25 minutes minimum), possible nutritional counseling (average 30 minutes), evaluation (5 minutes minimum), and documentation (average 15 minutes). The total time requirement was 1 hour 15 minutes minimum.
   - Metabolic support—services for patients requiring specialized nutritional support may include anthropometrical or metabolic cart measurements, thorough review of medical record, interpretation of assessment results, care plan, calculation of nutrient needs, documentation, daily monitoring, and reassessment. Total time requirements varied, but the minimum for each patient (average 15-day hospital stay) was 225 minutes.

   Personnel may be responsible for non–patient activities, such as supervising students, participating in quality management activities, engaging in research, supervising employees, and maintaining professional competence. These responsibilities should also be included in a listing of duties to determine productivity levels.

2. Identify time allocated per task. To complete this step, first determine procedures necessary to complete designated tasks and the amount of time required for each. Categories of activity as given above can be used as a guide, but both activities and times should be congruent with policies and standards of practice for each institution. For example, DeHoog (1985) determined that at University Hospital, Seattle, Washington, 0.80 relative

Exhibit 6–5 What To Include in Charting/Documentation

Charting provides legal documentation of nutrition care provided, timely written communication for nutritional status, comprehensive assessment regarding nutritional support or nutrition education for patients and/or family; written communication regarding nutritional goals of therapy and documentation for reimbursement.

All documentation should be written in the SOAP style:

Subjective
- Information provided by patient, family or caretaker.
- Significant nutritional history.
- Pertinent socio-economic, cultural information.
- Level of physical activity.
- Current dietary intake (in terms of nutrients).

Objective
- Factual, reproducible observations (i.e., anthropometrics and lab data).
- Height, weight and age.
- Desirable weight or a realistic goal.
- Current diet order.
- Nutritionally pertinent medications.

Assessment
- Interpretation of patient's status based on subjective and objective data.
- Evaluation of nutritional history.
- Estimation of nutritional requirements.
- Projected rate of weight gain or loss, if appropriate.
- Assessment of comprehension and motivation, if appropriate.
- Assessment of diet order.
- Anticipated problems and/or difficulties for patient compliance/adherence.

Plans
- Diagnostic
  - Suggestions for gaining further useful subjective or objective data.
  - Further workup, data gathering, consultations, etc.
- Therapeutic
  - Goal of nutritional therapy.
  - Recommendations for nutritional care.
  - Any referrals.
  - Follow-up.
- Patient education
  - Brief description of specific written or verbal instruction provided.
  - Description of specific recommendations (i.e., vitamin/mineral supplementation, supplemental feedings, etc.).

Within 24–48 hours post screening a nutritional assessment and care plan should be documented on the appropriate progress note. Note may include:

1. Objective data
   - Age
   - Height (include growth standard for age if less than 18 years)
   - IBW; % of IBW
   - Usual weight (or pre-illness weight); % of usual weight
   - Current weight
   - Diet history
   - Current diet prescription
2. Evaluation of pertinent clinical laboratory data
3. Assessment of functioning/nonfunctional gastrointestinal tract
   Indicating evidence of maldigestion and/or malabsorption (nausea, vomiting, diarrhea, fistula obstruction, severe Protein Calorie Malnutrition).
4. Evaluation of factors which may affect nutrient intake, digestion, absorption.
   - Medications
   - Previous surgeries
   - On-going treatment modalities
   - Chronic disease processes
5. Statement of goal of nutritional therapy and/or outcome, for example to:
   - Stop weight loss
   - Increase anabolism
   - Prevent negative nitrogen balance
   - Promote wound healing and weight maintenance
   - Minimize nitrogen wasting
   - Maintain body weight and protein stores
6. Statement of evaluation of appropriateness of current diet order.
7. Documentation of nutritional needs:
   - Estimated Basal Energy Expenditure
   - Estimated Energy needs; Kcal/kg
   - Estimated Protein needs; gm/kg IBW
   - Vitamin/mineral supplement, if appropriate
   - Estimated fluid needs
8. Development/documentation of a nutritional care plan
   Indicating type of nutritional support and its implementation (dietary modifications/delivery system oral, enteral or parenteral).
9. Follow-up documentation
   Statement of progress of nutritional status weekly to include:
   - Nutrient Intake Analysis of 3 days, if appropriate
   - Patient's weight, weight patterns
   - Significant I & O changes
   - Pertinent lab changes
   - Any changes in the nutrition prescription
   - Status: where the patient is in meeting goals
   - Statement of goal of therapy and/or outcome
   - Documentation of patient's reaction/tolerance of dietary modification
10. Statement of patient education, if appropriate, to include:
    - Patient and/or significant other has been instructed on prescribed needs
    - Statement of expected adherence to prescribed diet
    - Patients discharged on medications with food/drug interactions
    - Statement of an expected outcome/goal (for reimbursement purposes)
11. Discharge documentation to include:
    - Summary of nutritional therapies
    - Outcome of nutritional therapies
    - Pertinent information on weights, labs, intake
    - Statement on expected "progress"
    - Statement on discharge summary submitted for follow-up

Source: Reprinted from *Nutritional Assessment and Support*, 4th ed., by A. Grant and S. De Hoog, pp. 218–220, with permission of Susan De Hoog, © 1991.

Exhibit 6–6 Example of Problem, Intervention, and Evaluation (PIE) System of Documentation

---

6/1/93 **Problem:** Eight days of dysphasia secondary to cranial nerve dysfunction; violent coughing when attempting to take liquids. Aspiration pneumonia. Nutritional assessment reveals weight loss (10 kg in 2 weeks) and visceral protein depletion (albumin 3.1 g/dL; iron-binding capacity 215 µg/dL).
Intervention: Goal is protein repletion and weight/fat maintenance. Enteral feeding required: 92 g of protein per day, 1380 kcal of nonprotein per day, and 1746 mL of fluid per day through J tube. Begin with 35 mL/h standard isotonic formula first 24 hours. Flush tube with 60 mL of water 4 × per day. Monitor for tolerance.
6/2/93 **Intervention:** Increase standard isotonic formula to 60 mL/h. Flush tube with 60 mL of water 4 × per day.
Evaluation: No signs of intolerance to tube feeding.
6/3/93 **Intervention:** Increase formula to 65 mL/h and add 25 g of modular protein and 360 mL of extra water. Also flush tube with 35 mL of water 4 × per day.
Evaluation: Passed three watery stools 6/2/93. Monitor closely.
6/4/93 **Evaluation:** Two watery stools 6/3/93. Additional 1-kg weight loss.
Intervention: Continue enteral feeding at goal rate. Begin discharge planning and education.

---

value units (RVU) (0.1 RVU = 10 minutes) was required to complete one initial nutrition assessment for a high-risk patient based on the following activities:

Procedure	RVU
Screening	0.05
Data gathering/chart review	0.15
Nutritional history/diet history evaluations	0.20
Developing nutrition care plan	0.15
Charting	0.05
Consultation (M.D., R.N., others)	0.20
TOTAL = 0.80 RVU, or 1 hour and 20 minutes	

This type of analysis requires predetermination of a classification system to differentiate time needed to provide services for patients at various acuity levels. DeHoog developed the form shown in Appendix D-2 to use in determining time requirements for a particular institution.

3. Develop ratios or standard times needed for individuals in different positions (dietitians, technicians, aides/clerks) to complete designated tasks. These standards are then used to monitor productivity.

Monitoring Clinical Productivity

Once established, productivity levels of clinical staff members must be monitored. This is accomplished by having each dietitian keep a record (collected weekly or monthly) of specified activities such as those described by Ford and Fairchild (1990):

- patient admissions (Adm)
- discharges (out >3 days, out <3 days)
- transfers (Tx)
- number of patients classified
- number of initial assessments (IA)
- number of metabolic profiles (MP)
- number of follow-up notes (SOAP or F/U)
- number of charted diet instruction

The forms used to collect this information are shown in Exhibit 6–7 and Exhibit 6–8.

Using information generated in monthly reports, it is possible to assess both individual and departmental productivity. Results provide a basis for justifying clinical staff as well as for goal setting. For example, Huyck and McNamara (1987) found that in their 1984 study only 22 percent of clinical staff time was spent with patients; each nutrition assessment averaged 48 minutes, and inpatient nutrition counseling required 62 minutes. These data can be compared with institutional norms, as well as results at other institutions, to determine the need to streamline activities, set new priorities, redistribute workloads, or make other changes deemed appropriate.

Another way to calculate productivity levels is to use institutional data. Table 6–1 is an example of staffing and workload information compiled into a monthly productivity report for the clinical nutrition area at Ohio State University Medical Center. Two primary factors that influence the workload of the clinical staff are the number of inpatient days and the number of outpatient clinic visits. These are referred to as variable standards, since they change according to patient census. The department of nutrition and dietetics worked with the department of strategic planning to develop equations to determine the impact these statistics have on clinical workload and their effect on the number of consultations, diet instructions, and so on.

Each month these variable standards are added to a predetermined fixed standard. The fixed standard reflects job tasks that do not fluctuate according to the census. Providing student education and attending discharge rounds are examples

Exhibit 6–7 Monthly Clinical Operations Statistics Sheet for Admissions, Transfers, and Discharges

Service: _____

Date	Day	Admissions and Transfers	Out > 3	Out < 3
	Tu			
	W			
	Th			
	F			
	Sa			
	Su			
	M			
	Tu			
	W			

Source: Reprinted from Ford, D.A., and Fairchild, M.M., Managing inpatient clinical nutrition services: A comprehensive program assures accountability and success, *Journal of the American Dietetic Association*, Vol. 90, pp. 695–702, with permission of the American Dietetic Association, © 1990.

of tasks included in the fixed standard. The result of these calculations reveals how many hours the staff has "earned" to provide clinical nutrition services. Earned hours are divided by the number of hours the staff actually worked during the same time frame. This determines the "productivity rate." The desirable range for the productivity rate is between 0.95 and 1.05. Productivity rate is one of the determining factors used when the clinical area is filling vacancies, requesting temporary employees, planning or proposing new programs, or making other hiring decisions.

Productivity can be monitored using a departmental computerized system (Leyshock and Tracy 1992). The system in place at Westmoreland Hospital, Greenburg, Pennsylvania, uses a matrix correlating patient acuity and time needed to provide appropriate care. Data can be categorized and printed in numerous formats; these are used for problem identification and analysis.

For example, one set of monthly statistics uncovered two important problem areas. First, dietitian time was insufficient to provide prescribed care for the increased workload on Mondays and Fridays. This finding led to greater staff coverage on high-volume days. Second, one of the reasons the workload was so high on Monday was because high-risk patients were often admitted on weekends. This triggered a feasibility study and justification for dietitian availability on weekends for high-risk patient consultations and follow-ups. Monthly reports in this system are used to plan effective workload distributions, to ensure that staff time is used appropriately, and to track quality improvement indicators.

STAFFING PATTERNS

Staffing needs are directly linked with activities required to identify patients at nutritional risk and provide appropriate nutritional intervention procedures. Positions for both professional and support personnel are justified by demonstrating their required roles in maintaining departmental standards of practice (Ford and Fairchild 1990).

Two different approaches may be used to develop staffing models; both are based on productivity requirements.

1. Staffing analysis can be conducted by using volume data for numbers of patients within various classifications and the written policy and procedure manual as

Exhibit 6–8 Monthly Clinical Nutrition Productivity Statistics Sheet

Service: _____

Date	Visits	IA						MP		SOAP or F/U	Charted DI	Kcal Count	Kardex Class.
		I	II	IIIA	IIIB	IIIC	IIID	IV	III	IV				

Source: Reprinted from Ford, D.A., and Fairchild, M.M., Managing inpatient clinical nutrition services: A comprehensive program assures accountability and success, *Journal of the American Dietetic Association*, Vol. 90, pp. 695–702, with permission of the American Dietetic Association, © 1990.

Table 6–1 Institutional Productivity Report

	Clinical Units of Service[1]				
Work Unit	August	September	October	Current Month November*	Current YTD
Variable Standard[2]	.1170	.1170	.1170	.1170	.1170
Fixed Standard[3]	1555.27	1555.27	1555.27	1555.27	7776.35
Total Work Unit Volume[4]	18980.00	17398.00	18039.00	17544.00	88979.00
Productivity Rate[5]	1.03	0.97	0.95	1.15	1.01
Earned FTE	21.37	21.00	20.75	21.10	20.85
Earned Hours[6]	3775.93	3590.83	3665.83	3607.91	18186.89
Total Productivity Hours[7]	3655.83	3672.50	3834.80	3119.50	17986.03
Regular Hours	3515.46	3463.10	3687.20	2881.00	17041.35
Overpercent Hours	49.04	74.40	44.80	66.10	297.78
Overtime Hours	−2.76	28.30	−1.60	19.30	57.52
Total Worked Hours	3561.74	3565.80	3730.40	2966.40	17396.65
Contract Hours	0.00	0.00	0.00	0.00	0.00
Comp Hours	94.09	106.70	104.40	153.10	589.38

[1]Unit of service: patient days (derived from inpatient days and outpatient nutrition clinic visits).
[2]Variable standard: 0.1170 hours of clinical staff time spent per unit of service, varies according to monthly census data.
[3]Fixed standard: 1,555.27 hours of clinical staff time per month spent on routine day-to-day operations that are not significantly affected by patient census.
[4]Total work unit volume: sum of units of service or patient days.
[5]Productivity rate = earned hours ÷ total productivity hours
    $1.15 = 3607.91 ÷ 3119.50$
[6]Earned hours = variable standard × total work unit volume + fixed standard
    $3607.91 = 0.1170 × 17544 + 1555.27$
[7]Total productivity hours: sum of all hours worked.
*Calculations based on figures from the month of November.

Courtesy of The Department of Nutrition & Dietetics and The Department of Strategic Planning, The Ohio State University Medical Center, Columbus, Ohio.

---

the basis for standards of quality care expected at the institution (Huyck and McNamara 1987; Ford and Fairchild 1990). A time utilization study can reveal both the time required to complete designated tasks and deficiencies in performance that may be due to inadequate staffing. Results may lead to a reorganization of duties, clarification of standards and/or policies and procedures, or justification of new positions. For example, after initiating an integrated system of nutrition services, the number of dietetic technicians at Yale-New Haven Hospital was increased from three to seven (Ford and Fairchild 1990).

2. Joint Commission criteria can be used as the basis for defining productivity standards and determining clinical staffing needs (McManners and Barina 1984). Again, results may lead to adjustments in dietitian schedules, procedural changes, and justification of new staff positions.

The often-quoted "usual" dietitian-to-patient ratio of 1 to 100 was established more than a half-century ago (MacEachern 1935, 469–471), a time before many of today's standards, procedures, complex patient care systems, and mandated regulations were in place. Thus it is inappropriate to base staffing needs on such a norm. The American Dietetic Association's *Clinical Dietetic Staffing Kit* (1982) can be used as a guide to determine total full-time equivalents (FTEs) needed to provide services in line with current institutional standards of quality care. Since this dated staffing kit does not take into account acuity levels, internal techniques must be used to address recent advances in dietetic practice.

CRITERIA-BASED PERFORMANCE STANDARDS

Assignment of specific tasks and responsibilities based on policies and procedures, job descriptions, time requirements, and productivity standards can be translated into objective standards of performance. These standards can be used as the basis for quality management and monitoring activities, performance appraisals, programs to enhance professional growth and responsibility, and measures to ensure competence in providing patient care (Ford and Fairchild 1990). Criteria-based performance standards clearly delineate what should be done, how long it should take, desired processes, and expected outcomes.

MONITORING AND EVALUATION SYSTEM

Quality management by any name (quality assurance, quality improvement, total quality management) is a structured system for monitoring and evaluating performance in light of established standards and for taking steps to correct any observed variances. Methods for development and use of such a system are the topic of this manual. Procedures, guidelines, and examples of system components are given for the Joint Commission's quality improvement process, since this approach is recommended in current accreditation manuals.

SUPPORTIVE ENVIRONMENT

Quality improvement can most easily occur in departments in which both management and employees share a common vision and work toward achievement of the same goals. Clinical nutrition managers are challenged, therefore, to create and maintain an environment in which personnel are

- encouraged to take personal responsibility for the quality of their actions
- empowered to make decisions, especially in those matters that directly affect their work
- recognized and rewarded for their accomplishments, particularly as they relate to quality improvement
- kept informed of quality issues, performance, and progress
- involved in setting standards and solving problems related to substandard performance
- supported in their efforts to achieve personal goals congruent with organizational goals and to seek self-fulfillment through their work.

It is obvious that excellence in dietetics is more than setting up a system to measure quality practice; it is an integrated approach to total departmental effectiveness. Such effectiveness requires both commitment and transformational leadership, attributes that are discussed elsewhere in this manual.

REFERENCES

American Dietetic Association. 1982. *Clinical dietetic staffing kit.* Chicago.

American Dietetic Association. 1986a. *Productivity management for nutrition care systems.* Chicago.

American Dietetic Association. 1986b. *Standards of practice: A practitioner's guide to implementation.* Chicago.

American Dietetic Association. 1992. Economic benefits of nutrition services: Executive summary of the legislative platform of the American Dietetic Association. Chicago.

American Society for Parenteral and Enteral Nutrition. 1988. Standards for nutrition support: Hospitalized patients. *Nutrition in Clinical Practice* 3:28–31.

American Society for Parenteral and Enteral Nutrition. 1989. Standards for nutrition support: Hospitalized pediatric patients. *Nutrition Clinical Practice* 4:33–37.

American Society for Parenteral and Enteral Nutrition. 1982. Standards for home nutrition support. *Nutrition Clinical Practice* 7:65–69.

Brown, M.D.M., and L.W. Hoover. 1990. Productivity measurement in food service: Past accomplishments—a future alternative. *Journal of the American Dietetic Association* 90:973–981.

Brylinski, C. 1993. Strategic planning: An overview. *Clinical Nutrition Management Newsletter* 12, no. 1:1–4.

Butterworth, C.E. 1974. The skeleton in the hospital closet. *Nutrition Today* 9, no. 2:4–8.

Campbell, S.M., and V.A. Behm. 1988. The dietetic team: A step toward progressive clinical dietetic practice. *Dietetic Currents* 15, no. 4:17–22.

Chafin, S. 1990. Acuity-based productivity standards. *Clinical Nutrition Management Newsletter* 9, no. 2:1–2. June.

Christensen, K.S., and K.M. Gstundtner. 1985. Hospital-wide screening improves basis for nutrition intervention. *Journal of the American Dietetic Association* 85:704–706.

DeHoog, S. 1985. Identifying patients at nutritional risk and determining clinical productivity: Essentials for an effective nutrition care program. *Journal of the American Dietetic Association* 85:1620–1622.

DeHoog, S. 1988. Nutritional screening and assessment in a university hospital. In *Nutrition screening and assessment as components of hospital admission*, ed. J.D. Gussler, 2–8. The eighth Ross roundtable on medical issues. Columbus, Ohio: Ross Laboratories.

Fischer, K.H., and M. Olmsted. 1981. *Standards of dietetic care: A foundation for quality assurance.* Loma Linda, Calif: Seventh-Day Adventist Dietetic Association.

Foltz, M.B., et al. 1993. Nutrition screening and assessment : Current practices and dietitians' leadership roles. *Journal of the American Dietetic Association* 93:1388–1395.

Ford, D.A., and M.M. Fairchild. 1990. Managing inpatient clinical nutrition services: A comprehensive program assures accountability and success. *Journal of the American Dietetic Association* 90:695–702.

Ford, D.A., et al. 1990. *The Yale-New Haven Hospital (YNHH) nutritional classification and assessment system.* New Haven, Conn: Yale-New Haven Hospital.

Grant, A., and S. DeHoog. 1991. *Nutritional assessment and support.* 4th ed. Seattle, Wash: Authors.

Gilmore, S.A., et al. 1993. Standards of practice criteria: Consultant dietitians in health care facilities. *Journal of the American Dietetic Association* 93:305–308.

Grossbauer, S. 1993. Nutrition screening for Medicare reimbursements. *FoodService Director* 6, no. 4:154.

Hunt, D.R., et al. 1985. A simple nutrition screening procedure for hospital patients. *Journal of the American Dietetic Association* 85:332–335.

Huyck, N.I., and P.M. McNamara. 1987. Monitoring accountability of a clinical nutrition service. *Journal of the American Dietetic Association* 87:620–623.

Huyck, N.I., and M.M. Rowe. 1990. *Managing clinical nutrition services.* Gaithersburg, Md: Aspen Publishers, Inc.

Kamath, S.K., et al. 1986. Hospital malnutrition: A 33-hospital screening study. *Journal of the American Dietetic Association* 86:203–206.

Klopfenstein, J.D. 1993. Positioning clinical nutrition services for a dynamic future. *Topics in Clinical Nutrition* 8, no. 3:33–38.

Leyshock, P.J., and D.L. Tracy. 1992. Using a computerized system to monitor clinical dietetic productivity. *Topics in Clinical Nutrition* 7, no. 2:69–77.

Lutton, S.E. 1990. From the chair. *Clinical Nutrition Management Newsletter* 9, no. 2:2. June.

Lutton, S.E., et al. 1985. Levels of patient nutrition care for use in clinical decision making. *Journal of the American Dietetic Association* 85:849–852.

MacEachern, M.T. 1935. *Hospital organization and management.* Chicago: Physicians' Record Co.

McManners, M.H., and S.A. Barina. 1984. Productivity in clinical dietetics. *Journal of the American Dietetic Association* 84:1035–1041.

Menashian, L. 1992. *Standards of practice: Guidelines for the practice of clinical dietetics.* Fresno, Calif: Fresno Community Hospital and Medical Center.

Nutrition interventions manual for professionals caring for older Americans: Executive summary. 1992. Washington, DC: Greer, Margolis, Mitchell, Grunwald & Associates, Inc.

Schiller, M.R., et al. 1991. *Handbook for clinical nutrition services management.* Gaithersburg, Md: Aspen Publishers, Inc.

Scollard, T.A. 1993. Diagnostic charting: Increased identification of malnutrition. *Clinical Nutrition Management Newsletter* 12, no. 4:1–4

Smith, P., and A. Smith. 1988a. *Screening for hospital malnutrition: Implications for the management and marketing of nutritional care.* Research monograph no. 1. Deerfield, Ill: Clintech Nutrition Co.

Smith, P., and A. Smith. 1988b. *Superior nutritional care cuts hospital costs.* Chicago: Nutritional Care Management Institute.

Smith, P., et al. 1991. *Nutritional care cuts private-pay hospital days.* Chicago: Nutritional Care Management Institute.

White, J.V., et al. 1992. Nutrition screening initiative: Development and implementation of the public awareness checklist and screening tools. *Journal of the American Dietetic Association* 92:163–167.

Chapter 7

Improving Quality through Group Processes

A shared tenet of all quality programs is a management style that encourages those who perform the work to be involved in making decisions about the products and services they produce. This implies the use of group problem-solving and process improvement techniques. In order to have a successful quality management program, it is important to know and be able to use the skills that facilitate group interactions. This chapter focuses on tools and techniques that enhance group efforts.

Effective, efficient work groups do not just happen. Successful work groups share several characteristics, including the following:

- a clear understanding of purpose
- a composition of members that has been carefully selected
- education and training in the quality tools that assist in group problem solving

THE QUALITY MISSION STATEMENT

The development of a quality mission statement is advocated by many experts in the field. The purpose of the mission statement is to state publicly the organization's commitment to its quality initiative. According to James (1990), the statement should include the following:

- a commitment to constancy of purpose
- a dedication to continuous improvement
- a focus on customers (both internal and external)
- an understanding of the products and services provided
- measurement systems that evaluate quality

At the same time, the quality mission statement should be easily understood, applicable to all employees within the organization, and simple enough to be easily memorized (Orme and Parsons 1992).

Quality mission statements generally originate at the top management levels of the hospital—either the board of trustees or the executive management group. There is nothing wrong, however, with the department's adapting the organization's statement. Doing so creates ownership and a sense of integration with the hospital as a whole.

If the quality initiative begins at the department level, development of a quality mission statement is a good first step. It is important to keep in mind, however, that the departmental statement may need to be revised if and when the larger organization implements a program.

Likewise, the development of a mission statement by individual work groups is often recommended. Creating the statement provides the opportunity for group members to align their mission to that of the department and/or hospital, to explore and commit to their reason for being, and to begin the process of defining their identity as a team.

PROBLEM-SOLVING WORK GROUPS

The use of teams or work groups to solve problems is an integral part of quality programs. There are different ways of implementing this aspect of the program, however. The method of choice is best made at the organization or department level. Two specific variations of problem-solving teams are explored here: quality circles and task forces.

Quality Circles

A quality circle, for our purposes, is defined as an ongoing group of employees who systematically work together to identify and solve problems that affect their work. The problems that are addressed by these groups can include such issues as safety, morale, productivity, and costs. Quality circles are used extensively in the American manufacturing setting.

Task Forces

A task force is described as a structured group of employees who join together to solve a specific problem systematically. The issues that are addressed by task forces are comparable to those solved by quality circles. Moore and Miller-Kovach (1988) have described the use of this method as it applies to the provision of nutrition services.

Common Characteristics

Quality circles and task forces have several characteristics in common. In fact, they are more alike than dissimilar. Common characteristics include the following:

- a formalized structure
- use of a systematic problem-solving process
- predominantly voluntary participation
- management-supported meetings
- similar group size

Both quality circles and task forces have a formalized group structure. Membership is designed to focus on those employees who do the work. Indeed, the majority of groups are made up of individuals who have specific, direct knowledge of the work processes and tasks that are being addressed. Each participant has a role to fill, and these roles are clearly defined for all members. There are some minor differences in specific roles between task forces and quality circles that are addressed later.

Quality circles and task forces also share a belief in using problem-solving processes. Integral to each group's functioning is some kind of problem-solving process that is used systematically to identify, analyze, and develop solutions for the problem being addressed. The specific process varies from group to group and organization to organization. The specific process used can be equally effective in either a quality circle environment or a task force environment.

Membership is usually voluntary for both quality circles and task forces. Voluntary involvement at the first-line employee level is universal. There are some exceptions, however, for other group roles. For example, managerial involvement in quality circles and task forces may not be optional. In some organizations, first-line supervisors are expected to lead groups, thus making this a key job responsibility. In other organizations, managerial representation from a specific hospital function may be critical to the scope of the issue being addressed.

Facilitators, who play an important role in both task forces and quality circles, have advanced training in group processes and problem-solving techniques. Although the decision to become a facilitator may be voluntary, there is certainly an organizational expectation that facilitators will actively participate in groups after the provision of the specialized training.

In order to function successfully, management support is needed for both quality circles and task forces. Support of the initiatives is often demonstrated by providing time away from the primary job, as well as a place to meet. Of course, management commitment to the outcomes of the group's efforts is also critical.

Quality circles and task forces also share a similarity in size. In order to function effectively, the group's size needs to be limited no matter how the process is accomplished. For both types of structures, the range of membership is three to fifteen associates. On average, however, groups consisting of eight to ten members is the norm.

Differentiating Characteristics

In addition to sharing several commonalities, quality circles and task forces also have differences. These differences, although not large in number, are striking. They include

- the nature of the issues addressed
- the group's life cycle
- specific member roles
- the authority to act

The first dissimilarity between quality circles and task forces is in the nature of the problems that are solved. In general, quality circles are free—within established boundaries—to select the problems that they want to address. An integral part of the quality circle process is identification and selection of issues. This process can take the initial six weeks of a quality circle's existence (Crocker et al. 1984).

Task forces, in contrast, are convened to solve a specific problem or task that has already been identified in another way. This does not mean that the problem has been exclusively identified by management. Rather, the issue to be addressed can come from a variety of sources, including patient satisfaction data, an employee suggestion program, or an internal focus group. It is the manager's role, however, to determine that the task force is the forum for solving the issue and to initiate the process of putting together a team.

Quality circles and task forces also differ in the longevity of the group's existence. Quality circles are, by definition, ongoing. The same group of people meet regularly to train, refine problem-solving skills, and work on opportunities for improvement. The group develops an identity and is integrated into the formal structure of the organization's quality program.

Conversely, the life cycle of a task force is limited to the task or problem at hand. Once the issue for which the task force was convened is addressed, the group dissolves. The members, as a group, form only a temporary identity. Because of their limited nature, task forces are unlikely to be chartered by a quality council, develop self-directed mission statements, or select a unique name that is organizationally recognized.

Defined roles for selected group members are another differentiating characteristic between quality circles and task forces. In the traditional quality circle structure, there is both a leader and a facilitator. The leader is often the manager of the work group and, as such, is not elected by the group to the position. The leader is charged with facilitating the group's regular meetings and ensuring that the agenda is maintained. This aspect of work is often a part of the manager's job description. In the traditional quality circle, the role of the leader is to coordinate and facilitate, not to manage or control.

The quality circle's facilitator functions primarily as a consultant to the group. He or she provides ongoing training and guidance on an as-needed basis. The facilitator may also serve as a liaison between the quality circle and the organization's quality council. Because of the consultant nature of the facilitator, he or she may be involved with several quality circles at the same time (Crocker et al. 1984).

In task forces, the functions that are provided by the quality circle leader are assumed by the facilitator. Rather than being a consultant, the facilitator is an active member of the task force. It is the facilitator's job to serve as an unbiased coordinator of the group process. The facilitator does not contribute to the decisions made by the group, but rather maintains the structure and process needed to ensure effective interactions among group members. By filling this role with the facilitator, the manager in the task force can then become an active, equal member in the problem-solving process. The manager's role is to bring the organizational perspective to the work group without exerting undue control or authority over the outcome.

This differentiation between facilitative and managerial roles leads to the final distinction between task forces and quality circles—the scope of authority. In the usual quality circle structure, the product that is produced is recommendations for action. These recommendations are presented to the organization's management team as a written report and/or oral presentation. The process followed and the data used to develop the recommendations are an integral part of the group's report or presentation. It is then up to the organization's management to approve, disapprove, or alter the recommendations. Following management action, the quality circle may or may not be directly involved with the implementation of the recommendations (Crocker et al. 1984).

This is very different from the scope of authority given most task forces. For these groups, the authority to implement solutions is the norm. The structure of the task force process is designed to ensure that the group has the resources and information needed to solve the problem. When the manager who is responsible for implementing the solution becomes a group member, he or she abdicates the ability to veto the consensus solution reached by the group (Moore and Miller-Kovach 1988). By forming and joining the group, the manager commits to the implementation of the solution.

Advantages of Quality Circles

The structure and nature of quality circles leads to several advantages over task forces. Three are discussed here.

The first advantage is a result of the ongoing nature of quality circles. Because quality circle group members meet regularly, the opportunities for comfortable relationships are encouraged. In the proper setting, this assists in establishing high levels of trust, mutual respect, and cooperation among group members. These attributes are very important to successful group problem solving. Furthermore, the stability of the membership means that, as new issues are identified for exploration and analysis, time is not used to identify roles, establish rapport, and organize the structure of the group.

The second advantage of quality circles involves training. In organizations that establish quality circles, training and skill development is readily handled within the context of the group. This practice enhances the likelihood that members will have comparable skills, familiarity with quality tools, and consensus-building methods. This promotes homogeneity within the group, making the ties among members stronger.

The final advantage deals with the selection of issues to be addressed. Because opportunities for improvement are selected by the group members, there may be a greater sense of ownership. The circle members are likely to limit their efforts to those issues that matter most to them. Successful efforts can lead to a synergy between overall job satisfaction and a positive forward movement for the group.

Advantages of Task Forces

Task forces can have advantages over quality circles. Five potential pluses are explored below.

The first advantage of task forces deals with membership commitment. Because participation in a task force is finite, organizational members are free to volunteer for issues that are important to them without making the long-term commitment needed for quality circle membership. This advantage can mean a great deal in hospitals in which professional personnel are specialized and unlikely to be willing to devote large portions of time on a long-term basis to a group without knowing exactly what the group will be doing.

As a corollary to this issue of finite time, the defined nature of the issue to be addressed by a task force adds an implicit structure to the group that may not be as readily achieved with quality circles. There are no questions about the mission of the task force and the problem to be solved. This allows the task force members to get right down to work on problem solving, which is appealing to some people.

A distinct advantage of task forces is the ability to maintain work group momentum throughout the process. Because the group has a defined mission and a designated end point, there is a reduced risk that complacency will set in and that the phenomenon of meeting for the sake of meeting will occur. Task forces have a clear direction and an incentive to complete the work so as to fulfill one's commitment to the group.

Another advantage to task forces lies in the lack of ongoing group relationships. Because group membership changes—depending on the issue being addressed—creative, independent thinking is encouraged, as opposed to the group-defined roles that can be established in quality circles. Addressing the designated problem is the focal point of a task force; maintaining and enhancing relationships among group members is secondary.

The final advantage to task forces has to do with the predetermined knowledge that solutions will be implemented. This facet can maximize participant motivation to optimize the solution, because the success or failure of the outcome falls directly back to the group. While ownership in quality circles is established through issue selection, guaranteed solution implementation serves a similar role for task forces.

Making the Selection

In addition to advantages, both quality circles and task forces have disadvantages as well. By and large, what is a strength for one is a weakness for the other. Taking the advantages and disadvantages of these group processes into account, decisions must be made at the organization or department level as to which is preferred. Of course, hybrids of the two are also possible.

The critical factor in selecting a group problem-solving method is that departments or organizations select a model with which they are comfortable. Whether the model is a quality circle or a task force, it is the success of group outcomes that makes or breaks a quality program.

Specific action steps to implement both quality circles and task forces are shown as Exhibits 7–1 and 7–2.

Brainstorming

Brainstorming is a technique that is an integral tool in most group efforts. It is used to expand the thinking of group members so that all facets of an issue are considered. Without brainstorming or other creative methods, group efforts are likely to be limited to a variation of the "way things have always been done." Brainstorming encourages the generation of ideas about the ways things "could be done," whether within the realm of possibility or not.

When used effectively, brainstorming allows the greatest flow of ideas to be generated in the least amount of time. The goal of a brainstorming session is to produce the maximal quantity of ideas about a given topic or issue. Again, the objective is the quantity, not the quality, of ideas presented.

There are several variations of brainstorming that can be used (Crocker et al. 1984). Most frequently, brainstorming sessions are conducted in a single group meeting where individuals state their ideas to the rest of the team. The verbalization of ideas can be free flowing or structured. In structured brainstorming, team members present ideas in a round robin fashion, with each person taking his or her turn in sequence. Free-flow brainstorming encourages members to speak spontaneously; there is no taking turns or raising of hands.

As stated above, brainstorming is a mainstay of group problem solving. It is used to generate potential opportunities for improvement, possible causes to problems, probable solutions, and implementation strategies. An example of the use of

Exhibit 7–1 Quality Circle Action Steps

1. Form the circle
   - Recruit group members.
   - Select or appoint a leader.
   - Select a name.
   - Provide initial training.
2. Define the circle's mission, goals, and objectives.
   - Include measurements (e.g., reduce waste by 5 percent).
   - In addition to direct work issues, include education and training skills, quality of work life, team recognition, and/or any other factors that are important to the group.
3. Select a problem or opportunity for improvement.
   - Use brainstorming, the modified Delphi technique, or the nominal group technique.
   - Evaluate against the mission, goals, and objectives.
4. Identify possible causes that contribute to the problem.
   - Use cause-and-effect diagrams, flowcharts.
5. Investigate selected causes.
   - Use check sheets, Pareto charts, run charts, or histograms.
6. Select possible solutions.
   - Use brainstorming, the modified Delphi technique, the nominal group technique, or force field analysis.
7. Investigate solutions.
   - Use cause-and-effect diagrams, flowcharts.
   - Use check sheets, Pareto charts, run charts, histograms.
8. Develop recommendations.
   - Base recommendations on data.
9. Write a report for and/or make a presentation to management.
10. Participate in implementation as directed by the management team.
11. Ensure effectiveness of the solution.
    - Use a control chart.
12. Select a new problem and begin the process again.

Exhibit 7-2 Task Force Action Steps

---
1. Select a problem or opportunity for improvement.
2. Select a facilitator.
3. Form the task force.
4. Identify possible causes that contribute to the problem.
   - Use cause-and-effect diagrams, flowcharts.
5. Investigate selected causes.
   - Use check sheets, Pareto charts, run charts, histograms.
6. Select possible solutions.
   - Use brainstorming, the modified Delphi technique, the nominal group technique, or force field analysis.
7. Investigate solutions.
   - Use cause-and-effect diagrams, flowcharts.
   - Use check sheets, Pareto charts, run charts, histograms.
8. Select a solution.
   - Base the solution on data.
9. Implement the solution.
10. Ensure effectiveness of the solution.
    - Use a control chart.
11. Disband.
---

brainstorming with respect to a clinical nutrition service is presented as Exhibit 7-3.

Conducting a Brainstorming Session

Brainstorming sessions are best conducted by a facilitator who does not participate in the generation of ideas. The facilitative steps to conducting a brainstorming session include the following:

- Develop a clear statement about the issue or topic that is being brainstormed. Questions, concerns, or clarifications about the subject should be addressed before beginning the generation of ideas. It is a good idea to write the statement where it is visible for the group members to see throughout the session.
- Decide how the flow of ideas should be presented—whether participants are to take sequential turns presenting ideas or to speak out as an idea comes to mind. Inform the participants about what is expected of them before the session begins. If the stated behavior is not followed, remind group members as needed as the session progresses.
- Write down all of the ideas as they are presented so that all group members can see them. Whenever possible, write the ideas in the exact words of the speaker. Flipcharts and spacious blackboards are useful tools for capturing the ideas; overheads tend to be limiting because, if several ideas are generated, it is impossible for the participants to view all of the ideas at one time.
- Limit reactions, both verbal and nonverbal, about the ideas being presented. As a facilitator, it is all right to offer positive, supportive feedback to the person presenting an idea because this encourages further participation. This should not be confused, however, with giving feedback about the idea itself. Other group members should be told to refrain from showing any reaction to the ideas as they are being voiced. Any group member who does not follow this guideline should be immediately dealt with.
- Encourage group members to explore variations of the ideas that are generated. If spontaneity is not present, ask some questions to get the group going.

Most brainstorming sessions are self-limiting. For most topics, sessions can be expected to last 5 to 15 minutes (*The Memory Jogger* 1988). It is very difficult to break a brainstorming session into parts, so allowing sufficient time to complete the process in a single meeting is important.

The final role of the facilitator is to keep track of all of the ideas that come out of brainstorming sessions. Indeed, the unedited list of ideas should be transcribed into the meeting minutes or kept with the team's records. It is not unusual for the group to return to the ideas that came out of a brainstorming session as it works its way through the rest of the quality process. Likewise, quality circles often return to their initial list of issues when they have completed their work in one area and are ready to move on to another.

Critical Success Factors

There are two critical success factors for brainstorming. The first is the selection of the facilitator. It is extremely important that the group have a strong facilitator who has experience in managing brainstorming sessions. Balancing the need for creativity and maintaining manageability within the group is much more difficult than it may seem.

The facilitator needs to encourage and foster creativity within the group. At the same time, however, the facilitator is charged with ensuring that the participants stick to the issue, do not interrupt each other, are open to all of the ideas being generated, and share the floor equally. In order to guide the

Exhibit 7-3 Potential Causes for Incomplete Calorie Counts That Came Out of a Brainstorming Session

---
- The patient's menu is lost.
- The person who is supposed to check the tray is too busy.
- The person who is supposed to check the tray is not properly trained.
- The tray is missing foods.
- The menu is not clearly marked for the calorie count.
- Nurses remove the trays from the room before the information is collected.
- Patient families circumvent the process.
- The patient sabotages the process because he or she doesn't want us to know what is being eaten.
- The patient is off the floor during mealtimes.
- The technician never gets the menu.
- The technician misplaces the menu before it is calculated.
---

group without alienating individuals, the facilitator must have extraordinary interpersonal communication skills. Likewise, the facilitator must also be able to transcribe accurately the unedited ideas that are being verbalized quickly.

The second critical success factor is to not let the brainstorming session become more than it is. One must be extremely careful to ensure that the ideas that come out of a brainstorming session are treated as ideas and not facts. No idea should be acted upon until there is measurable information that validates it.

The Pluses and Minuses of Brainstorming

The use of some form of brainstorming as part of the quality improvement process has several advantages. Perhaps most importantly, brainstorming serves as an effective team-building tool because it strengthens cooperative ties among group members.

Spontaneous brainstorming sessions, characterized by a free flow of ideas, maximize creativity because group members can state their thoughts as they come to mind. Sequential brainstorming, however, has the advantage of equalizing the contributions of individual group members. Because sequential brainstorming involves taking turns to share ideas, quieter team members or those who are reluctant to share thoughts spontaneously are more likely to participate fully.

Brainstorming, when done improperly, can have several problems. For example, free-flowing sessions can be easily dominated by a few group members. While this situation may be valuable when the group's most creative members are developing a synergy of ideas, there is also the possibility that the session will be dominated by those individuals who simply like to hear themselves talk.

Another brainstorming problem occurs when, despite efforts to the contrary, negative reactions to ideas are expressed by session participants. The damage of this situation is compounded when the reactions reflect feelings about the person who had the idea, as opposed to the idea itself. When this happens, good ideas are lost and the self-esteem of individual group members can be hurt.

It is not unusual for individual participants to remove themselves psychologically from the brainstorming process once they have contributed what they perceive as being the best idea. In these cases, the person is likely to begin formulating an argument about why his or her idea should be used instead of creating more ideas on the topic. This, of course, defeats the purpose of the exercise.

Finally, the more effective the brainstorming session, the more energy and excitement is created in the group. Often there is a tendency to want to act on the ideas that are generated as a result of the brainstorming process. It takes discipline to harness the excitement of the group to the next step of gathering pertinent data.

Although there are potential problems with brainstorming, the disadvantages are minimal and can be managed readily by an effective facilitator or group leader.

ORGANIZING IDEAS

A brainstorming session can produce a plethora of ideas in a matter of minutes. The sheer magnitude of the number of ideas can create a feeling of panic. This need not be the case, however, because there are several techniques that can be used to sort and prioritize the outcomes of brainstorming.

If brainstorming is used for problem identification, careful attention must be paid to the idea selection if the advantages of using a group process are to be maintained. All too often, a group can brainstorm a number of creative opportunities for improvement, only to have the problem of choice selected for them by an outsider. In most cases, the outsider is a manager.

This is not to say that idea selection by decree is always a bad idea. Indeed, there are several advantages to doing so, including the following:

- The group can begin to work on a specific idea immediately after its formation (i.e., rapid progression through the process is more likely to hold members' attention).
- Problems known to the organization's upper management can be addressed first.
- Ideas that are known to lend themselves to a structured problem-solving process can be selected to assist the group's experience with new techniques.

The lesson to be learned from this discussion is to decide how the idea will be selected before progressing to brainstorming. Once the group has generated its own ideas, it will want to retain ownership for those ideas throughout the process.

Two techniques that efficiently process the products of a brainstorming session are described here—the nominal group process and the modified Delphi technique. Simply put, both of these tools translate the words of ideas into numerical equivalents that can then be dealt with through simple mathematics. The product of the process is the selection of a single idea that will be investigated by the group.

The Nominal Group Technique

Of the tools described, the nominal group technique is the simpler of the two. There are six steps to completing the process. These steps are summarized as Exhibit 7–4 and are described below.

The nominal group technique is best managed by the group's facilitator. It begins by writing out the ideas that were generated by brainstorming in succinct statements that can be readily viewed by all group members as the process unfolds. Use of a large blackboard or flipchart pages taped to walls is a

Exhibit 7–4 Using the Nominal Group Technique To Select an Idea

---
It is the role of the group's facilitator to:

1. Write idea statements out where all group members can see them. Identify each idea with a sequential alphabet letter; count the total of ideas listed (number of ideas = n).
2. Have each group member list the alphabetic letter identifiers vertically on a sheet of paper.
3. While asking the question, "Of all the ideas listed, which one do you think is the most important?," instruct each participant to write the number, n, next to the identifier that corresponds with his or her answer.
4. While asking the question, "Of all the remaining ideas listed, which do you think is the next most important?," instruct each participant to write the next sequential number (e.g., n-1, n-2, n-3, etc.) next to the letter that corresponds with his or her answer.
5. On a master form, list the letters vertically and the corresponding ratings horizontally for each idea.
6. Sum each horizontal row. The lettered row with the highest numerical total yields the idea on which the group should focus its efforts.

---

good way to accomplish this task. Once the statements are written, they are sequentially identified by a letter in alphabetical order, with A being the first, B being second, and so on. If there are more than 26 ideas, the alphabet is resumed by lettering ideas as AA, BB, etc. This done, a count of the total number of ideas is taken.

Each group member is then provided with a sheet of paper and asked to list the ideas, by letter, in a vertical row down the left-hand side of the page. Once everyone has completed this task, the prioritization begins.

Having counted the number of ideas listed, the facilitator asks each participant to put that number next to the letter of the one idea that he or she thinks is most important. The need to pick the idea based on importance should be emphasized by the facilitator.

The facilitator continues this process until all of the ideas have been sequenced. In other words, the facilitator descends through the numbers (i.e., the most important idea equals the number of ideas available, the next most important is the number of ideas minus 1, and so on) until reaching the number *1*. The facilitator ends the exercise by instructing the group to put a *1* by the last remaining letter. This idea represents the least important idea to that team member.

Duplicating the list generated earlier by individuals, the facilitator creates a master form to tally the team's scores. Each participant reads off his or her ratings (e.g., A = 6, B = 1, C = 3) for the facilitator, who transcribes them onto the master form. After all of the team members have contributed their listings, the scores are summed for each lettered idea.

The idea with the highest numerical total is chosen. An example of the tabulations associated with the nominal group technique are shown as Table 7–1.

Table 7–1 Calculations Used in the Nominal Group Technique*

Idea	Joe	Mary	Betty	Bob	Fred	Total
A	1	6	4	3	2	16
B	2	5	5	2	3	17
C	3	4	6	1	4	18
D	4	3	1	4	5	17
E	5	2	2	5	6	20
F	6	1	3	6	1	17

*Based on the above sorting, idea E would be selected.

The Modified Delphi Technique

The modified Delphi technique is a little more complex than the nominal group process. While the nominal group technique prioritizes ideas by importance, the modified Delphi technique focuses on both importance and personal preference. This technique also encourages interaction and discussion among team members as part of the process.

The general steps for conducting the modified Delphi technique are summarized as Exhibit 7–5. Like the nominal group process, the modified Delphi technique is best managed by a skilled facilitator.

The process begins with an open discussion about the ideas being considered. Taking the ideas that were generated by brainstorming, the group openly and candidly discusses the importance and merits of each idea individually. If there is consensus among group members, ideas can be dropped, reworded, or combined as a result of the open discussion. At this point, each idea should also be evaluated to ensure that it

- will meet the group's objective
- is doable within the scope and control of the group
- can be adequately evaluated through the use of available or readily attainable data
- can be addressed within an acceptable time frame.

The ideas that remain at the conclusion of the open discussion are written out, identified by letter (as for the nominal group technique), and tallied.

During the next step, each individual within the group rank orders the ideas based on his or her preference, using *1* for the most preferred idea and n (i.e., the number of total ideas) as the least preferred idea. It is important for the facilitator to stress that the rank ordering is to be based on personal preference for the idea, not the level of importance.

Next, each group member assigns a factor of importance to each of the ideas. Usually an importance factor of 1 (most important) to 10 (not at all important) is used. The same importance factor can be assigned to any number of ideas. In other

Exhibit 7-5 Using the Modified Delphi Technique for Idea Selection

---
It is the role of the group's facilitator to:

1. Encourage a group discussion about each of the ideas that came out of the brainstorming session.
2. List each idea for evaluation.
3. Instruct group members to rank order each idea in order of personal preference, with *1* being the most preferred.
4. Using a pre-established scale of importance, with *1* being most important, have each group member assign a level of importance for each idea.
5. Have each member multiply his or her preference score by the level of importance to yield a single number for each idea.
6. Compile each member's scores and total the sum for each idea. The idea with the lowest total score is the one selected.
---

words, two ideas can be given an importance rating of 1, but they cannot share a preference rating.

This done, the importance and preference factors are now combined. Either the group facilitator or each group member multiplies the individual idea's preference ranking and the importance factor together. The product of this multiplication becomes the idea's total.

To complete the process, individual member totals are then summed for each idea. The idea with the lowest sum is chosen. An example of the modified Delphi technique tabulations is shown as Table 7-2.

Critical Success Factors

Two critical factors must be in place for either the nominal group or modified Delphi techniques to be successful.

First, all group members must feel free to assign scores with which they are individually comfortable. In other words, there cannot be peer pressure applied to entice team members to "vote" for an idea based on whose idea it is. If there are concerns about this happening in an open forum, then those steps that involve scoring and summarizing can be done privately on paper and turned into the facilitator for presentation of the outcomes.

The second critical factor is that group members must understand the ideas sufficiently to be able to make informed choices. Toward this end, the discussion period must be used to educate and inform. Hidden agendas and political persuasion between and among group members during the discussion and scoring steps are a sure way to undermine the process. Having a skilled facilitator who can manage the process is key to its success.

Although not crucial, it is best to prioritize immediately after the conclusion of the brainstorming session. This minimizes confusion about what was meant by a statement, the inclination to add or change statements, and the inclination of members to lobby for personal choices.

Additional Considerations

The nominal group technique and the modified Delphi technique share several advantages. First and foremost, the processes ensure that every group member has an equal voice in the idea selection. The group member who speaks the loudest or most often has no more voice in selecting the idea than the member who chooses to limit his or her involvement to listening to others. This equality does a great deal to encourage a feeling of group ownership for solving the issue.

Another advantage is that members are forced to choose between ideas in a structured manner. This avoids the need for consensus, which is nearly impossible when faced with these kinds of choices. Thus the tools are an efficient way of sorting ideas that leads to a group decision without sacrificing time.

Finally, because both methods translate ideas into numbers, evaluation of outcomes can be calculated easily. This process introduces objectivity into an arena that is fraught with subjectivity. Furthermore, when there are many ideas or a large number of participants, the scores can be transferred easily to a spreadsheet software program for tabulation.

The modified Delphi technique has the unique advantage of recognizing both preference and importance in choosing an idea to investigate. The nominal group technique limits

Table 7-2 Calculations Used in the Modified Delphi Group Technique*

Idea	Team Member															Total
	Joe			Mary			Betty			Bob			Fred			
	P	I	T	P	I	T	P	I	T	P	I	T	P	I	T	
A	1	2	2	6	5	30	4	7	28	3	2	6	2	2	4	70
B	2	1	2	5	5	25	5	7	35	2	1	2	3	3	9	73
C	3	4	12	4	3	12	6	7	42	1	1	1	4	4	16	83
D	4	5	20	3	4	12	1	1	1	4	4	16	5	9	45	94
E	5	7	35	2	1	2	2	2	4	5	4	20	6	10	60	121
F	6	7	42	1	1	1	3	3	9	6	5	30	1	2	2	84

*P, Preference score; I, importance score (1 very important; 10 not important); T, total. Based on the above scoring, idea *A* would be selected.

prioritization to the variable of importance only. However, the nominal group technique is more amenable to dealing with large numbers of ideas. Because only one factor is rated, members have less to deal with in rank-ordering ideas. Also, the group can choose to limit the number of ideas that they rank. One method of accomplishing this objective is to limit ranking to the "one half plus one" rule (*The Memory Jogger* 1988). This rule states that you rank only "$n/2 + 1$" of the ideas. For example, if there was a list of 30 ideas, members would rank-order importance from 16 to 1, leaving no score by the lowest 14 ideas. This is often a good technique because group members can spend a great deal of time rank-ordering ideas that, in the end, will have no chance of being selected.

The nominal group technique and the modified Delphi technique are not foolproof, however. Because they are based on numbers, outcomes can yield several ideas that either tie or are extremely close in their totals. It becomes very difficult to say that an idea that scores 0.5 point higher or lower than another is clearly the "best" idea for investigation. This disadvantage can be reduced, however, by singling out the tied or top ideas and repeating the process. Doing so spreads the scores between the top ideas so that differentiation between the ideas is amplified.

Sorting out the ideas that are generated by a group can be an arduous process. Use of the nominal group technique, the modified Delphi technique, or a hybrid of the two techniques as a tool, however, structures the process into a manageable task. At the same time, the outcome embodies the spirit of group decision making.

CAUSE-AND-EFFECT DIAGRAMS

Quality management relies on comprehensive evaluation of problems. The cause-and-effect diagram is one of the most widely used quality tools by which to structure the potential roots of a work process that can be made better.

The cause-and-effect diagram is also known by other names. In the quality literature, especially publications geared toward the manufacturing sector, the cause-and-effect diagram is referred to as the Ishikawa diagram. This identifier is associated with the tool because a man with the surname of Ishikawa developed it when he was working with the Japanese steel industry in the 1940s. The diagram is also called a fishbone analysis. This name derives from the diagram's resemblance to a fish skeleton when it is completed. Regardless of what it is called, the cause-and-effect diagram is an integral tool in most quality management programs.

The cause-and-effect diagram is a method for evaluating process dispersion. In other words, the diagram defines the relationships between events (causes) and outcomes (effects). The construction of a cause-and-effect diagram organizes and graphically shows a group's ideas about what may be the roots of a problem at work. Cause-and-effect diagrams, along with flowcharts, are the most frequently used quality tools to analyze work processes and identify opportunities for improvement.

Cause-and-effect diagrams lend themselves to work group efforts. It is possible for an individual to create a diagram, but the tool is best suited to group settings. In their traditional format, cause-and-effect diagrams are used to facilitate brainstorming sessions.

Cause-and-effect diagrams can be used to evaluate the full range of clinical nutrition products and services. In addition to the analysis of problems relating to the provision of support services and clinical products, cause-and-effect diagrams can be helpful in assessing problematical patient outcomes associated with the provision of medical services. For example, progressive weight loss as a function of hospital stay can be investigated by using this tool.

Developing a Cause-and-Effect Diagram

Construction of a cause-and-effect diagram has five major steps. The diagram takes up a lot of space, and it is important to keep it intact and visible to the group's members. Expansive blackboards or walls filled with blank paper make a good medium for the development of the diagram.

The first step in developing a cause-and-effect diagram is to write the problem statement at the center of the right-hand edge of a large space. The problem statement is a short, concise phrase of the outcome or effect that is to be evaluated. Examples of problem statements include the following: patients are not being screened for nutrition risk in accordance with departmental policy; calorie counts are inaccurate; and patients are leaving the hospital without drug-nutrient interaction information.

In developing the problem statement, it is important that all of the group members agree on both the words and the meaning of the effect being evaluated before proceeding. When the statement is defined and written at the head of the diagram, a straight arrow is drawn from the center of the left-hand edge of the writing surface to the effect.

The second step in developing a cause-and-effect diagram is to determine the major cause categories for the problem being studied. In traditional cause-and-effect diagrams, four defined main causes are used: people, methods, equipment, and materials. Other major causation areas can be developed as needed. For example, major headings for a problem associated with clinical nutrition care may define patients, practitioners, procedures, and environment as major sources. Write the major cause sources above and below the central arrow and then draw connecting arrows. The "skeleton" of a basic cause-and-effect diagram is shown as Figure 7–1.

The third step is to brainstorm specific causes under each major area. This is usually accomplished by considering each major heading separately and then brainstorming all of the potential causes that could lead to the effect. As relationships between causes and/or subcauses are identified, they are de-

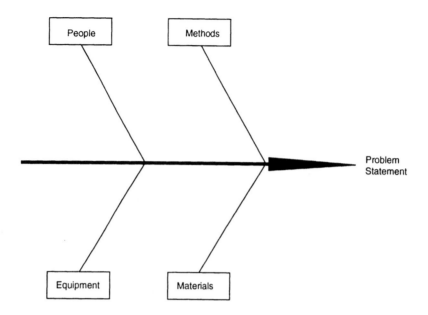

Figure 7–1 Basic Cause-and-Effect Diagram

picted graphically by adding another "bone" to the "fish." It is acceptable to repeat ideas under different major topic areas. For example, inadequate training could be a cause that leads to problems in both the people and methods areas.

The fourth step is to post the diagram where it can be viewed for several days by team members and others in the work area (if appropriate). This step allows for individual reflection of both the identified causes and their relationships. During this time, members should be encouraged to comment or add to the diagram.

Finally, the group selects from the diagram those causes that are most likely to have the greatest impact on the outcome and/or those that the group can actively investigate and act on. This step is taken only after the diagram is reviewed as a group for completeness. The causes to be investigated can be selected by vote or by using one of the organizational methods described in the brainstorming section. Cause selection should always be done in a fair, structured manner, however.

If the group is unable to select the causes that it believes need to be acted on, the process should be halted temporarily. Each team member is then assigned causes to investigate and reports his or her findings back to the group. Those causes and subcauses that will be targeted for action are circled. Action plans to verify their importance in the problem being addressed are then developed and carried out. A completed cause-and-effect diagram is shown as Figure 7–2.

It is a good idea to save cause-and-effect diagrams for future reference. If the initial causes that were selected by the group are found not to be major contributors, or further enhancements to the process are desired at a later date, the diagram can be used as a starting point.

Perks and Pitfalls of Fishbone Diagrams

Cause-and-effect diagrams are an important tool, but there are pitfalls that must be avoided to ensure their success. First, the effect needs to be evaluated in the largest possible context. This is true even if the identified cause cannot be changed by the work group. For example, consider a hospital that depends on nursing to collect the data for calorie counts and also has a nurse staffing shortage. Although a lack of nursing cooperation is a viable cause of the problem of incomplete calorie counts, the nursing shortage is probably a bigger cause. Including both causes in the diagram assists the group's efforts in identifying the cause on which to take action.

It is also critical that the construction of a cause-and-effect diagram be spearheaded by an effective, experienced facilitator. Listening to ideas during brainstorming, determining cause placement on the diagram, clarifying the speakers' meaning, and writing the causes succinctly present a real challenge. Moreover, it must be remembered that the construction of a cause-and-effect diagram is intended to be a tool in the discussion of the problem. If not handled skillfully, the diagram can easily become the center of attention. This tendency can be avoided with the use of a good facilitator.

The construction of a cause-and-effect diagram can leave the group believing that they have analyzed the problem objectively. Indeed, as also happens in the use of other quality tools such as brainstorming, it is tempting for the group to want to act immediately on the identified causes. This is a mistake, because the diagram is constructed on the basis of subjective information and opinion, not verified fact. To ensure the integrity of the quality improvement process, it is impor-

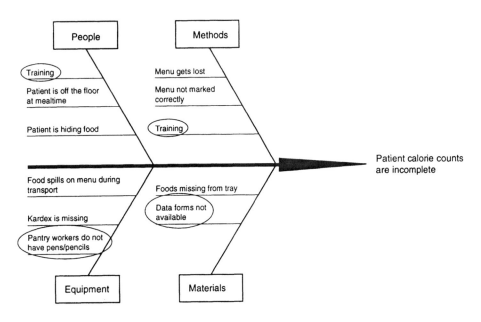

Figure 7–2 Cause-and-Effect Diagram for Incomplete Calorie Counts

tant that the potential causes be validated with reliable data prior to taking action.

There are several direct and indirect advantages of using cause-and-effect diagrams as a tool in a quality management program. Perhaps the biggest advantage is the diagram's visual recognition that the problems encountered in the course of work are multifaceted and multidimensional. When viewed in the context of a cause-and-effect diagram, the often-used management approach of determining which single person is responsible for a problem becomes ludicrous. The diagrams also do a great service in dispelling the myth that a problem can be fixed with a single solution.

Cause-and-effect diagrams are also advantageous because the development process can be readily learned and applied by almost all workers in an infinite number of settings. They are also inexpensive.

Cause-and-effect diagrams are preferred to simple brainstorming because they show graphically the sequential hierarchy of causes and subcauses to the effect. This, in turn, represents the relative influence of the cause to the effect. This cannot be accomplished with traditional idea-generation techniques. The tool is not without disadvantages, however. For example, it is sometimes difficult to capture accurately the appropriate location of a cause as the diagram is being constructed. At other times, major cause areas may not be easily identifiable at the onset of the brainstorming session. When this occurs, a variant method of construction can be used. As potential causes for the effect are brainstormed, each is written on a single card or adhesive memo note. After the brainstorming is complete, the diagram can be constructed by grouping the causes and defining relationships after the fact.

Finally, team members should always remember that form needs to follow function in the use of all quality tools, including cause-and-effect diagrams. Although it is easy to get into debates about whether the diagram was properly constructed, the actual placement of causes is relatively unimportant. If the diagram captures the ideas of the group that put it together, and those individuals understand its meaning, then its usefulness is assured.

Cause-and-effect diagrams are one of the key quality tools. They are used to evaluate systematically the potential causes that result in a negative outcome. At the same time, the diagrams present visually the relationships between causes and subcauses.

Traditional quality programs create cause-and-effect diagrams as part of a brainstorming session in a work group meeting. However, cause-and-effect diagrams can also be developed by using creative, alternative approaches in the work setting. For example, the diagram can be developed in a free-form fashion by posting the basic skeleton in a visible area and inviting members of the work group to contribute on their own. Likewise, the cause-and-effect diagram can be used to evaluate successful operations by stating the effect as a positive outcome and brainstorming causes for the success.

FLOWCHARTS

Of all of the tools that are available to assist in quality improvement efforts, none is more valuable or more often used than the flowchart. Indeed, flowcharts can be an integral resource during every step of the quality improvement cycle, including process understanding, problem identification, analysis, solution development, change implementation, and monitoring for sustained gains.

A flowchart is a visual representation of the series of steps and handoffs that make up a work process. It is a graph of the flow of work, including the relationships, decisions, and thought processes that go into making a product or service. A generic representation of a flowchart is shown as Figure 7–3.

Flowcharts are one of the most important quality tools to master because they demonstrate one's understanding of the importance of process in system improvement. Creating a flowchart is usually one of the first steps that is taken in any quality improvement initiative.

Flowchart Formats

Flowcharts can be constructed by using a variety of formats. Some quality programs emphasize the use of specific symbols in the construction of a flowchart. For example, ovals or circles often mark the beginning and end of the process being evaluated. Rectangles may represent an activity step within the larger process, and diamonds are used to indicate decision points. In virtually all flowcharts, arrows are the symbol employed to show the direction in which the work is flowing. Not all flowchart symbols are prescriptive, however, and some work groups prefer to limit their symbolic representations to boxes or circles and arrows.

Depending on the situation, consistent use of symbols may or may not be important. If your department is doing organizational quality improvement, the use of consistent symbols may be important to assist in the understanding of your flowcharts throughout the organization. If, however, the flowchart's use is limited to a problem-solving team or to a department's internal use, then there is no need to be overly concerned with using specific symbols. What is important, however, is that the people who are putting the flowchart together and those who will be using it in the course of work understand what it means.

Flowchart Applications

Flowcharts are especially helpful in improving support services and clinical products. These services and products, which form the infrastructure on which medical services are based, have a significant bearing on patient care. Support services and clinical products are usually multidimensional and cross-functional. Flowcharting the series of activities that lead to the products and services that patients receive is critical to understanding the impact that systems have on the provision of quality medical care.

As stated before, flowcharts have many uses. They are often an integral part of

- precisely defining the steps involved in completing a process as it is presently performed
- developing a common ground among suppliers, customers, and those performing the work so that mutual understanding of the process is enhanced
- analyzing an existing process to identify problems or opportunities for improvement
- Diagramming an action plan to implement a change that has many steps, involves several people, or simply is complex
- projecting the impact of potential changes in an existing process
- providing clear documentation of the performance of work for internal and external review
- educating others about changes that are being made in a process
- communicating work processes to others who are unfamiliar with the steps involved.

Some potential applications for the use of flowcharts in the provision of clinical nutrition products and services are listed as Exhibit 7–6. An example of a nutrition flowchart is shown as Figure 7–4.

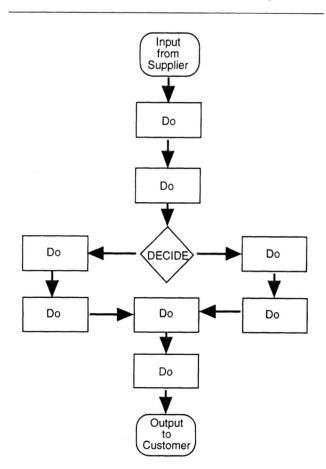

Figure 7–3 General Format of a Flowchart

Exhibit 7–6 Potential Applications for Flowcharts in Nutrition Services

- Diet change processing
- Screening for nutrition risk
- Tray assembly
- Tray delivery
- Selective menu processing
- Patient counseling
- Appointment scheduling
- Development of nutrition support recommendations
- Medical record documentation

Constructing the Flowchart

There are five basic steps that need to be performed whenever a flowchart is developed. The first is to identify the person or team members who will be responsible for the task. Some experts recommend that the flowchart be drafted by an individual who is familiar with the process being studied and then reviewed by others. Others recommend that the flowchart be constructed as a group project. The most appropriate method is probably situation- and time-dependent. However it is drafted, input regarding the flowchart should be obtained from knowledgeable suppliers and customers, as well as the people who are performing the activities, to ensure that the final product provides a clear, accurate, objective diagram that represents all points of view.

The second step is to establish clearly the process to be charted. This includes a definition of the beginning and end points as well as the inclusion or exclusion of activities that are peripheral to the process being charted. Once there is clarity and consensus on this point, the actual flowchart development can begin.

It is recommended that one start a flowchart at an end point and then work forward or backward. It really does not matter whether it is the beginning or the end. Exhibit 3–1, which defines your roles as a customer and supplier, may be helpful in defining these initial and ending points.

The next step is to list in sequential order every action that is taken in the production of the product or service. Invariably, there are steps where a decision has to be made and the next action is dependent on the outcome of that decision. When this occurs, it is recommended that each of the decision outcomes be taken separately. This is accomplished by selecting one of the choices and following it through to its conclusion. When this has been done, the same process is completed for the second choice. This sequence is repeated until all of the possible outcomes have been charted.

The final step is to review the completed flowchart for completeness and accuracy. It is important to ask as many questions as possible and develop as many scenarios as feasible to ensure that all the activities and possible decision points have been covered.

Maximizing Effectiveness

Several details should be noted in order to maximize the effective use of flowcharts. One factor involves the selection of participants in the development of the tool. It is critical that the right people be involved. All too often, the development team is limited to department members who share a homogenous view of the work flow. Almost all processes include extradepartmental suppliers and customers whose viewpoints must be considered if an accurate portrayal of the flow is to be made. When developing a flowchart, it is better to err on the side of getting too much input from diverse sources than to limit feedback to a select few.

Another detail involves the amount of space allocated to a flowchart's development. Because complexity can develop quickly with even the simplest of processes, it is important to leave more room than is anticipated. Taping sheets of flipchart paper to a wall is an often-recommended method of constructing a flowchart. This technique is especially appropriate if the chart is being developed in a group setting. All parts of the chart should be visible at the same time; fragmenting the activities on separate pieces of paper or overheads can lead to errors, missed steps, and overlooked opportunities for improvement.

Like space, adequate time must be devoted to the development of the flowchart. It is often recommended that the charting process be broken into several sessions to allow time to reflect on its development.

While the procedure of developing a flowchart is important, it is not a good idea to spend too much time worrying about using the correct symbol. This is especially important during the time that the steps of the process are being sequentially delineated. If a prescriptive coding of the steps is needed, it can be fine-tuned during the final stages of the flowchart's development. Whenever possible, it is a good idea to try to keep the flowchart as simple as possible while retaining the integrity of the process that is being mapped. A flowchart should be accurate and clear. It does not have to be pretty.

When the flowchart is complete, review it to ensure that every loop is completed. In other words, all actions and decision paths must lead ultimately to the final product or service. If there is a step that has an output that does not go anywhere, there is either insufficient knowledge about the process on the part of the group that is putting it together or a big problem in the process.

When a flowchart is being constructed, it is a good idea to avoid making suggestions or judgments about the process that is being evaluated. Analysis of the flowchart needs to be treated as a separate exercise. As a corollary to this, the process that the flowchart represents should be clearly known, understood, and marked. Flowcharts can represent what is, what could be, or what should be. A single flowchart should never mix these three possibilities together, however.

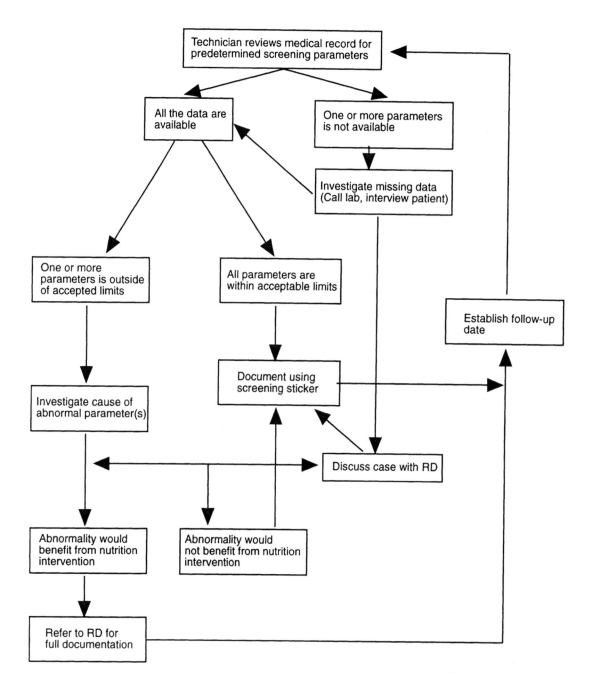

Figure 7–4 Flowchart for Nutrition Screening. Courtesy of the Department of Nutrition Services, Cleveland Clinic Foundation, Cleveland, Ohio.

Benefits of Flowcharts

The benefits of using a flowchart are numerous. Flowcharts

- are an inexpensive means of gathering valuable information
- assist those who do the work to understand their part in providing patient care, thus encouraging pride and commitment to the process
- enhance the ability for all of the process participants to identify opportunities for improvement
- encourage communication on a common ground among suppliers, process performers, and customers
- serve as a valuable staff orientation tool because new employees can be "walked" through the process in a classroom situation, which helps alleviate the activity

isolation that often occurs with simple on-the-job training

In summary, flowcharts fill a critical role in the quest for quality improvement. By specifying the scope of work, the importance of an existing process becomes clear. Because efficient work means that every step in a process has a purpose, there should be a readily recognized, tangible value added to the process as the result of each step contained in the flowchart. If this is not the case, it becomes readily apparent when conducting the exercise. The same holds true for the unveiling of redundancies, missed communications, and the identification of "cracks" in the system. Finally, flowcharts can assist in enhancing the flow and clarity of communication about a process throughout the organization.

FORCE FIELD ANALYSIS

Most of a group's time is spent identifying and investigating opportunities for improvement in work processes. The outcome of this work is a solution or solutions. Force field analysis is a useful quality tool that assists in evaluating the probable success of a proposed solution.

A force field analysis yields a visual description of the reasons why a single solution will both succeed and fail. In other words, the analysis provides a concise illustration of the overall impact of a potential solution. A sample of a generic force field analysis is shown as Exhibit 7–7.

Why Use Force Field Analysis?

There are several advantages to using force field analysis as part of the group's effort to evaluate a solution or solutions to the problem being investigated. First, the analysis consolidates the data that have been collected about a possible solution into a single graph. At the same time, however, it expands the evaluation beyond the data to include qualitative statements that can have a significant bearing on the success or failure of a possible solution.

For example, let us assume that a work group of dietitians is looking at ways to improve the quality of the nutrition assessments performed on obese patients at their hospital. One of the solutions that is being considered is the purchase of fat-fold calipers for each clinician in the department. A supporting assumption for this solution is that, with the use of calipers, the percentage of obese patients on whom body composition measurements can be included in the assessment will go from the current rate of 15 percent to 80 percent. Another supporting reason is that 90 percent of the physicians surveyed would like to have body composition measurements as part of the assessment. A resisting assumption that does not have specific data but does affect the likelihood for implementation of the potential solution is that funding for capital equipment purchases is severely restricted. By graphing all of the assumptions as part of a force field analysis, both qualitative and quantitative statements can be included.

Another advantage of force field analysis is its incorporation of the relative importance for each of the supporting and resisting assumptions. By adding this dimension to the evaluation, the quality of the decision reached is better than that which is the result of merely listing the advantages and disadvantages of a potential solution.

Finally, performing a force field analysis requires that the group take the time to examine fully the impact that a proposed solution may have. By virtue of completing the analysis, the group must examine all of the variables that affect its solutions. It demands that the group look at the big picture. Just as the construction of flowcharts and cause-and-effect diagrams assists in the group's understanding of the problem, force field analysis performs the same function for solutions.

Creating the Graph

There are six steps to constructing a force field analysis graph:

1. Brainstorm all of the assumptions that would favor and work against the implementation of a solution.
2. As a group, rate the relative importance of every assumption on the brainstormed list.
3. Chart the results by listing the supporting assumptions in a vertical column on the left-hand side of a large piece of paper. Resisting assumptions are listed in a vertical column on the right-hand side.
4. Draw a straight, vertical line down the center of the paper. This line represents the crossover point between supporting and resisting assumptions.
5. To each side of the center line, write the numerical indicators of relative importance.
6. For each of the assumptions, draw a line from its predetermined level of importance to the center line.

The result is a graph that visually depicts the number and cumulative importance of the opposing and supporting arguments for the solution being evaluated.

Once the graph is constructed, the interpretation is fairly straightforward. Indeed, if all of the assumptions and relative importance ratings are valid, then potential success of a proposed solution can be predicted with some accuracy. So long as the cumulative space of the supporting assumptions is greater than that of the resisting assumptions, the improvement will probably succeed. The reverse, of course, is also true.

Making It Work

There are two critical success factors to keep in mind when interpreting the force field analysis graphs. First, the graph

Exhibit 7–7 General Format of a Force Field Analysis

Supporting Assumptions:	Relative Importance: 10 5 1	Relative Importance: 1 5 10	Resisting Assumptions:
One			One
Two			Two
Three			Three
Four			Four
Five			Five
Six			Six
Seven			Seven
Eight			Eight
Nine			Nine

must be complete and accurate to provide meaningful information. Therefore, the knowledge and skill level of the group determine the usefulness of the tool. Second, it is important to remember that a force field analysis is based on assumptions, not facts. And although the assumptions may be data driven, the relative importance of the assumption is not. Therefore, the interpretation must not be construed as an objective finding in the overall evaluative process.

REFERENCES

Crocker, O. et al. 1984. *Quality Circles: A guide to participation and productivity.* New York: Facts on File Publications.

James, B.C. 1990. Implementing continuous quality improvement. *Trustee* 43, no. 4:26.

The Memory Jogger. 1988. Methuen, Mass: GOAL/QPC.

Moore, C., and K. Miller-Kovach. 1988. Task force: A management technique producing quality decisions and employee commitment. *Journal of the American Dietetic Association* 88, no. 1:52–55.

Orme, C.N., and R.J. Parsons. 1992. Customer information and the quality improvement process: Developing a customer information system. *Hospital and Health Services Administration* 37, no. 2:197–212.

Chapter 8

Quantitative Quality Management Tools

A common element in all quality management programs is the need to base decisions on objective data. Indeed, measuring work processes and making improvements that can be evaluated numerically are key facets of quality. One need not, however, become a master statistician to manage a quality program.

There is a myriad of quantitative measurement tools that can be incorporated into a quality program. It is beyond the scope of this book to include them all. Rather, this chapter delineates the measurement tools that are most commonly used in newly established quality programs, including check sheets, Pareto charts, histograms, run charts, control charts in benchmarking and Scatter diagrams. Other tools can be learned and used as one's familiarity with process improvement grows.

CHECK SHEETS

Because one of the overriding principles of quality management is that all analyses must be data driven, assessing whether subjective opinions have a basis in fact is integral to the quality process. To accomplish this, there are quality tools that collect data efficiently.

Check sheets are a very simple, yet effective, way to gather information about a process. They are neither complicated nor complex. Check sheets are most often used as a starting point for getting information to answer the questions, "How often, how long, or how many?" By gathering specific information from a defined set of sample observations, it is possible to detect patterns and the frequency of many problems. Examples of the use of check sheets in dietetics are presented as Exhibit 8–1. The possibilities for check sheets in nutrition services are endless, as are the data they provide.

Successful use of check sheets requires the following elements:

- The situation evaluated is clearly defined and understood by everyone who is collecting data.
- The situation evaluated reflects the problem or question that is being asked.
- The sample used is representative of the situation being studied, and, if there is a great deal of variability in the sample, defined subsets have been identified and grouped accordingly.
- The sample size or data collection period is adequate to get enough information to make a valid assessment.
- The data collectors are gathering information consistently and with an open mind; biases must be recognized and dealt with before the study.

Exhibit 8–1 Application of Check Sheets to Nutrition Services

- How often are renal diets ordered?
- How many consultations for outpatient services do we receive?
- How many patients are NPO (nothing by mouth) for more than five days?
- How often are weights and heights missing from the admission notes?
- How long is the response time from consultation order to completion?

To ensure simplicity, the most effective check sheets are limited to distinct variables for the situation being evaluated. If data are desired for more than one event or many subsets of variables within a single event, it is better to conduct more data-collection periods than to try to obtain additional information on a single check sheet.

This concept of simplicity is illustrated with Exhibit 8–2. A dietetics department is interested in investigating why nutrition risk screens are not consistently completed within 24 hours of patients' admissions. Through brainstorming, the group has compiled a list of possible explanations. As shown by Exhibit 8–2, the explanations can be classified as major causes and subcauses. As stated above, the initial check sheet should be limited to the variables of unavailable charts, unavailable patients, lack of information, and other. If, based on the findings obtained from this check sheet, further information is desired about subcauses, then another check sheet can be developed. In other words, if the major reason that screens are not done within 24 hours is found to be insufficient information, then another check sheet that looks at the frequency of unavailable heights, missing weights, and problems associated with laboratory results can be developed.

Developing a Check Sheet

To construct a check sheet, the events being observed are listed in a vertical column down the left-hand side of the form and the observation period is listed across the top. It is important to leave ample room for data entry. A row at the bottom of the form and a column at the right-hand margin can be allocated to include totals if the individual who is collecting the data will be doing manual calculations. Exhibit 8–3 provides an example of a possible check sheet to assess the frequency of diet orders.

It is sometimes helpful to develop check sheets using a computer software program that has both spreadsheet and graphic capabilities. The check sheet can be printed directly off the file. The file then can be used to combine the check sheet data obtained from the various observers, compute the sums, and be converted into a Pareto chart or histogram for group analysis and decision making.

Check sheets can also serve as a valuable tool for ongoing monitoring of clinical nutrition systems. When used in this manner, it is often helpful to illustrate visually the sequential data sets in a run chart, another quality tool. Likewise, multiple observation sets lend themselves to incorporation into a computer spreadsheet or database. Retrospective analysis of sequential check sheets is invaluable in determining trends and practice patterns and in ensuring that quality gains are maintained.

PARETO CHARTS

Pareto charts are a quality tool that serves as an integral part of group problem identification and decision making. A Pareto chart is a bar graph that visually illustrates the relative importance of problems or causes of problems. In Pareto charts, the bars are traditionally presented in a vertical format and in descending rank order to ease interpretation. The generic Pareto chart format is shown as Figure 8–1.

Exhibit 8–2 List of Possible Reasons for a Problem: Basis for Developing a Check Sheet

---

Problem: Nutritional risk screens are not performed within 24 hours of patients' admissions

Charts are not available.
  Patient and the chart are out of the ward.
  Other providers are using the chart.
  Charts are not returned to chart rack but are left throughout the area, making them difficult to locate.

Patients are not available.
  Patient is out of the ward.
  Patient has visitors.
  Patient asked not to be disturbed.
  Patient is busy with other staff.

Screening information is not available.
  Height is not recorded in the chart.
  Weight is not recorded in the chart.
  Lab data are not available— physician did not order.
  Lab data are not available—blood was drawn, but results have not been posted.

Exhibit 8–3 Check Sheet To Determine Frequency of Diet Orders

Diet Order	Day of Month						
	1	2	3	4	5	6	Total
NPO							
Clear Liquids							
Full Liquids							
General							
Bland							
Soft							
Mechanical Soft							
Pediatric							
Calorie Controlled							
High Calorie							
Renal							
No Added Salt							
2 Grams of Sodium							
No Added Sugar							
Diabetic							
Puréed							
AHA Phase I							
Total							

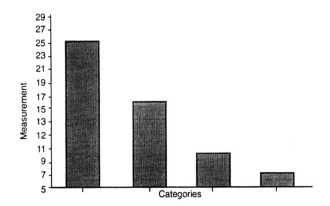

Figure 8–1 Pareto Chart

Exhibit 8–4 Pareto Chart Applications in Nutrition Services

- Development of diet instruction materials (i.e., spend the most resources on those that are used most often)
- Sources of patient dissatisfaction
- Types and frequency of diet orders used in the hospital
- Causes for incomplete nutrition screens (e.g., missing weights, lab data not ordered)
- Where tray accuracy errors occur (menu checking, trayline, in delivery)

The most common use of the Pareto chart is to organize data and assist in making decisions about how energies will be directed to achieve the greatest gains. In other words, Pareto charts help the group decide what problem or, if that has already been identified, what cause of a problem should be worked on. The charts are also a useful tool to monitor gains.

This quality tool is named for and is based on the Pareto principle. This principle states that relatively few factors have the greatest impact on any process. This same phenomenon is often referred to as the "80/20 rule." Exhibit 8–4 identifies some potential uses for Pareto charts in the provision of nutrition products and services.

Many Pareto charts start out as check sheets. Indeed, translation of the data collected from a check sheet into a Pareto chart is a logical "next step" for many groups as they delve into resolution of a problem process. As stated earlier, check sheets are used to gather information that answers the questions, "How often, how long, or how many?" While check sheets serve as the data-collection form to detect patterns and the frequency of many problems, Pareto charts display the data in a visual medium that enhances analysis.

With advance planning, one can combine the check sheet and Pareto chart into a single computer file. The advantages to this system include reduced chance for error and time savings in the development, transcription, and translation of data involving a single event or a series of observation events.

Graphing a Pareto Chart

Pareto charts can be done manually on graph paper or maintained as a computer file with ready transference to a graphic printout. A Pareto chart is developed in the following steps:

1. Decide what problems or causes of a specific problem are to be targeted for evaluation. These become the categories for the chart.
2. Determine the unit of measure that best represents the impact of what is being evaluated. Cost, numerical frequency, and percentage are the most common measurements used.
3. Gather the information needed to construct the chart. Existing data resources may be available. If not, data should be collected using a check sheet with the categories from above.
4. Once the data are collected, sort them by category from highest to lowest.
5. Prepare the graph by marking the vertical axis with the unit measurement and appropriate numerical intervals.
6. List and plot the categories from left to right, with the first bar being the one with the greatest unit measurement.
7. If there are several categories with inconsequential unit measurements, they can be combined and shown as "other." An "other" category should always be listed at the far right of the horizontal axis, regardless of the unit measure.

Success Factors for the Pareto Chart

The successful use of Pareto charts requires careful attention to three specific areas. First, it is important that the unit of measurement be appropriate for the problem or process being investigated. This step can have a significant bearing on the outcome.

The importance of unit measurement selection is illustrated in Figure 8–2. For this example, let us assume that there is a group of dietitians who are interested in looking at enteral product usage in their hospital. The first Pareto chart illustrates product usage by the number of cases purchased per year. The second Pareto chart shows this same information on a cost-per-year basis. These charts demonstrate that frequency and cost do not always correlate and that the unit of measurement chosen for the Pareto chart must be carefully determined by the impetus for looking at the data in the first place. In some cases, it may be wise to illustrate the data in several ways so that the group can gain a better understanding of how the category affects the problem or process as a whole.

In addition to careful selection of the measurement unit, it is critical that the chart be clearly marked with all pertinent data.

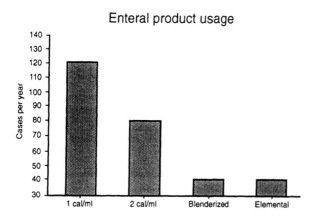

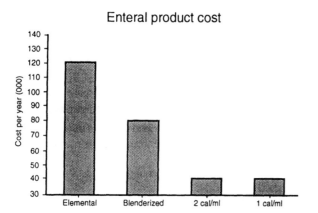

Figure 8–2 The Importance of Measurement Selection in Pareto Charts

This would include the measurement unit used for the vertical axis, the categories developed for the horizontal axis, and the amount of time the chart covers. The need to identify clearly the last component can be demonstrated by looking back at Figure 8–2. In evaluating the cost of enteral feeding products, it would be important to know whether the expenditures shown on the Pareto chart were those incurred in the course of a month or those incurred annually.

Finally, it is important to interpret the Pareto chart carefully. A common mistake in evaluating problems with Pareto charts is to get caught up in the graphics and not look at the data in a larger sense. For this reason, common sense should always supersede the chart.

Most of the time, the category that has the highest bar measurement is the most worthy of group efforts. There are exceptions to the rule, however. For example, Figure 8–3 graphs tray errors by source. Although the Pareto chart demonstrates that the trayline was the most frequent error source, further evaluation may reveal that the kinds of errors made at this level (e.g., missed spoon, no napkin) are less likely to affect patients' care negatively as those made in menu checking (e.g., missed fluid restriction, high protein content for a renal

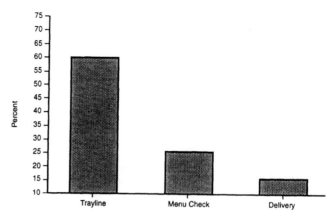

Figure 8–3 Sources of Tray Errors

diet). Clearly, in this case, menu checking should be the causative agent that is addressed if the group's objective is to improve the clinical nutrition course of hospitalization.

Pareto charts are a valuable tool when a group is deciding what problem or part of a process to tackle. By expressing raw data in a comprehensible format, Pareto charts assist in the group's ability to make decisions and direct efforts to the factors that will reap the greatest gains. The use of Pareto charts is not limited to the beginning of a process evaluation, however. Replicating the Pareto chart after putting a change into effect is helpful in evaluating the degree of success and, if done repeatedly, ensuring that gains are maintained.

HISTOGRAMS

A histogram is a type of bar graph that shows the distribution and frequency of numerical data. Histograms illustrate trends and patterns that may not be readily identified from a table of numbers or a check sheet. Like Pareto charts, most histograms have their origin in check sheets and are concerned with the range of findings from a single period of observation. As such, histograms are generally used to analyze problems rather than monitor gains.

On a histogram, the horizontal axis represents the sequential continuum of ranges that the data fall into. The first step in constructing a histogram involves deciding the categories (or bars) that are to be included on this axis, which is done as follows:

1. Count the number of measurements (data points) to be included for the graph.
2. Calculate the range for the data set. This is the highest value minus the lowest value.
3. Using Table 8–1 as a guide, determine the number of categories to be included.
4. Determine the interval width for the categories. This is done by dividing the range by the number of categories. Round off as needed.

Table 8–1 Determining the Number of Bars in a Histogram

No. of Observations	No. of Categories (Bars)
< 50	5–7
50–100	6–10
100–250	7–12
>250	10–20

Source: Adapted from *The Memory Jogger*, by Goal/QPC, p. 39, with permission of Goal/QPC, © 1988.

5. Construct the categories by starting with the lowest data point and adding the interval width to it. This becomes the first category. The interval width is added to each consecutive number until a category that includes the largest data point is achieved.

Plotting the Graph

Histograms can be constructed manually on graph paper or generated from a spreadsheet software program that has graphics capabilities. Using the categories from above, the horizontal axis can be plotted. The frequency measurement, usually expressed as the actual number of occurrences, is plotted on the vertical axis. This done, the number of data points for each category is determined and the corresponding number is plotted on the graph.

When constructing a histogram, it is important that the bars be connected. In most bar graphs, including Pareto charts, the bars represent distinct findings. As such, the bars are not connected. Histograms, however, chart a continuum of data. Connecting the bars illustrates this important difference (Spoeri 1991). The general format of a histogram is shown as Figure 8–4.

When completed, the histogram provides information about both the distribution and variation in the work process being evaluated. The distribution is the general shape of the bars as they are spread across the graph. All histograms do not have a

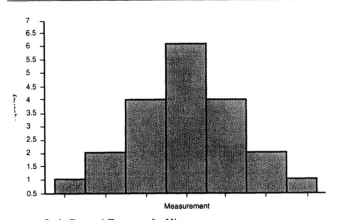

ure 8–4 General Format of a Histogram

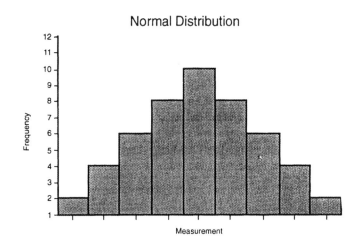

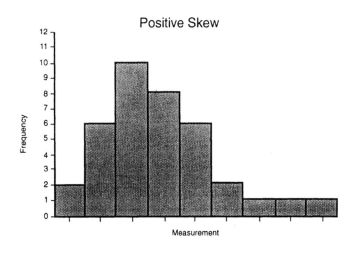

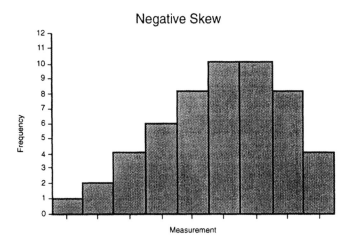

Figure 8–5 Histogram Distributions

normal distribution, also known as a bell shape. Rather, many work processes have a natural skew. Distributions commonly observed in histograms are shown as Figure 8–5.

Process variability is depicted by the "spread" of the data over the graph. When the data are limited to a few categories with a high frequency in each, the process has a low variability. In other words, the range for the data set is limited. When the range of the data is large and the frequency for single categories is small, the process has large variability. Examples of histograms with differing variability are shown as Figure 8–6.

Interpreting a Histogram

Histograms can provide work groups with a great deal of information on which to make process improvements. In general, processes that show a great deal of variability are ripe for improvement. Likewise, evaluation of current findings against desired service standards enhances objective evaluation on which to base changes.

For example, there is a nutrition department that has a goal of initiating 90 percent of outpatient instructions within ten minutes of the patient's arrival for their appointment. A work group looking at this process develops a histogram that depicts the current practice (Figure 8–7). Clearly, to meet its goal, the work group must work to reduce both the time it takes to get patients seen and the variability in existing practices.

When evaluating a histogram it is important to know that, although many work processes have a natural skew, the skew should remain consistent during multiple-sampling episodes. In other words, if an initial set of data has a positive skew and an improvement to the work process is made, the next data set should have the same positive skew with a smaller variation.

Occasionally, a histogram will show a distribution with more than one peak. Multiple, distinct peaks indicate that the reason for variation in the process is coming from more than one source.

When constructing a histogram, it is important to pick your category expression carefully to be sure that the histogram is providing information about the question that you are trying to answer. For example, Figure 8–8 shows two histograms that present data about 50 patients who recently completed a weight management program. The top histogram shows the patients' weight in pounds and the bottom graph depicts weight as a percentage of the goal weight established at the beginning of the program. Although both are valid examples of properly constructed histograms, the bottom one is preferred if the objective is to demonstrate the effectiveness of the weight management program.

Like Pareto charts, the use of histograms is not limited to the beginning of a process evaluation. Repeating the histo-

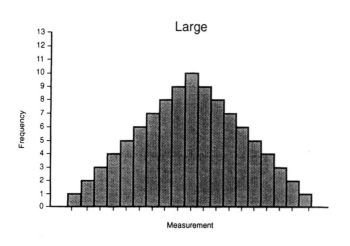

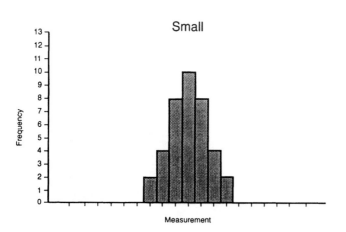

Figure 8–6 Histogram Variability

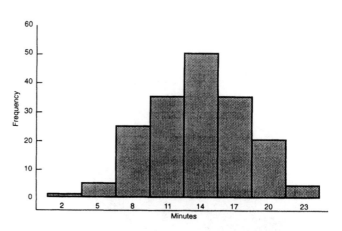

Figure 8–7 Histogram Case Study ($n = 175$): Wait Time To See Dietitian

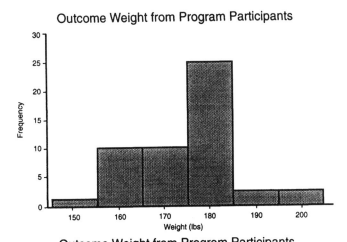

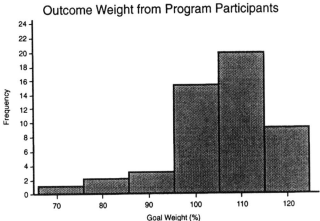

Figure 8–8 The Importance of Selecting Measurement

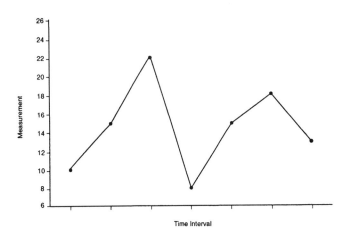

Figure 8–9 Run Chart

gram after putting a change into effect is helpful in evaluating the degree of success and, if done repeatedly, ensuring that gains are maintained.

RUN CHARTS

The run chart, which is also known as a trend chart, is another quality tool that is easy to construct. The run chart provides a visual representation of how a measured event is changing over time. It is a line graph on which each point represents one observation period. The average of the observation periods can also be visually represented on the chart. An example of a run chart is shown as Figure 8–9.

Run charts are used to determine the presence of trends in how a measured event is being performed. In quality management, run charts are invaluable for monitoring performance over time. One of the most important uses of run charts is to confirm that gains made as a change in a process are maintained. When graphically illustrated in combination with the average, data can be assessed in relation to a static point. This can be helpful, especially when a large number of data points are included.

Constructing a Run Chart

Run charts are not difficult to construct. Like Pareto charts, they can be done manually on graph paper. Software packages with graphic capabilities are also an alternative. Maintaining the run chart measurements in a computerized file is an easy way to keep concise, consistent information as well as have ongoing chart printouts. The steps to constructing a run chart are as follows:

1. Determine the trend to be tracked.
2. Establish the most appropriate measurement expression and place it on the vertical axis.
3. Decide on the time interval and place it on the horizontal axis.
4. Collect the data.
5. Plot each measurement period as a point on the graph.
6. Connect the points.
7. Calculate the average of the points and draw it as a horizontal line through the graph (optional).

There are four critical success factors when using run charts. First, all of the data points must be plotted in chronological order. Charting measurements in chronological order is critical, because the purpose of a run chart is to evaluate trends over time. If the measurements are not represented in this manner, trend analysis is meaningless. Although not critical, graphing data in consistent time intervals assists in visualizing what is happening in the workplace. This point is illustrated in Figure 8–10. The top graph would, without close scrutiny, lead one to believe that there was a precipitous decrease in the rate of menu entry errors. By keeping time intervals constant, as shown on the bottom graph, the data now show a steady downward trend in the error rate.

Second, data must be expressed in a consistent manner. Common expressions include percentage and number ob-

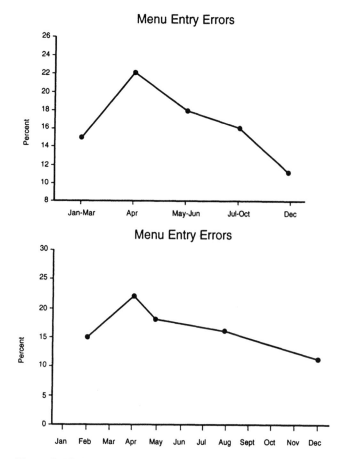

Figure 8–10 The Importance of Time in Run Chart

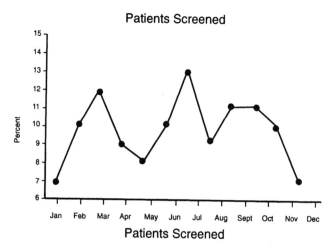

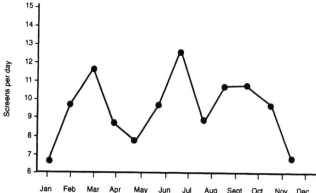

Figure 8–11 The Use of Measurements in Run Charts

served. The method of expressing the measurement should be determined by evaluating the purpose for developing the run chart. For example, Figure 8–11 shows two run charts that are used to determine trends that affect screening patients for nutrition risk. The first run chart identifies the number of patients that need to be screened in relation to the number of total patients in the hospital. Because of its purpose, the measurement for this run chart is expressed as a percentage. The second run chart tracks the number of nutrition screens performed per day. The measurement used in this case is the actual number.

Third, be sure to mark the chart axes clearly with respect to the measurement units and time intervals used. To provide meaningful information to many people, it is important that there be clarity as to what is being charted over what period of time. Although it is always important, this factor takes on special meaning when the run chart is to be displayed in the work area.

Finally, data must be interpreted cautiously. The single greatest error in the use of run charts is undue significance given to a single data point. This is inappropriate. The purpose of run charts is to determine trends. Therefore, action should not be taken nor judgments made until or unless a clear trend is established. This requires careful, thoughtful interpretation.

Evaluating Trends from a Run Chart

When evaluating a run chart for trends, it is expected that data points will fall above and below the average line. If the trend is holding steady, one can expect the number of points above and below the average line to be about the same. If, however, the run chart has nine or more consecutive points on either side of the average line, this means that something significant is happening. Likewise, if there are six or more sequential points that are increasing or decreasing without any reversals, then something significant is happening.

Run charts, although able to pick up significant trends, do not provide insight into the cause of the change. They are wonderful, however, for pointing out trends before they may be obvious. This facet enhances the ability to correct problems before a crisis occurs.

Negative trends that are revealed by a run chart require investigation. Problems in a work process, departmental interfaces, and/or other causative agents can be determined and removed by using other quality tools. Marking the time when changes are made on an ongoing run chart can assist in tracking improvements. Keeping the run chart going is an effective means of ensuring that gains are maintained.

Positive trends also need investigation because they indicate that something has changed for the better. The effective use of other quality tools assists in determining the source of the improvement, recognizing those who are responsible for it, and incorporating it into the permanent structure of the work process. Continuation of the run chart, with the upward-adjusted average, illustrates the overall improvement in the task measured.

Run charts can also be an effective tool for communicating information to work groups and motivating continued performance. By posting run charts of key work processes prominently in the work area, individuals can be kept abreast of performance. Upward trends can inspire effort, and the positive effects of improvements in work processes can be readily seen by everyone involved. Used in this manner, run charts are a simple, yet visually powerful, quality tool.

CONTROL CHARTS

Control charts are a variant of run charts. In addition to showing trend information over time, control charts contain calculated control limits. These limits indicate, on a statistical basis, what the boundaries of variability are for a work process. That done, the process can then be evaluated within the boundaries. Like run charts, control charts are a valuable tool for evaluating trends in existing work processes. Control charts are particularly valuable for monitoring the effect of changes made to a process as the result of a quality improvement initiative.

In the language of quality management, data points that are within the statistical boundaries are called "in control," whereas those outside the boundaries are referred to as "out of control." The line that defines the upper boundary is termed the upper control limit (UCL). The lower boundary is called the lower control limit (LCL). All control charts have UCLs; LCLs may or may not be used. Lower limits are not included in control charts when either the lower boundary is not relevant to the process being examined or the calculated LCL is a negative number that is impossible to achieve.

The theoretical basis for control charts lies in the premise that work processes have natural variation. For example, when a marinara sauce is made according to its recipe, the actual sodium content may vary between 140 and 160 milligrams per cup. This natural variation is due to the slight differences in the sodium content of the ingredients being used. If, however, a sample of the sauce came back from the lab with a sodium level of 250 milligrams per cup, then this would indicate that something extraordinary has occurred. Control charts use statistical data that assist in differentiating between natural variations in a work process and those that are due to some external factor. Put simply, control charts are a quality tool that measures when the process is at fault and when other factors are responsible for a problem.

Developing a Control Chart

There are several kinds of control charts. Different formulas and charts can be constructed based on the data that are being used. For example, a "P" chart calculates the control limits for proportional defects with qualitative characteristics. A comprehensive description of the types of control charts and the formulas needed to calculate the limits is beyond the scope of this text. *The Memory Jogger* (1988) provides a concise summary of this material.

For most clinical applications, control charts can be constructed using upper and lower limits of two standard deviations (Spoeri 1992). When using this method, the UCL is equal to the mean plus two standard deviations; the LCL, if used, is the mean minus two standard deviations.

The construction of a control chart begins with the run chart. If not already done, the mean (i.e., average) of the data contained in the run chart is calculated. Using the data and the mean, the standard deviation is then calculated (Exhibit 8–5). The standard deviation is then doubled and added to the mean to establish the UCL and doubled and subtracted from the mean to determine the LCL. Although the standard deviation can be calculated manually, the process is made much simpler and faster by using a spreadsheet software package. Additional advantages of using a computer include the ability to display the control chart graphically directly from the data and to add to and revise the chart as more observation periods are conducted. The control chart is plotted by adding the mean, UCL, and LCL (if appropriate) to the run chart. Figure 8–12 depicts the basic format of a control chart.

Interpreting a Control Chart

Data points that are between the plotted limits are "in control"; they are within the natural variation of the process as it exists in its current form. Any data points that are outside the limits are "out of control," meaning that some external force or action was introduced to the normal process. The chef's adding salt to the marinara sauce when the recipe did not call for it is an example of an external force. Natural variation can be improved by changing the work process. Occurrences that are "out of control" can be eliminated only by finding the source of the problem and removing it.

When interpreting a control chart, it is important to remember that control limits should not be confused with specifications. Specifications are established goals that have been set for a process based on the customer's need. For example, a nutrition department may have a specification that dietary consultations be acknowledged within 24 hours of receipt. A control chart that was constructed to evaluate this process could have a UCL well within (or outside of) the 24-hour mark.

Control charts provide a great deal of information to a work group, since both natural and unnatural variations in a work

Exhibit 8–5 How To Calculate Standard Deviation

Use the following formula:

$$s = \sqrt{\frac{\sum X^2 - \frac{(\sum X)^2}{n}}{n-1}}$$

Where X = a raw score
$\sum X^2$ = each raw score squared then added together
$(\sum X^2)$ = all the raw scores summed then the total squared

Example: Number of outpatient counseling sessions per week:

Raw scores Week		Raw score squared
Week 1	52	2704
Week 2	57	3249
Week 3	72	5184
Week 4	39	1521
Week 5	42	1764
Week 6	55	3025
Week 7	30	900
Week 8	64	4096

Sum = 411   Sum of squared raw scores 22,443
Average = 51
Sum squared = 168,921
n = 8

$$\text{Thus } s = \sqrt{\frac{22{,}443 - \frac{168{,}921}{8}}{n-1}} = \sqrt{\frac{22{,}443 - 21{,}115}{7}}$$

$$= \sqrt{\frac{1{,}328}{7}} = 189.7 = 190$$

$$s = \sqrt{190} = 13.7 = 14$$

Control chart upper limit = 14 × 2 = 28 + 51 = 79
Control chart lower limit = 51 − 28 = 23

---

process can be changed. By knowing the scope of a process' variability, the frequency of external influences, and the level of exactness required to meet the needs of the customer, informed decisions about improvement efforts can be made. This concept is best illustrated by using an example.

A CASE STUDY

A work group is concerned about the clinical outcome of patients with congestive heart failure. The physicians are having difficulty stabilizing the patients; frequent changes in medications are needed. Although including a 2-liter fluid restriction as part of the diet order is standard protocol for these patients, there is a concern that the restrictions are not being followed.

Figure 8–13 is a control chart that shows the actual average fluid intake of 20 patients who were on a 2-liter fluid restriction. Using a check sheet, the intake of each patient was recorded for a period of two weeks. The average daily intake was plotted on the chart. The mean and UCL (i.e., mean plus two standard deviations) were calculated from the average daily intake data. In addition, the specification of 2 liters was included on the chart to assist with evaluation.

Figure 8–13 shows that, on average, patients on 2-liter fluid restrictions drank less than their allowance. There was, however, a great deal of variability in the amount of fluid consumed. Indeed, the UCL of 3.5 liters suggests that, under the current process of administering fluid restrictions, a patient would have to have more than 1.5 liters over his or her allowance for the process to be considered "out of control." The control chart indicates that there is a significant problem in the process that is currently in place; it does not support the notion that a few patient care providers are ignoring the restriction.

Based on this finding, the work group can investigate reasons why the process has such a large variability. Perhaps nursing staff are unfamiliar with what foods are counted as fluids, or patients are not being adequately instructed about the need for limiting intake. By evaluating the existing process, investigating causes, and implementing solutions, the work group can reduce the variability of the system. Repeating the check sheet process and developing a new control chart will prove (or disprove) the effectiveness of their efforts.

BENCHMARKING

Benchmarking is the search for best practices that lead to superior performance and the measurement of practices, products, and services against these high standards (Camp 1989). The benchmarking process fits well with quality improvement efforts because its basic premise is one of opportunity, not past history. It can be used as a tool to set realistic long-term goals and as a tool to measure the effectiveness of changes made to meet these goals (American Hospital Association 1992).

"The first step in benchmarking is to understand your own process in detail" (Main 1992). These processes then can be compared internally to those of other departments or externally to those of other organizations. An example of the former would be comparing the nutrition services department procedures for scheduling outpatients with those of the hospital's other allied health departments. Comparing staff dietitians' productivity rates among a group of community hospitals is an example of external benchmarking. Benchmarking need not be limited to the health care industry. For example, the department may want to examine the billing practices of catalog companies in order to improve its own billing systems. Much more may be learned from retail companies with years of experience in billing clients than from other hospitals that may also just be getting up to speed in this area.

Quantitative Quality Management Tools 93

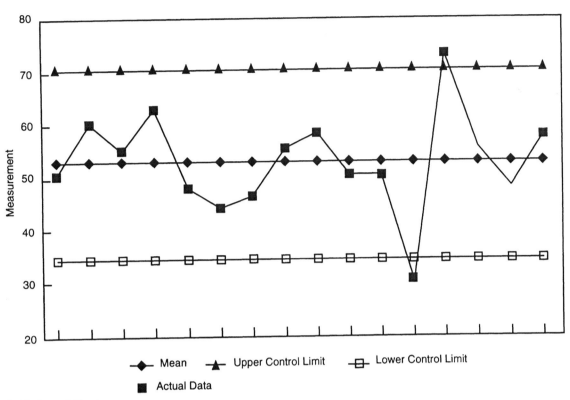

Figure 8-12 Control Chart

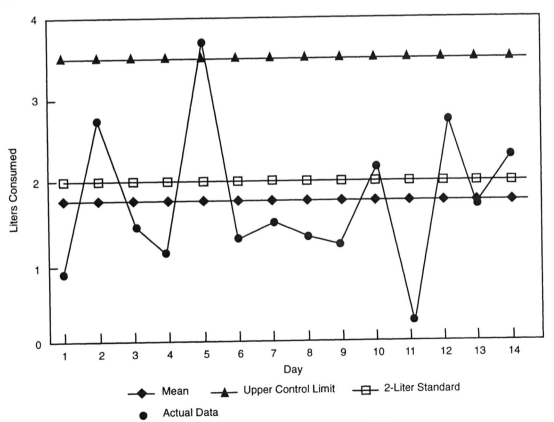

Figure 8-13 Sample Control Chart: Actual Intakes of 20 Patients on 2-Liter Fluid Restriction

Mecon Associates (University Hospital Consortium/Mecon 1992) describe a step-by-step process for conducting operational benchmarking:

1. Identify the area and criteria to be benchmarked.
2. Find benchmarking candidates.
3. Plan and conduct survey or site visits.
4. Develop action plan.
5. Follow up.

Benchmarking is conducted so that participants can compare their practices with those of the industry as a whole, its best performers, and closely matched peers within the group. Survey items addressing clinical nutrition practices may include

- total number of nutrition consultations completed within a specified time frame
- dietitian-to-patient ratio
- total billed revenues
- case mix index
- labor dollar expenses (University Hospital consortium/Mecon 1993)
- patient satisfaction with clinical nutrition services.

Statistical, organizational, and operational characteristics of each organization and department are used to determine which participants are most similar as a whole, and might best serve as a peer comparison.

The benchmarking process must be conducted ethically and legally. Confidentiality is respected, and use of the information is confined to the limits agreed upon in advance. In order for the process to be the most effective, all participants must be willing to exchange information mutually, honestly, and cooperatively (University Hospital Consortium/Mecon 1992).

If done effectively and continuously, benchmarking provides several benefits that can be used to the department's or organization's advantage. These include

- creatively incorporating the best practices into operations
- stimulating and motivating staff and providing opportunity for professional growth
- encouraging receptiveness to change (Camp 1989).

The focus should be on comparing current performance with what could be achieved if improvements identified through benchmarking were implemented (Staff 1992).

SCATTER DIAGRAMS

Scatter diagrams are used to show the relationship between two qualities or variables, one on the x-axis and the other on the y-axis. Variations can be seen at a glance.

An example of a scatter diagram is shown in Figure 8–14. In this example, the dietitians were reviewing data regarding late

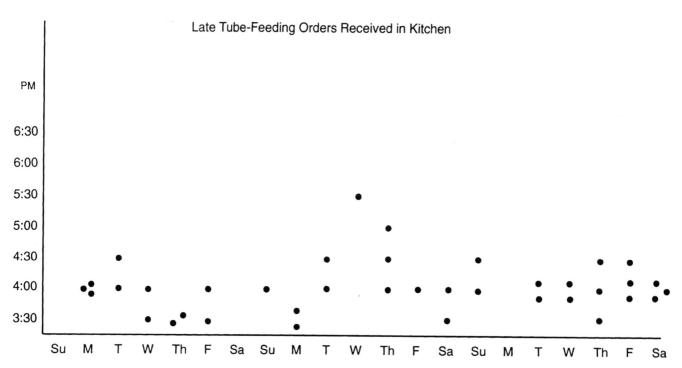

Figure 8–14 Scatter Diagram

tube-feeding orders. The scatter diagram was developed by using days of the week for the x-axis and time of day for the y-axis.

A scatter diagram is created by putting a point on the graph at the appropriate point for each occurrence being studied. In the example shown in Figure 8–4, tube-feeding order slips were used as the data source. For each day, every late tube-feeding order received in the kitchen was recorded by putting a point on the graph. On the first Monday, three late orders were received, all around 4:00 P.M. The second Wednesday, an order was received at 5:30 P.M. By analyzing data in this fashion, patterns of activity are easily grasped. Scatter diagrams can also be used to illustrate such things as the relationship between tray accuracy and volume of meals served (Jackson 1992, 142), hospital length of stay and incidence of nutritional risk by level, and patient satisfaction with dietary counseling in relation to number of scheduled counseling sessions attended.

REFERENCES

American Hospital Association. 1992. *Benchmarking: Comparing your hospital's performance against the "best."* Teleconference. Chicago: American Hospital Association.

Camp, C. 1989. *Benchmarking: The search for industry best practices that lead to superior performance.* Milwaukee, Wis: ASQC Quality Press.

Jackson, R. 1992. *Continuous quality improvement for nutrition care.* Amelia Island, Fl: American Nutri-Tech, Inc.

Main, J. 1992. How to steal the best ideas around. *Fortune*, October 19, 102–106.

The memory jogger. 1988. Methuen, Mass: Goal/QPC.

Spoeri, R.K. 1991. The inspection of data. In: *Quantitative methods in quality management: A guide for practitioners.* Chicago: American Hospital Publishing, Inc.

Spoeri, R.K. 1992. *Developing quality indicators: The inspection of data.* Teleconference. Chicago: American Hospital Association. May 28.

Staff. 1992. Clinical benchmarking project demonstrating potential of shared information. *University Hospital Consortium News*, September/October, 4.

University Hospital Consortium/Mecon. 1992. *Case study operational benchmarking for managers.* In-service. San Ramon, Calif: Mecon Associates.

University Hospital Consortium/Mecon. 1993. Benchmarking task force minutes. Oak Brook, Ill.

Part III

Monitoring and Evaluation

Chapter 9

Departmental Planning for Continuous Quality Improvement

Quality assessment and improvement begin with creative planning, the key to any successful program (Marshik-Gustafson et al. 1981). Planning is at the heart of any quality management program, such as the ten-step process presented in Chapter 4. Implementation phases of the process will fall short of expectations if insufficient rigor is applied during the planning steps. Hopkins (1990) noted that many times the failure to implement effective quality review systems can be traced directly to a misunderstanding about the clear differentiation between planning and implementation tasks associated with quality improvement initiatives.

This chapter contains background information and examples of comprehensive planning for the quality review process:

- outlining roles and responsibilities
- developing a scope of care
- defining important aspects of care

Also included is a discussion of working with cross-functional teams to make nutrition an integral part of total patient care as defined by other departments.

ASSIGN ROLES AND RESPONSIBILITIES

As described elsewhere in this manual, departmental mission and goal statements, organization, levels of acuity and priority systems, steps of the nutrition care process, and responsibilities of dietetic team members are essential facets of the framework for quality nutrition services. Documents, policies, and procedures related to these elements of nutrition services should be in place before a quality management system is designed.

The first step in planning an effective quality review system is the delineation of responsibilities for quality-related activities. Tasks should be outlined for each person involved in the quality process, as shown in Exhibit 9–1.

The department director provides overall leadership for the development and implementation of the quality management process. This responsibility includes the following:

- working with the dietetic or multidisciplinary teams to define scope of practice, important aspects of care, patient populations, major disorders requiring nutritional intervention, and quality indicators
- reviewing annually the responsibility for monitors and audits, the scope of practice, the aspects of care and their weighted value, the indicators, and data-collection methodology (Hopkins 1990)
- scheduling routine quality reviews
- motivating and training staff for continuous quality improvement
- role-modeling commitment to quality
- overseeing quality review activities

Some hospitals devote a full-time equivalent (FTE) to quality management activities. In these cases the quality manage-

Exhibit 9–1 Roles and Responsibilities for Quality Monitoring and Evaluation

---

A. Director, Clinical Nutrition, is responsible for:
- development and implementation of the Quality Management Program;
- evaluation of all medical staff, nursing, and parent complaints regarding clinical nutrition care;
- reporting quality management activities within the department to the Vice-President, Support Services, on a monthly basis, and the hospital's Quality Management Department on a quarterly basis; and
- reporting the annual assessment of effectiveness of the department's quality management plan.

B. Assistant Director, Clinical Nutrition, is responsible for:
- any quality management activities requested by the Director;
- monthly data collection and analysis;
- monthly reporting of quality management data to clinical nutrition staff; and
- assisting with the implementation of actions to improve care when appropriate.

C. Pediatric Clinical Nutritionist is responsible for:
- completion of Meal Rounds surveys; and
- any other quality management activities requested by the Director and Assistant Director.

D. Special Nutrition Technician is responsible for:
- any quality management activities requested by the Director and Assistant Director.

E. Diet Assistant Team Leader is responsible for:
- assisting with data collection; and
- any quality management activities requested by the Director and Assistant Director.

F. Department Secretary is responsible for:
- typing and distribution of all reports and minutes of department meetings concerning the quality management program.

Courtesy of Alice E. Smith, Director, Clinical Dietetics, The Children's Memorial Hospital, Chicago, Illinois.

ment dietitian is responsible for coordinating the data-collection process within the department, including development of data-collection tools and protocols (Gardner 1990).

Often the chief clinical dietitian or clinical nutrition manager is responsible for clinical nutrition quality management and chairs the departmental quality committee. Other personnel participate in the quality process by consistent delivery of first-rate performance of their job responsibilities, constant vigilance for possible shortfalls in service, membership on quality circles or task forces, and conduction of quality monitoring and evaluation activities as assigned.

DELINEATE THE SCOPE OF SERVICES

This step forces a department to list "what is done in the department" to address the needs of inpatients, outpatients, or other "customers." Each department must delineate its scope of services as a basis for identifying those important aspects of care that will be monitored and evaluated. The defined scope of care should include major patient population groups, disorders requiring nutritional intervention, teams that include nutrition services, and procedures or types of care that are commonly provided by the department. The scope of services provided by a department of nutrition services in a major research hospital in which food services are provided under contract by an outside company will be very different from the scope of services offered by the dietary department of a small community hospital. Exhibits 9–2 and 9–3 illustrate the scope of services for two different hospitals.

A fundamental principle of continuous quality improvement is attention to customer needs. Who the customer is should be clearly evident in defining the scope of services. For example, note that Children's Hospital (Chicago) gives attention to both the family and the hospitalized child (Exhibit 9–2). Southwest General Hospital (Exhibit 9–3) highlights an array of groups for whom food services are provided; the scope of care also delineates units or programs in which dietitians are involved.

Quality care is the bottom line in any hospital quality management program. Without standards and guidelines, wide variations in practice suggest that some patients receive unnecessary care, while others fail to receive needed services (Leape 1990).

Thus it is important to have standards or guidelines to undergird services delineated in the scope of care. As described in Chapter 6, guidelines or standards are "standardized specifications for care developed by a formal process that in-

Exhibit 9–2 Scope of Care: Children's Memorial Hospital, Chicago

---

The Department of Clinical Nutrition offers professional services to hospitalized patients and families and to ambulatory patients and their families. Services include the following:

- screening of new admissions for nutritional risk;
- assessment of at-risk patients to evaluate nutritional status and recommend nutritional support;
- assisting patients and families with menu selection;
- nutrient analysis of fluids and solids consumed;
- special nutrition instruction to prepare the child and family for discharge and home care; and
- monitoring of nutrition support and care plans for appropriateness, accuracy, and effectiveness.

Professional services are provided by registered dietitians in accordance with a physician's written order, at the request of appropriate medical and nursing personnel, or at the initiative of the registered dietitian. Under the supervision of the registered dietitians, the Special Nutrition Technicians prepare and distribute specialty infant formulas, tube feedings, and supplements to inpatients.

Other professional services include in-service education for medical and nursing personnel, training for dietetic students, community education, and clinical research.

Courtesy of Alice E. Smith, Director, Clinical Dietetics, The Children's Memorial Hospital, Chicago, Illinois.

Exhibit 9-3 Scope of Care: Southwest General Hospital

> The Nutrition Services Department provides the following services:
> - quality meals for patients, employees, physicians, visitors, and volunteers;
> - nourishments and supplements;
> - tube feedings;
> - modified diets;
> - nutrition counseling and education (inpatient and outpatient); and
> - nutritional assessments.
>
> The Nutrition Services Department provides nutritional care and services to patients on medical-surgical, cardiac step-down units, intensive and coronary care units, obstetric unit, pediatric unit, mental health unit, and a chemical dependency unit. The department is also involved in the stroke, diabetic, dysphagia, and cardiac rehabilitation programs, hospice and acute renal dialysis programs. A clinical dietitian is assigned to each of these patient units and is an active team member of each program.
>
> Courtesy of Rhonda Billman, Clinical Manager, Nutritional Services, Southwest General Hospital, Middleburg Heights, Ohio.

Exhibit 9-4 Characteristics of Useful Guidelines and Standards

> - Comprehensive—Includes all situations and indications for use of the procedure.
> - Specific—Clearly describes exact processes or nutrient needs; clearly differentiates one set of circumstances from another.
> - Detailed—Uses features such as laboratory or clinical data to distinguish between those who will benefit from certain aspects of nutritional care and those who will not.
> - Inclusive—Includes all major relevant factors that must be taken into consideration in the decision to recommend or carry out nutritional procedures.
> - Manageable—Presentation format, structure, and content are sufficiently simple that the guidelines can be used in routine practice.
>
> *Source:* Copyright 1990 by the Joint Commission on Accreditation of Healthcare Organizations, Oakbrook Terrace, Illinois. Reprinted from the *Quality Review Bulletin*, Vol. 16, with permission.

corporates the best scientific evidence of effectiveness with expert opinion" (Leape 1990). As the terms indicate, *guidelines* and *standards* are "recommendations" for care of individual patients. *Standards of Practice,* published by the American Dietetic Association (1986), lacks the specificity needed for general use by practitioners. Characteristics of useful guidelines and standards are outlined in Exhibit 9-4.

Departmental standards of care or standards of practice are integral to any quality assurance plan. Such standards, adapted to each institution, can serve as the basis for developing components and implementing a quality review. If the department does not have standards, they can be developed as different pieces of the quality management program are put in place.

Standards should be developed for problems and procedures most frequently encountered in practice. Such standards eliminate some of the inconsistency and ambiguity that creep in when several individuals perform similar tasks without the benefit of protocols or guidelines. For example, in the clinical area, standards of care or protocols should be developed to guide departmental processes of

- screening for nutritional risk
- nutritional assessment and reassessment
- assisted feeding
- monitoring nutritional intakes
- monitoring changes in nutritional status
- calorie counts
- patient education
- documenting nutritional care
- discharge planning
- food-drug interaction teaching
- nothing by mouth (NPO) or clear liquid diet orders (Huyck and Rowe 1990).

As Huyck and Rowe pointed out, standards of care should include the following components as illustrated in Exhibit 9-5:

- population—patients to be involved
- objective—rationale for the standard
- process—how it will be done
- personnel—who will do it
- time frame—when it will be done
- monitoring—how the work will be evaluated
- documentation—how the process and outcome will be communicated.

Standards of practice may also be developed for disease-specific groups or selected populations. The sample standards of practice given in Exhibit 9-6 includes guidelines for the nutrition care process and its application for individual patients. Depending on the case mix of the hospital, standards of practice may be created for a variety of situations, such as methods of feeding (i.e., tube feedings), diagnoses (i.e., renal disease), hospital service area (i.e., outpatient radiation therapy), illness stage (i.e., rehabilitation or cardiac step-down services), or life stage (i.e., infants or elderly).

As described in Chapter 6, standards may be adapted from other sources but should be individualized to suit specific needs, practices, and resources at your hospital. If published standards are adopted and revised for institutional use, copyright laws demand that permission be obtained and credit be given to the original source when printing revised standards.

IDENTIFY IMPORTANT ASPECTS OF CARE

Important aspects of care can be considered as groupings of interrelated processes or activities that most affect the quality of care and services delivered in a health care organization. This step of planning for quality management requires the de-

Exhibit 9-5 Standard of Nutrition Care for Rehabilitation Unit Patients

---

Objective

To provide a plan of nutritional care to bring the patient to the highest level of independence while providing adequate nutrition.

Personnel

Clinical dietitian with referral to the dietetic technician at the time of implementation when appropriate.

Time Frame

Within 48 hours of admission to the unit.

Process

Review medical record.
Interview patient.
Document height, weight, %IBW, nutritional status, feeding problems, nutrient needs, dietary modifications.
Document nutrition care plan for nutrient sources, education needs, plans for referral.
Attend patient care conference.
Implement care plan.

Monitor

Nutritional status
Feeding skills
Understanding
Documentation
Changes in nutritional status.
Progress toward independence.
Recommendations.

NOTE: %IBW, percentage of ideal body weight.

Source: Reprinted from *Managing Clinical Nutrition Services*, by N.I. Huyck and M.M. Rowe, p. 123, Aspen Publishers, Inc., ©1990.

---

partment to identify high-volume, high-risk, high-cost, and/or problem-prone activities or processes generally considered within their scope of practice. This exercise is essential so that quality monitoring and evaluation focuses on department activities having the greatest impact on patient care.

Practitioners sometimes confuse aspects of care and indicators (Hopkins 1990). Aspects of care encompass those few elements that most affect the quality of care and services delivered in the health care organization. Indicators, on the other hand, are measurable variables related to a structure, process, or outcome of care. Important departmental functions or aspects of care are target areas that can be analyzed to reflect the quality of services provided; indicators are measuring devices used to help identify specific performance issues that should undergo intense review to assess quality of care.

When several individuals are involved in delineating important aspects of care, the brainstorming process may be used to generate ideas. When the draft list of ideas has been generated, a Delphi technique may be used to decide which elements will remain in the final list. Also refer to Chapter 7 for further discussion of brainstorming and use of the Delphi technique.

Staff may provide input for key aspects of care in other ways, too. For example, at the Ohio State University Medical Center, dietitians and dietetic technicians were surveyed to ascertain what each perceived as main functions, primary customers, and important customer expectations. As shown in Exhibit 9-7, perceptions were diverse and formed the basis for further discussion and priority setting.

After reaching consensus on the list, review each aspect of care to determine whether it is high volume, high risk, high cost, problem prone, or all of these. This last step is critical to help select those aspects of care that will receive priority in monitoring and evaluation. Exhibit 9-8 provides a format that may be used in generating the list of important aspects of care.

Exhibits 9-9, 9-10, and 9-11 illustrate important aspects of care as defined by three different hospitals.

Important aspects of care should be reviewed annually and revised if appropriate. For example, Children's Hospital (Chicago) initially delineated four aspects of care:

1. evaluation and assessment of patients' nutritional needs
2. development of nutrition care plans and goals
3. periodic assessment of the effects of nutrition therapy
4. education/consultation of patient and family.

However, the hospital plans to reduce the number of important aspects of care to simplify the quality management process. The Joint Commission requires that at least two indicators be developed for each aspect of care. Twelve indicators are currently monitored, a number deemed excessive given the sample size required for evaluation of each indicator. Instead of numerous departmental indicators, nutrition services plans to shift its emphasis to cross-functional (multidisciplinary) indicators.

Patient and Customer Satisfaction: An Important Aspect of Care

In most quality management systems, customers are placed at the center of attention. Thus patient satisfaction must be addressed as an important aspect of nutrition care. Components of satisfaction and methods for assessing patient satisfaction are discussed at length in Chapter 3.

Long-term care marketing expert Becky Branan notes that customers "include not only residents and their families, but also facility staff, physicians, hospital discharge planners, federal and state surveyors, volunteers, and the community" (Tishman 1992). Correspondingly, dietetic professionals must take a broad view of who comprises its customer base, and take steps to ensure that each is given appropriate focus in the list of activities that will comprise the basis for quality improvement activities.

Exhibit 9–6 Standard of Practice: Enteral and Parenteral Nutrition

	Initiation To Occur within 24 Hours of Order for PN or EN	
	Enteral Nutrition	*Parenteral Nutrition*
1. Complete assessment elements 1–5.		
2. Determine caloric and protein needs: BEE × 1.25 × stress factor; or Kg wt. × 30, 35, or 40.		
3. Determine formula or TPN appropriate for patient; or Assess appropriateness of formula or TPN composition as ordered.	1. Select appropriate formula. If another is ordered and inappropriate, request change. 2. Start TF at 300 mOsm (1/2 full strength). 3. Start rate at 50–75/hr. If another rate or concentration is ordered and inappropriate, request change.	1. Determine appropriate percentages of amino acids, dextrose, and lipid as well as their rate of infusion. 2. Assess the above with what is ordered and, if appropriate, request change to better meet patient's needs.
4. Include document in medical record.		
5. Assess labs available and request labs needed.	1. Should have albumin available. 2. Transferrin is best to follow for progress. Should be ordered by Tuesday AM. 3. Follow other labs as appropriate.	1. Follow TPN protocol. 2. Follow other labs as appropriate.

Parenteral and Enteral Nutrition Monitoring

	Frequency	*Enteral*	*Frequency*	*Parenteral*
1. Check formula or TPN composition and compare to that ordered for previous day.	Daily (M–S)	If no change made or needed, continue with monitoring. If change made, look for rationale. If change made and not necessary, then speak with MD and request change.	Daily (M–S)	
2. Check rate and concentration of TF and rate of TPN and compare to that ordered for previous day.	Daily (M–S)	If increased, then calculate cal., pro., and cc free water to be given. If decreased, then calculate the above and determine the rationale for change.	Daily (M–S)	(No need to calculate fluid balance unless patient fluid balance is an issue.)
3. Check weight of patient.	2 ×/week	Check as available. Wt. gain ideally less than 2 lbs/week.	2 ×/week	
4. Check I and O sheet for ml. of TF/TPN/lipid infused previous day.		When building patient up to ideal rate and concentration, do this daily and calculate intake. When patient stable, do this 2 ×/week unless changes occur.	Daily (M–S)	
5. Check labs for nutritionally pertinent indicators and tolerance.	Daily (M–S)	Check nutritional parameters as available. Check hydration parameters (BUN, Na, etc.).	Daily (M–S)	Follow TPN protocol.
6. Check with RN for tolerance.	Daily (M–S)	Check stool, residual, bloating, etc.	Daily (M–S)	
7. Document.				

Note: BEE, basal energy expenditure; TPN, total parenteral nutrition; TF, tube feeding; I and O, intake and output; BUN, blood urea nitrogen.

Source: Reprinted from *Managing Clinical Nutrition Services*, by N.I. Huyck and M.M. Rowe, p. 119, Aspen Publishers, Inc., © 1990.

Exhibit 9–7 Form Used for Staff Input on Department Key Functions and Customers

---

Area: Clinical Nutrition

Circle one: Dietitian   Dietetic Technician

1. Identify your area's main functions, products, and services. What does your department do? (please list up to 5)
   Summary of results:
   - Nutritional assessments
   - In- and out-patient education. That is, educate patient and family on prescribed diet and home enteral feeding.
   - Provision and documentation of nutritional therapies such as tube feedings and oral supplements

2. Identify the customers who experience those main functions, products, and services. Who and what areas get your work? Please include areas inside and outside your department.
   Summary of results:
   - In- and out-patients and families
   - Physicians
   - Nurses and other hospital staff
   - Dietary clerks
   - Students
   - General community population

3. In your opinion, what are the most important customer expectations?
   Summary of results:
   - Patient education that is effective. That is, patient understands discharge diet or diet counseling.
   - Expect a palatable, hot meal
   - Receive understanding and compassion
   - Optimal nutritional care
   - Timely follow-through

Courtesy of the Ohio State University Medical Center, Columbus, Ohio.

CROSS-FUNCTIONAL INITIATIVES

Nutrition services are often integral to quality services provided by others within the institution (Hard 1991). The 1994 Joint Commission accreditation manual reflects the philosophy that key health care functions often necessitate collaboration between several different departments (1994 Accreditation Manual 1991). The new approach focuses on "key processes" rather than departments. For example, "Care of the Patient" may include such processes as

- patient rights
- admissions
- patient evaluation
- nutritional care
- nonoperative and other invasive procedures
- patient and family education
- continuity of care.

Exhibit 9–8 Aspects of Care

List the major aspects of care and/or services provided by your department/service. For each item, assign attributes of high volume, high risk, and/or problem prone, as appropriate.

Aspects/Services	Attributes*
1	
2	
3	
4	
5	
6	
7	
8	
9	
10	
11	
12	

*HV = high volume   HR = high risk   pp = problem prone

Courtesy of Hospital Council of Southern California, © 1990.

This new organization of content is designed to reflect the importance of cross-functional groups in delivery of high-quality care.

Nutrition support team quality plans already include patient-focused nutritional outcomes. Owens et al. (1989) reported a plan that included nutritional assessment, nitrogen balance, and total iron-binding capacity as quality indicators, all facets of care normally monitored by dietitians who care for patients receiving parenteral nutrition. Exhibit 9–12 shows important aspects of care and related indicators delineated by another nutrition support team (Powers et al. 1991). The aspects of care described are clearly cross-functional and need

Exhibit 9–9 Important Aspects of Care: Henry Ford Hospital

The following functions, considered important aspects of care, are monitored and evaluated:
  a. Evaluation and assessment of patients' nutritional needs.
  b. Development of nutrition plans and goals.
  c. Periodic assessment of the effects of nutritional therapy.
  d. Counseling and evaluation of patients/families.
  e. Quality control of food services.
  f. Environmental assessment of sanitation, infection control, safety.

Courtesy of Hildreth A. Macy, Associate Director, Food and Nutrition Services, Henry Ford Hospital, Detroit, Michigan.

Exhibit 9-10 Important Aspects of Care: Holyoke Hospital, Inc.

Important aspects of patient care and service have been identified and these are the basis for the department's quality assurance program. They are as follows:
- evaluation of the patient's nutritional needs
- appropriateness of the diet order to diagnosis
- appropriateness of enteral/parenteral nutrition
- appropriateness of NPO or Clear Liquid Diet
- assessment of patient's mealtime intake
- assessment of effects of nutritional therapy
- change in patient status
- education and understanding of the patient

Courtesy of Marcia Durell, Chief Clinical Dietitian, Holyoke Hospital, Inc., Holyoke, Massachusetts.

---

cooperation of a dietitian for proper implementation. This delineation also gives clear indications or guidelines for both enteral and parenteral nutrition.

Examples of cross-functional activities are also contained in nursing literature. For example, Standards of Oncology Nursing Practice (Maiskowski and Rostad 1990) include nutrition as one of 11 high-incidence problem areas common to patients with cancer. This illustrates how nutrition care processes can be easily integrated into the nursing process of data collection, goals, planning, intervention, evaluation, and reassessment (McGuffin and Mariani 1990).

The trend is for nurses to develop unit-specific quality assurance plans (Leary 1990). Unit-specific plans allow health care professionals to individualize quality analysis to include the unique needs of patients on that service. For example, on a geriatric medical unit elderly patients are debilitated; they often suffer from nutritional problems, and skin care is a high-volume, high-risk, and problem-prone aspect of care. The prevalence of pressure ulcers is clearly a measure of quality care—and the nutritional status of the patient is a key factor in maintaining skin integrity. Thus, dietitians are important members of cross-functional teams, and they should be in-

Exhibit 9-11 Important Aspects of Care: Southwest General Hospital

---

1. Evaluation and assessment of patients' nutritional needs (high volume and high risk).
2. Patient and family nutrition education and counseling (high volume and high risk).
3. Safe food storage (high volume and high risk).
4. Food preparation, quality, and distribution (problem prone, high volume, and high risk).
5. Distribution and nutritional monitoring of enteric tube feedings (high risk and problem prone).

Courtesy of Rhonda Billman, Clinical Manager, Southwest General Hospital, Middleburg Heights, Ohio.

Exhibit 9-12 Important Aspects of Care Delineated by a Nutritional Support Team

---

- Nutritional assessment and consultation
- Indications for enteral and parenteral nutrition support
- Provision of optimal nutrition support including the attainment of nutritional goals and the prevention, detection, and management of complications

Source: Reprinted from Powers, T., Deckard, M., Stark, N., Cowan, G.S., A Nutrition Support Team Quality Assurance Plan. *Nutrition in Clinical Practice,* Vol. 6, pp. 151–156, with permission of the American Society for Parenteral and Enteral Nutrition, ©1991.

---

volved in the development of unit-specific, patient-focused plans. Cross-functional teams are discussed further in Chapter 17.

Another example of dietetics-nursing collaboration is highlighted in a comparison of the mealtime care given to patients by nurses using two different meal-delivery systems (Carr and Mitchell 1991). A research study revealed that the use of meal-delivery systems designed to free nurses from "non-nursing" functions may have an adverse effect on nurses' involvement in other aspects of mealtimes, such as checking patients' well-being, observing feeding difficulties, providing assistance during mealtimes, and observing the contents of discarded dishes. A cross-functional discussion of issues related to patients' mealtimes may uncover unsuspected problem areas that could have an impact on the overall quality of patient care.

QUALITY MANAGEMENT PLAN

These three steps (assign responsibility, delineate scope of care, identify important aspects of care) form the basis for a quality improvement plan. Many hospitals have a standard format that must be used by every department in the institution. Generally, the plan includes

- responsibilities
- scope of care
- important aspects of care or monitoring
- quality improvement goals
- monitoring and evaluation plan
- plan for reporting, analysis, and communication of quality management activities.

If you do not currently have a plan, it can be started by listing responsibilities, scope of care, and important aspects of care as described in this chapter. Other components can be added as developed by the quality coordinator with input from various work groups. Examples of sample plans are in Appendixes A-1 to A-3.

REFERENCES

1994 Accreditation manual for hospitals—first peek. 1991. *Briefings on JCAHO* 2:10. December.

American Dietetic Association Council on Practice Quality Assurance Committee. 1986. *Standards of practice: A practitioner's guide to implementation.* Chicago: American Dietetic Association.

Carr, E.K., and J.R.A. Mitchell. 1991. A comparison of the mealtime care given to patients by nurses using two different meal delivery systems. *International Journal of Nursing Studies* 28:19–25.

Gardner, S. 1990. Q and A on QA. *Clinical Nutrition Management Newsletter* 9:6–8. December.

Hard, R. 1991. Food service, pharmacy team up for nutrition. *Hospitals* 65:46–48. June 5.

Hopkins, J.L. 1990. Planning for M & E. *QRC Advisor* 6:1,5–7. September.

Huyck, N.I., and M.M. Rowe. 1990. *Managing clinical nutrition services.* Gaithersburg, Md: Aspen Publishers, Inc.

Leape, L.L. 1990. Practice guidelines and standards: An overview. *Quality Review Bulletin*, 16:42–49.

Leary, C.B. 1991. Use of nursing process to develop unit specific quality assurance plans. *Journal of Nursing Quality Assurance* 4, no. 2:1–6.

Maiskowski, C., and M. Rostad. 1990. Implementing the ANA/ONS standards of oncology nursing practice. *Journal of Nursing Quality Assurance* 4, no. 3:15–23.

Marshik-Gustafson, J., et al. 1981. Planning is the key to successful QA programs. *Hospitals* 55:67–73. June 5.

McGuffin, B., and M. Mariani. 1990. Clinical nursing standards: Toward a synthesis. *Journal of Nursing Quality Assurance* 4, no. 3:35–45.

Owens, J. P., et al. 1989. Concurrent quality assurance for a nutrition-support service. *American Journal of Hospital Pharmacy* 46:2469–2476.

Powers, T., et al. 1991. A nutrition support team quality assurance plan. *Nutrition in Clinical Practice* 6:151–156.

Tishman, E. 1992. Total quality management: The bridge to customer satisfaction. *Provider* 18, no. 10:30–42.

Chapter 10

Quality Indicators, Criteria, and Monitors

The final step in planning quality care is development of indicators, criteria, thresholds or triggers for evaluation, and monitors. This step will proceed with greater ease if the first three steps (assigning responsibility, delineating scope of care, and identifying important aspects of care) are carefully developed. This chapter provides an overview of quality indicators, guidelines for indicator development, suggestions for criteria establishment, and selection of means to trigger intensive evaluation.

PRACTICE GUIDELINES PRODUCE QUALITY INDICATORS

The work of health care professionals results in improved health of patients; the work of dietitians leads to better nutritional status of clients in the hospital, at home, or in the community. Understanding the work process and how it can be improved is fundamental to organizationwide advances in quality care (Batalden 1991).

Practice guidelines set forth what is considered the preferred method of carrying out various aspects of the dietitian's work. These guidelines are based on consensus expert opinion and, to a lesser extent in dietetics, clear scientific evidence of effectiveness (Marder 1990). Practice guidelines give direction on how various tasks should be accomplished. The American Dietetic Association has published guidelines for several areas of dietetic practice (see Appendix 10-A).

Performance indicators provide the tools by which to measure the performance of the health care organization or individuals who work there (Marder 1990). Practice guidelines describe what should be done; indicators measure either how often or how well the guideline is used (process) or whether the expected result is achieved (outcome). Practice guidelines are the basis for good indicators, and well-developed indicators provide information to evaluate the effectiveness of guidelines.

WHAT IS A QUALITY INDICATOR?

An indicator is a measurable variable "that can be used as a guide to monitor and evaluate the quality of important patient care and support service activities" (Characteristics 1989). Performance indicators are quantitative tools used to monitor and evaluate the quality of important governance, management, clinical, and support functions that affect patient outcomes (Angaran 1991).

In themselves, indicators do not directly measure quality. Rather they serve as a screen or flag to identify potential problem areas. When performance related to an indicator appears out of line, it prompts those charged with monitoring the activity to take a closer look at performance in that defined area of practice. Thorough assessment of the target area not only forms the basis for assessing the quality of care provided by the institution, but also suggests possible approaches to quality improvement.

ATTRIBUTES OF INDICATORS

Indicators are quantitative measures. They must have the capacity for statement in numerical terms, usually as a ratio of occurrences to the total number of events in the defined universe. Take the following indicator, for example: "Outpatient clinic patients receiving diet instruction for fat-controlled diets achieve a significant decrease in serum cholesterol levels within 6 months of instruction" (Seidel and Mitchell 1991). This indicator is quantitative. It is possible to determine cholesterol levels of all patients six months after counseling. A monitoring of the indicator requires a simple tally of patients who achieved significant reduction in cholesterol levels in relation to all patients counseled. For purposes of illustration, say 365 individuals received counseling in a three-month period and medical records show that 93 patients had a significant decrease in blood cholesterol levels. The ratio would be:

$$\frac{\text{No. patients with significant drop in cholesterol levels}}{\text{No. patients who received fat-controlled diet instructions}}$$

In numerical terms the ratio is 93/365, or 25 percent. As we shall see later, not all cases are reviewed when performance is monitored. Rather, a specified random sample is selected to represent the total population.

Indicators address factors identified with quality care. In the past, quality assurance focused on problem areas that were sometimes unrelated to true quality issues. Indicators, however, are linked directly with one or more factors associated with quality care as outlined in the Joint Commission's *Primer on Indicator Development and Application* (1990):

- Accessibility—How easy is it for patients to get the care they need?
- Timeliness—How long do patients have to wait to get services or an appointment? How long do they spend in the waiting room?
- Effectiveness of care—To what extent is care provided in the manner dictated by current standards and guidelines?
- Efficacy of care—To what degree does the service meet the need for which it was intended?
- Appropriateness of care—To what extent does the service provided match the needs of the patient?
- Efficiency of care—To what degree did the service meet the needs of the patient with minimum effort, expense, or waste?
- Continuity of care—To what extent was patient care coordinated effectively among health care professionals and across institutions, organizations, and time?
- Privacy of care—Were patient rights protected with regard to information?
- Confidentiality of care—Did health professionals refrain from disclosing privileged information as warranted by circumstances of each case?
- Patient and family participation in care—To what extent was the patient (or family) involved in the decision-making process with regard to health matters?
- Safety of care environment—Were standards of safety and sanitation maintained?
- Supportiveness of care environment—To what degree were facilities, equipment, space, medical products, and comfort products/services available to the patient when needed?

It is not difficult to develop indicators related to the factors of quality listed above. Take, for example, the first indicator of quality (accessibility). What percentage of patients at your institution who have one or more nutritional risk factors receive appropriate nutritional intervention? Or, considering timeliness, what percentage of patients who receive emergency diet instructions in the hospital return to the outpatient nutrition department for more extensive nutrition counseling? Regarding appropriateness of care, what percentage of specialized nutrition support (tube feedings and parenteral nutrition) is warranted for the particular case? Or, considering effectiveness, what percentage of patients on NPO (nothing by mouth) or oral feedings meet institutional criteria for enteral or parenteral feedings? Note that the indicators are stated in quantitative terms, and they can be measured easily.

Indicators must be valid. The primary purpose of an indicator is to spell out in measurable terms the level of quality expected. This is particularly important when qualitative measures, such as patient satisfaction, are used to define quality. Validity of an indicator is the degree to which the indicator accomplishes its purpose. For example, in dietetics one valid indicator of quality nutrition services would be patient satisfaction with meals or dietary counseling.

Indicators are meaningful. That is, each indicator should mark a key process. Each must detect an instance of a potential problem associated with an important aspect of quality of care. Each indicator should have a direct and identifiable relationship with quality or it should not be included in the list of items regularly monitored. Also, indicators must be both sensitive and specific. That is, the indicator must allow detection of all cases in which actual problems exist. However, the indicator ought not be so general that it includes cases that are problem free. For example, in determining the quality of nutritional care, medical record entries have greater significance than do departmental notes or Kardex files. It would be inappropriate to have an indicator related to Kardex notes while eliminating any reference to medical record documentation. Also, the indicators must pinpoint actual nutritional problems from documentation in the medical record, but they should not mistakenly identify nutritional problems if none exist.

Indicators flow from the important aspects of care. When indicators are developed, they should reflect a direct relationship to important aspects of care described in Chapter 9. Alternatively, every important aspect of care should have at least one indicator that can be used to monitor quality of services designated in that aspect of care. One important aspect of care and its affiliated indicators are shown in Exhibit 10–1.

Indicators address "key functions." As hospitals move toward greater accountability and cost efficiencies, increased emphasis is placed on those activities that are most crucial to high-quality patient care. Such functions "involve care provided in several hospital departments, and the indicators will thus reflect the performance of a spectrum of several clinical, managerial, and support processes" (New Indicators 1989). The two key functions first addressed were effective use of medications and the prevention, detection, and control of infections. Indicator development includes the following:

- development of a flowchart outlining the relevant processes and illustrating the relationships among involved departments and professional groups
- identification of aspects of the process that warrant attention, such as those important to good patient outcomes, processes that could be improved, or areas that are thought to present problems at many hospitals
- task force development of indicators for testing among pilot hospitals

Dietitians should be involved in multidisciplinary task forces for indicator development whenever possible. Not all key functions will include nutrition care services. However, dietitians can make important contributions to indicator development task forces by creatively and aggressively advocating use of nutrition-related processes and outcomes as important dimensions of quality patient care and disease prevention.

A process similar to the one used to define "key functions" might also be used within nutrition and dietetics departments. For example, a flowchart can be created to show the overall process for delivering nutrition services. This flowchart reveals key decision points, documentation, flow of information, and action steps. The flowchart quickly reveals important aspects of care, problem areas, and processes that might benefit from improvement. Thus decisions on which indicators to develop evolve from an analysis of the flowchart. Dietitians can use such an analysis to illustrate the need for nutrition and dietetic services within key functional areas.

TYPES OF INDICATORS

Indicators are often classified into categories, depending on what they measure. The following three categories are sometimes used:

1. Indicators Based on Seriousness

Sentinel Events

Sentinel events are subject to intensive review for each occurrence. A sentinel event indicator measures a serious, undesirable, and often avoidable occurrence. Each instance of the event is rigorously reviewed to determine what steps can be taken to prevent a repeat of the undesirable outcome or process. A few examples of sentinel events that would always receive follow-up assessment are

- food poisoning
- patients receiving the wrong enteral or parenteral feeding
- foreign objects found in food
- vermin observed in the food service area
- patients receiving spoiled food
- patients suffering respiratory arrest secondary to hypophosphatemia of refeeding (Skipper 1991)

Rate-Based Indicators

Rate-based indicators require further assessment only if trend data show significant deviation from desired performance levels. Most indicators are rate based, as illustrated in the indicators used at Holyoke Hospital, Holyoke, Massachusetts, shown in Exhibit 10–2. Rate-based indicators need further review only if data show a trend toward significant departure from desired levels of performance, or notable deviations from established norms. Rate-based indicators usually measure patient care processes or outcomes for which expected standards allow less than perfection (100 percent or 0 percent). For example, desired patient tray accuracy may be 97 percent, or number of acceptable meal complaints may be 5 percent.

2. Indicators Based on Care Delivery

Dimensions of quality care typically include outcomes, processes, and structures (Donabedian 1988). These aspects of care are graphically illustrated by the model shown in Fig-

---

Exhibit 10–1 Relationship of Important Aspects of Care to Clinical Indicators

Important aspect of care:
  Periodic assessment of the effects of nutritional therapy.
Indicators:
  1. The patient receiving specialty formula shall have a special nutrition lab profile which lists the current physician's order.
  2. Patient's trays prepared by the food service department shall be viewed for accuracy of implementation of the diet order.
  3. There shall be no incident reports indicating incorrect formulation of physicians' orders.
  4. Patients receiving specialty formula shall have appropriate weight change after implementation of recommended nutritional care plan.

Courtesy of Alice E. Smith, Director, Clinical Dietetics, The Children's Memorial Hospital, Chicago, Illinois.

Exhibit 10–2 1992 Quality Indicators Used at Holyoke Hospital, Inc.

---

1. Percent of patients not receiving any alimentation for 3/5 days or longer.
2.* Percent of patients on parenteral nutrition being overfed/underfed.
3.* Percent of patients on enteral nutrition having diarrhea.
4. Percent of diet orders not matching diagnosis.
5. Percent of staff who show inadequate understanding of Universal Precautions Policy and Procedure.
6. Percent of physicians not satisfied with services of clinical nutrition.
7. Percent of out-patients not satisfied with diet instruction.
8. Percent of Nutrition Assessment Sheets not following Policy and Procedure.

*Inter-disciplinary studies.

Courtesy of Marcia Durell, Chief Clinical Dietitian, Holyoke Hospital, Inc., Holyoke, Massachusetts.

---

ure 10–1 (Brown 1992). Components of the model are described below.

Outcome Indicators

Outcome indicators examine what happens to a patient as a result of dietetic intervention. These indicators seem to be the most difficult to develop but are the most valuable in demonstrating the effectiveness of nutrition services (Hopkins 1992). Outcomes describe the result of nutritional intervention: What happened to the patient in terms of palliation, control of disease or its indicators, elimination of the nutritional problem, or rehabilitation? Lohr (1988) gives a classic list of outcome measures known as "the five Ds":

- death
- disease (determined by physiological variables)
- disability (health status or functional measurement; for example, days hospitalized, days of disability, rate of complications)
- discomfort
- dissatisfaction.

To be truly valuable, outcome measures must be linked directly with nutrition care activities. Without such a correlation, the actual impact of nutrition services can never be determined. Until and unless there is a link between process and outcomes, outcome indicators fail the test of being valid measures of quality care (Lohr 1988).

Hopkins (1992) contends that the reason it is so difficult to come up with outcome indicators is that the right measures are often overlooked or poorly understood. Using the simple task of washing dishes, Hopkins shows how, by asking the right questions, appropriate measures might be volume, input, process, outcome, performance rate, benchmark, or documentation. For example, the question "How many dirty dishes were there?" results in a measure of volume. The question "Do the dishes meet the standard for clean?" measures outcome. Measuring "number of times dishes meet the clean standard/number of loads" gives a performance rate. Thus by understanding work as a process and asking the right questions, dietitians can go beyond process and documentation to measure outcomes.

The American Dietetic Association Council on Practice Quality Assurance Committee developed outcome-oriented indicators of nutrition care for the adult patient in areas of oncology, cardiovascular care, and surgery (Queen et al. 1993). The objectives for American Dietetic Association involvement in this process were as follows:

- Enhance delivery of quality care to the public through development and promotion of indicators that assess nutrition status and outcomes of care.
- Provide nutrition professionals with validated outcome-oriented clinical indicators for use in practice and quality assurance programs that lead to efficient and effective delivery of care.
- Provide indicators that the Joint Commission will adopt and incorporate into its clinical indicator development as part of the Joint Commission's *Agenda for Change* project.
- Form a resource base of technical experts who can assist dietitians in development and modification of indicators for a particular setting (Frankmann 1990).

These indicators were subjected to extensive field testing (Dougherty and Frankmann 1991) and are given in Appendix B. As this project continues, dietitians can expect to see the development of other disease-specific indicators and greater standardization in practice guidelines from one hospital to the next.

In another project, the nutrition committee at Bridgeport Hospital, Bridgeport, Connecticut (Bernstein 1992), defined three outcome measures for the adequacy of nutrition based on laboratory values:

1. Serum albumin >3.0 g/dL
2. Serum prealbumin (PAB) (transthyretin) >11.0 mg/dL
3. PAB increase of at least 2.0 mg/dL in 1 week.

Failure to achieve these values is taken to indicate either inadequate nutrition support or inadequate physiological response.

Other outcome criteria for nutrition services are patient satisfaction surveys and measures of patient understanding of diet instructions. Outcome measures must track key nutritional processes or they will be useless in achieving continuous quality improvement (Reinertsen 1993).

Process Indicators

Process indicators measure one or more elements of care provided within the institution. For several years, process in-

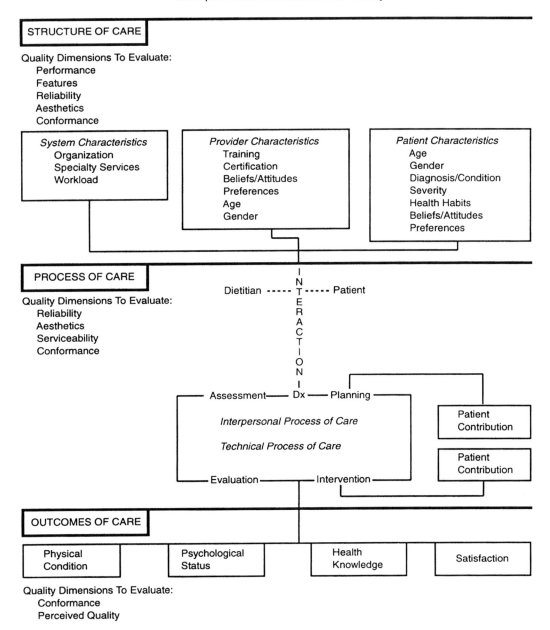

Figure 10–1 Conceptual Model of Quality Care Components. *Source:* Reprinted from Brown, D.S., A Conceptual Framework for Evaluation of Nursing Service Quality, *Journal of Nursing Care Quality*, Vol. 6, No. 4, pp. 66–74, Aspen Publishers, Inc., © 1992.

dicators dominated quality assurance studies conducted by dietitians. A look at the American Dietetic Association's *Guidelines for Evaluating Dietetic Practice* (1976) shows almost complete reliance on process factors. Monitor of tray error rates (Dowling and Cotner 1988), and accuracy of prescribed sodium-restricted diets received by hospitalized patients (Smith and Fullen 1985) are examples of published quality assurance studies based on process indicators. Following are some other examples of indicators of the nutrition care process (Schiller et al. 1991):

- patients with nutritional risk factors who receive nutritional assessments
- patients who receive assessment of nutritional status within 48 hours of admission
- diet orders that are changed after recommendation by a clinical dietitian
- number of problems associated with administration of tube feeding, such as clogged tubes
- percentage of patients NPO for more than 3 days

- percentage of food consumption or percentage of food returned uneaten
- enteral feedings that are advanced properly.

As shown in Figure 10–1, process indicators can assess reliability, aesthetics, serviceability, and conformance with key aspects of the nutrition care process: assessment, planning, intervention, and evaluation. Both interpersonal and technical aspects of care can be monitored. Because both the process and outcomes of care depend on patient characteristics as well as their involvement in the process, patient contribution to planning and intervention may be considered in diagnosis-specific indicators or as exclusions for certain criteria.

Process indicators also can be used to catch noncompliance with departmental standards or policies and procedures. One problem with process indicators is that they are often unrelated to high-volume, high-risk, or problem-prone situations. Yet in many instances, processes are currently the best available measures of quality in nutrition and dietetic services, and they are often used as a basis for quality improvement programs. It is important, however, to be sure that any indicator reflects processes central to quality nutrition care. Guard against using indicators that monitor niceties or aspects of service at the fringe of quality, such as patient refreshment service or menu management.

Structure Indicators

Structure indicators allow assessment of health care settings such as equipment, facilities, providers, decision-making models, and organizational structures. They are often used in quality management programs, particularly in food service areas. These indicators relate to the availability and use of facilities, space, equipment, professional qualifications, and organizational structures. Examples of structure indicators are

- nutrition assessments and evaluation of nutritional status conducted by qualified personnel
- safety and/or sanitation factors that comply with standards
- percentage of clinical dietitian time spent on nonprofessional duties such as screening patients, checking menus, entering routine diet changes, and the like.

3. Indicators Indicating Direction of Performance

Desirable Indicators

Desirable indicators are used when expected levels of performance approach 100 percent. Many situations in dietetics call for compliance at or near the 100 percent level. Below is a partial list of indicators that fall into this category (Escott-Stump 1988):

- Diet orders are written in the patient's medical record.
- The screening tool is complete, logical, and accurate.
- The completed assessment or consultant's report is placed in the medical record when ordered by a physician.
- Formal diet instruction is given upon physician request.
- Relevant food and drug counseling is offered.
- Documentation is complete, accurate, and relevant.
- NPO orders and clear liquid diets are progressed as soon as possible within 72 hours.
- All patient nourishments are logical and appropriate for the diet order and the patient's medical condition.
- Needs of high-risk patients are given priority.
- Diet instructions and consultations are completed within 24 to 48 hours of admission or the physician's order.

Undesirable Indicators

On occasion it is preferred that situations do not arise. Thresholds for these indicators are set at 0 percent. In these cases, indicators are used to monitor activity for those rare instances and to keep the number as low as possible. Following is a list of possible undesirable indicators:

- patient complaints about a particular service, such as diet counseling, dietitian skills, or food service
- incident reports such as incorrect formulation of physicians' orders for enteral feedings or infant formula
- patients not receiving alimentation for five days or longer
- errors in test diets necessitating a delay in procedures or an extended hospital stay.

Combination Indicators

In practice, most written indicators relate to care delivery (outcome, process, structure), whereas the desirability or undesirability of the indicator is addressed in the wording or by setting high or low triggers for in-depth evaluation. The decision of when to conduct rigorous review for deviations is made during indicator development, when the timing of review is established; some indicators require investigation of every event (sentinel indicators), and other indicators call for analysis of trend data (rate-based indicators). Thus, although all types of indicators are used, core indicators relate to patient care whereas the seriousness and desirability of events are handled as ancillary aspects of indicator development.

Patient-focused indicators can be developed in cooperation with nurses or other health professionals. Interdisciplinary indicators usually center on the general management of a defined patient population and are outcome oriented. Williams (1991) suggested that an interdisciplinary indicator might be appropriate weight gain at discharge (defined as 10 percent or more of admission weight) for infants admitted to the hospital for failure to thrive. Such an indicator involves dietitians, nurses, and physicians in the management of such infants. To achieve the desired outcome, dietitians provide special consultation on caloric requirements and appropriate formula, whereas nurses assess the patient's response to feeding and monitor intake and weight gain. The physician provides general management of the infant's condition. Each professional on this interdisciplinary team has a role to play in achievement of the desired outcome.

HOW TO DEVELOP QUALITY INDICATORS

Numerous ideas can be gleaned from the literature and what other hospitals are doing, but each institution should develop its own quality management plan, including clinical indicators, criteria, data sources, triggers for evaluation, and monitors. Ideas or materials borrowed from others often need to be adapted to fit the particular circumstances of your institution. Following is a stepwise approach that can be used to develop indicators or revise a current quality assurance plan.

Before beginning the development of indicators, it would be helpful to review the characteristics set forth by the Joint Commission (1986) for departmental monitoring and evaluation activities. These activities should be

- comprehensive enough to cover all aspects of the department, as well as structures, processes, and outcomes
- planned, systematic, and ongoing
- based on indicators and criteria that are agreed upon by the department staff and approved by the facility quality committee
- accomplished by the routine collection and periodic evaluation of data
- structured to require appropriate actions to resolve identified problems or to search periodically for ways to improve care
- continuous, assuring that improvements in care and performance are sustained
- integrated with other departments to facilitate desired outcomes of patient-centered care and to coordinate monitoring and evaluation activities throughout the facility.

As these characteristics suggest, it may be helpful to keep in mind that indicators ought to be prepared for all aspects of the department or service, be approved by both the dietetics staff and the hospital quality committee, result in routine collection and evaluation of data, and lead to quality improvement. If quality improvement is a new concept for the dietetic staff, the department quality coordinator should regularly review these characteristics to be sure that the working committees are on target and to facilitate meshing of dietetic quality plans with activities throughout the hospital.

1. Decide Who Will Develop the Indicators

In departments of one or two dietitians, the task of writing indicators will fall almost totally on these individuals. Clinical nutrition and cross-functional patient-centered indicators may be formulated by clinical dietitians working alone or as part of a team. Writing indicators related to food service and sanitation may be a collaborative effort among clinical dietitians and administrative dietitians, food service manager(s), and supervisors.

In large departments there can be one or more indicator development task forces of two to four members, appointed by the appropriate quality assurance designee. The work of each task force should be delineated clearly to avoid overlap and confusion. As work progresses, each task force needs to communicate with the entire group so that consensus is achieved regarding indicators and other aspects of the quality management process.

2. Understand the Terminology

Before going further, be sure everyone understands the terminology in current use (American Dietetic Association 1993):

- *Clinical indicators*, like any quality indicator, are quantitative measures that can be used as a guide to monitor and evaluate the quality of important patient care and support service activities (Srp et al. 1991). Clinical indicators may be related to specific diseases, major clinical functions, or different nursing units. A few food and nutrition departments have clinical indicators directed toward specific diseases, but most use a more general approach, shaping clinical indicators around major clinical functions or participating in cross-functional quality initiatives.
- *Criteria* are objective limits of acceptable outcomes for a specific indicator, such as acceptable ranges for laboratory indices, standards of performance related to an indicator, or brief statements describing critical elements of an indicator.
- *Data sources* may describe either of two major resource sets. First, they may be references used to develop indicators, criteria, and thresholds, such as policies and procedures, research articles, textbooks, surveys, and samples of materials used at other institutions. Second, data sources may indicate potential sources for data collection, such as medical record, patient observation, patient or staff survey, patient interview, or environmental observation.
- *Thresholds* or *triggers* are expected levels of occurrence for defined criteria. Performance short of expected levels triggers an intense review of the quality indicator. When performance consistently exceeds the threshold, consid-

eration is given to improving the quality of care by increasing the threshold for desirable criteria or lowering the threshold for undesirable criteria.

- *Monitors* are data elements that will be collected on a routine basis as a way of determining whether criteria have been met.

Exhibits 10–3 and 10–4 provide different examples showing the relationships among indicators, criteria, data sources, thresholds, and monitors. These relationships can also be observed in the quality management plans shown in Appendix A–1 to A–4.

3. Adopt or Develop an Indicator Development Form

The use of a standard form provides structure to the decision-making process. Also, it gives rigor to the point-by-point consideration of indicator specifications. Completion of a form is also a good way to make sure all information is complete. A sample form is given in Exhibit 10–5.

The clinical indicator development form offered by the Joint Commission (Characteristics of Indicators 1989) is used by professional organization task forces (such as the American Society for Parenteral and Enteral Nutrition and the American Dietetic Association) to establish diagnosis-specific indicators that can be used across the respective professions. This form lends itself to the development of care-specific indicators but is difficult to use for many aspects of food service and general nutrition care.

Those involved in general (non–disease specific) indicator development should agree on one form and the same form should be used throughout the department. Clinical dietitians who develop care-specific or unit-specific indicators may use the same form or they may elect to use the Joint Commission form or a variation of it.

Exhibit 10–3 Southwest General Hospital Indicator Work Sheet

*Type of Indicator
S—Structure   O—Outcome
P—Process   R/M—Risk Management

DEPARTMENT NUTRITION SERVICES   DATE JULY 10, 1991

DATE	IMPORTANT ASPECT OF CARE	INDICATOR	OBJECTIVE OF INDICATOR	HOW MONITORED	BY WHOM	HOW OFTEN	TYPE*
APRIL 1991	Food preparation, quality & distribution	Quality & accuracy of food items served to patients meet dept. standards	1. To monitor the accuracy & quality of food served to patients 2. To monitor temp. of food served to pts.	Tray test evaluations	Assigned staff & employees	Weekly	S,P,O

Courtesy of Rhonda Billman, Clinical Manager, Nutrition Services, Southwest General Hospital, Middleburg Heights, Ohio.

4. Draft One or More Indicators

Determine which aspect(s) of care will be given priority—the one(s) for which indicators will be developed. To facilitate this process, identify the subsystems in the nutrition care process that indicators should monitor: screening and assessment, care planning, nutritional intervention, diet and nutrition counseling, evaluation, and documentation. For example, Winkler (1992) offers the following areas for possible indicator development for enteral nutrition:

- appropriateness of enteral nutrition
- timeliness of intervention
- appropriateness of prescribed formula
- adequacy of intake
- laboratory and bedside monitoring
- complications
- patient education
- discharge planning
- home nutrition support.

Dietitians can brainstorm similar lists for any patient population or important aspect of care. Any of the areas can be elaborated further, resulting in indicators to help identify areas requiring further review. Winkler (1992) suggested the following indicators under tube-feeding complications:

- tube feeding related to pulmonary aspiration
- hyperglycemia
- hypophosphatemia on long-term tube feeding
- tube malposition/occlusion
- nausea/vomiting
- diarrhea defined as greater than five stools per day.

Those working alone might select only one or two areas for indicator development; if task forces are used, they may be assigned one or more important aspects of care to address.

For each defined important aspect of care, write at least one indicator. The indicator may relate to process, clinical management, appropriateness, or outcome of care (Lehmann 1989). The indicator may be a current policy designed to pro-

---

Exhibit 10–4 Criteria Form and Related Data-Collection Sheets—Southwest General Hospital

QUALITY CRITERIA FORM

INDICATOR: Quality and accuracy of food items served to patients meet department standards.
SOURCE: Nutrition Services Policy & Procedure Manual
METHOD: Weekly test tray evaluations

Criteria	Threshold	Exception	Compliance
1. Temperature of food items on tray are within department standards as indicated on test tray evaluation form.	80%	None	A total score of 16 or greater for food temperatures on the test tray evaluation form with no food item receiving a score of zero.
2. Food items are correct portion size according to department standards.	90%	1. Specific amounts of food items are written on menu. 2. Small portion sticker is on menu.	A total score of 18 or greater for portion size on test tray evaluation form with no food items receiving a score of zero.
3. Appearance, quality, and taste and aroma of food items meet department standards.	80%	None	A total score of 16 or greater for appearance, quality, taste, aroma on test tray evaluation form with no food item receiving a score of zero.
4. Food items on tray are accurate and dishes are clean.	90%	Substitution card on tray for missing or substituted items.	A total score of 18 for section 2 of the test tray evaluation form.
5. Overall tray meets department standards.	85%	None	A total score greater than 120 (85%) on test tray evaluation form.

continues

Exhibit 10–4 Continued

DATA RETRIEVAL FORM

UNIT: _____
DATE: _____
RESPONSIBLE RECORDER: _____

1) + Met Criteria   3) 0 Variation
2) – Met Exception   4) J Justified Variation

Quality and accuracy of food items served to patients meet department standards. Date of test tray	Temperature	Portion size	Appearance/quality/taste/aroma	Accuracy and cleanliness of tray	Overall score					
TOTAL #/% COMPLIANCE										AVERAGE #/% COMPLIANCE
TOTAL #/% OF NON-COMPLIANCE										AVERAGE #/% NON-COMPLIANCE

continues

Aspen Publishers, Inc., 1994

Exhibit 10–4 Criteria Form and Related Data-Collection Sheets—Southwest General Hospital (continued)

QUALITY ASSESSMENT OF PATIENT TRAYS

MEAL: BREAKFAST
DATE: _____
DIET: _____

STANDARDS:

HOT CEREAL:	175–165 = 2,	164–155 = 1,	Less than 155 = 0
EGGS:	145–135 = 2,	134–125 = 1,	Less than 125 = 0
HOT ENTREE:	150–140 = 2,	139–130 = 1,	Less than 130 = 0
HOT MEAT:	150–140 = 2,	139–130 = 1,	Less than 130 = 0
COLD BEVERAGES:	40–50 = 2,	51–60 = 1	More than 60 = 0
HOT BEVERAGES:	185–175 = 2,	174–165 = 1,	Less than 165 = 0

TIME ASSEMBLED: _____ TIME EVALUATED: _____

Section 1

MENU ITEMS	HOT FOOD							COLD FOOD					TOTAL
Product	Juice	Fruit	Hot cereal	Egg	Hot entree	Hot meat	Toast	Pastry	Milk	Hot beverage	
1) Temperature of Food on Tray											
STANDARDS	40	50	175	145	150	150	CRISP	RM TEMP	40	185	
2) Temperature of Food Score	(2)	(2)	(2)	(2)	(2)	(2)	(2)	(2)	(2)	(2)	
3) Portion Size	(2)	(2)	(2)	(2)	(2)	(2)	(2)	(2)	(2)	(2)	
4) Appearance of Food	(2)	(2)	(2)	(2)	(2)	(2)	(2)	(2)	(2)	(2)	
5) Quality and Preparation	(2)	(2)	(2)	(2)	(2)	(2)	(2)	(2)	(2)	(2)	
6) Taste and Aroma	(2)	(2)	(2)	(2)	(2)	(2)	(2)	(2)	(2)	(2)	
7) Missing Items/Substitutions	(1)	(1)	(1)	(1)	(1)	(1)	(1)	(1)	(1)	(1)	

Section 2

8) Tray Completeness	Garnish ___ (1)	Condiments ___ (1)	Tray ___ (1)	Flatware ___ (1)	Plate Cover ___ (1)
& Cleanliness	Dishes ___ (1)	Napkins ___ (1)	Glassware ___ (1)	Soup Bowl ___ (1)	Coffee Pot ___ (1)

Sub Total: _____

Possible Points Section 1	100	ACTUAL ___
Possible Points Section 2	20	ACTUAL ___
Total Possible	120	ACTUAL ___ + 120 × 100 = ___ %

Overall Quality ___ Excellent (93–100%) ___ Satisfactory (85–92%) ___ Unsatisfactory (Below 85%)

LIST COMMENTS ON REVERSE SIDE:

Courtesy of Rhonda Billman, Clinical Manager, Nutrition Services, Southwest General Hospital, Middleburg Heights, Ohio.

Exhibit 10–5 Sample Indicator Development Form

Service/Department:

Aspect of Care:

Indicator:

Threshold:

Criteria:

Data Collection:
Time Period/Frequency:
(From _____ To _____)

Person(s) Responsible for Data Collection:

Describe Data Collection Method:
_____
_____
_____
_____
_____
_____

Courtesy of Hospital Council of Southern California, © 1990.

---

mote quality services. Examples provided earlier in this section can be used as a source for ideas.

Nurses who have written indicators note that care-specific indicators are better than general ones (Marder 1989; Myles 1989). In dietetics, however, both generic and disease-specific indicators are useful. Development of indicators for which data can be collected at a practical level requires concentration on specific clinical conditions so that data elements can be clearly defined. For example, it is easier to write clinical indicators for bone-marrow transplant patients for whom strict protocols are in place than it is to develop indicators for surgical patients who may represent numerous differences in severity of illness, assessment parameters, type of surgery, recommended dietary interventions, and nutrition counseling needs.

Most small hospitals confine their indicators to various aspects of the nutrition care process, even though these are more difficult to define. This is still the preferred approach because the volume of patients in any one area is usually not large enough to warrant development of indicators for each type of clinical condition. On the other hand, care-specific indicators may be preferred in large institutions where dietitians are assigned to units serving specific medical conditions.

Some practice groups of the American Dietetic Association have defined criteria for use by members. For example, dietitians in pediatric practice developed a comprehensive set of quality assurance criteria for selected pediatric conditions (Wooldridge and Spinozzi 1990). This published manual may be used as a model for indicator and criteria development by dietitians or other practice groups.

Indicator statements should be written in the plural (Characteristics of Indicators 1989). Indicators are events, not percentages, and the wording should reflect this. For example, an indicator might be written, "Patients at nutritional risk receive an in-depth nutritional assessment." Any terms that are ambiguous or need explanation should be defined. In the example given above, "nutritional risk" needs further explanation. If indicators are developed for a particular diagnosis, the exact *International Classification of Diseases,* 9th Revision—*Clinical Modification* (ICD-9-CM) codes should be included.

Answering a few questions will help assess the quality of the indicator before the process is continued.

- Why is this indicator important and what is it expected to measure? Note how the example in Exhibit 10–3 includes an objective for each indicator describing the importance and purpose of the indicator. To clarify your intent, state the purpose of the indicator you just developed.

- Does this indicator focus on a high-risk, high-volume, problem-prone aspect of care? If so, which of these does it address? If the indicator fails this test of importance for your facility, scratch out what you have done and start over. For instance, some of the indicators cited earlier relate to enteral and parenteral nutrition. A small hospital that rarely uses specialized nutrition support might better design its early indicators around quality issues more pertinent to daily practice.

- Does this indicator require multidisciplinary input? In other words, do other professionals provide patient care services that impinge on this indicator? If not, the indicator can be departmental in nature. If other groups are intimately involved in related activities, they should also participate in developing this idea as a cross-functional indicator. Some typical cross-functional indicators are those related to parenteral nutrition, height and weight information in medical records, discharge planning and education, and management of various medical conditions such as pressure sores, diabetes, or renal dialysis.

- Is the indicator a clear, objective statement that can be measured? Compare your indicator with examples given elsewhere in this manual to determine clarity, objectivity, and measurability.

Before continuing, a staff meeting should be held to debate and edit the proposed list of indicators. Changes may be made as appropriate.

For 1994 surveys, the Joint Commission requires a minimum of two indicators to monitor the quality of dietetic services. The number of indicators selected will depend on institutional size, experience, the number of professional staff, and availability of support systems for data collection and analysis.

Not all opportunities to improve care will surface as indicators in the planning process. In fact, numerous opportunities to improve care may not be reflected in any of the preplanned indicators. For example, if those who deliver patient trays harbor poor attitudes and display negative human relations skills, they often ignore opportunities to assist patients, such as helping them open food cartons, or fail to make corrections in errors overlooked during tray assembly. Caregiver attitudes toward patients have a significant impact on the quality of care provided as measured by patient satisfaction. Although improved human relations may offer an opportunity to improve care, they may not appear as an important aspect of care for the department, and thus will not surface as a quality indicator. This does not mean that negative employee attitudes should be ignored. Rather, they should be handled through supervisory efforts such as employee counseling, performance appraisals, adjusted work assignments, or resolution of underlying problems. Quality indicators address high-volume, high-risk, high-cost, problem-prone aspects of patient care; other operational strategies should be in place to facilitate the delivery of high-quality food and nutrition services.

5. Describe the Indicator Population

It is essential to plan what patient group or sample of events will be used to assess each quality indicator. First, begin the numerator with "number of patients" or "number of events." This is followed with wording directly from the indicator. For example, a numerator may be "number of patients on parenteral nutrition being underfed or overfed." The denominator also begins with the words "number of patients" or "number of events" and continues with a description of the larger population of patient cases or events to be studied. Using the example above, the denominator may read "number of patients on parenteral nutrition." In this case, 100 percent of patients on parenteral nutrition will be monitored; the regular report will show the percentage of patients on parenteral nutrition who are underfed or overfed.

6. Establish Thresholds or Triggers for Evaluation

A threshold for evaluation is the level or point that triggers intensive evaluation of care, although it is not necessarily the only datum that would lead to in-depth study. Thresholds signify expected standards of performance, expressed as the percentage of sampled events that meet specified criteria. A regular review of the percentage of cases meeting expected performance levels with notation of any unusual upward or downward trends offers a rapid and reliable method of identifying potential problems in key aspects of quality care.

Continuous quality improvement literature specifies that triggers be established for each *indicator*. A review of departmental samples such as those shown in Exhibit 10–4 shows that thresholds are sometimes set for criteria. Generally, the quality management process is exponentially more complex when thresholds are established for criteria, rather than for indicators.

Lehmann (1989), among others, cautions practitioners to "keep it simple." One way to facilitate simplicity is to ensure that indicators are measurable and to set thresholds for indicators, not for each criterion used to describe details of those indicators.

How are appropriate numerical thresholds established? Triggers may be based on professional standards and guidelines, clinical and quality management literature, the particular experience of a department, and comparative data from similar institutions (benchmarking). Many of the examples provided in this manual give thresholds of 0 percent and 100 percent. There is no allowance for lax performance, because high standards of quality have been set, exceptions are well defined, policies and procedures are in place, employees have been trained, and supervisors make it clear that they expect employees to adhere to established quality standards.

To set absolute thresholds (0 percent or 100 percent) for every indicator being monitored is to program the system for failure (Katz and Green 1992). Staff members will become frustrated with the intensity of review, since an investigation of every deviation is required. The system becomes unmanageable because enormous paperwork is required. Quality monitoring becomes a tedious and wearisome exercise rather than the vital tool it was intended to be. Thus it is neither smart nor practical to set standards at "perfection." Rather, thresholds of 20 percent or 80 percent, or even 30 percent and 70 percent, may be more reasonable at first. Thresholds are dynamic. They change as systems are stabilized and improvement is achieved. When performance is within the threshold range, that system should be studied to look for ways to improve the existing process. This is consistent with the philosophy of continuous quality improvement.

No one should be satisfied with mediocrity. Therefore, thresholds should be sufficiently rigorous to challenge both professional staff members and other employees to perform at their level best. At the same time, those responsible for quality management should recognize that human limitations, weaknesses in existing systems, and an inadequate departmental framework may hinder achievement of inappropriately high (or low) thresholds. In time, analysis of data should help pinpoint problem areas, address system inadequacies, facilitate problem resolution, and foster advancement to higher levels of quality care.

7. Determine Criteria That Characterize the Indicator

Criteria are objective limits of acceptable outcomes or specific parameters that define compliance with a particular indicator. Criteria may spell out acceptable ranges of laboratory values, notations in the medical record, content of certain forms, or reference to specific policies and procedures. Exhibits 10–6 and 10–7 illustrate the relationship between indicators and criteria. Powers et al. (1991) published detailed indicators and criteria for a nutrition support team.

Note that criteria in the quality management process currently advocated by the Joint Commission have an operational definition different from criteria previously used in patient care audits. In outmoded quality assurance programs, criteria were defined as "predetermined elements against which aspects of the quality of health care service may be compared.... Criteria should be relevant, understandable, measurable, behavioral, and achievable" (Walters and Crumley 1978). In the quality improvement framework, criteria are used to further define and embellish indicators, making it easier to ascertain which elements under review meet compliance specifications and which ones do not. Referring again to Exhibit 10–6, all screening forms containing the three specified pieces of information (criteria) would be judged in compliance with the indicator; others would be considered in noncompliance and would result in less than 100 percent performance.

Criteria for certain indicators may be already delineated in departmental policies and procedures. For example, an indicator may be developed indicating that a comprehensive nutritional assessment will be conducted on all patients found during nutrition screening to have one or more nutritional risk factors. Current procedures in the department may specify currently that a certain form will be completed and placed in the permanent medical record, or that certain parameters will be assessed and that the evaluation will be documented in the medical record. These current procedures can be used as the basis for criteria.

Exhibit 10–6 Example A Showing the Relationship between an Indicator and Its Related Criteria

---

Indicator:	Patients shall be appropriately screened for nutritional risk within 48 hours of admission.
Exclusions:	Patients admitted weekends and holidays are screened within 72 hours of admission; patients admitted to ICU and NICU who are automatically at risk.
Criteria:	a. The patient shall have screening status noted, dated and initialed on the screening form. b. The patient shall have weight and weight percentile documented on the screening form. c. The patient shall have height and height percentile documented on the screening form.

Courtesy of Alice E. Smith, Director, Clinical Dietetics, The Children's Memorial Hospital, Chicago, Illinois.

Exhibit 10–7 Example B Showing the Relationship between an Indicator and Its Related Criteria

---

Indicator:	Patients not receiving any alimentation for 3/5 days or longer.
Procedure:	Nutrition Assistants monitor all of their patients to ensure that they have not gone more than five (5) days NPO; three (3) if known to be at nutritional risk (in ICU/CCU, serum albumin level below "Adequate," inadequate PO intake prior to being NPO, pertinent diagnosis). If/when that occurs, they notify the registered dietitian (RD) or registered dietetic technician (DTR) responsible for that patient.
Criteria:	a. The patient is on peripheral or central hyperalimentation. b. There is documentation that the patient is terminal, has multiple system failure, or refuses nutritional support.
Action:	If neither of the criteria exist, the problem will be brought to the attention of the physician. The timeliness and quality of the response will be monitored and reported quarterly.

Courtesy of Marcia Durell, Chief Clinical Dietitian, Holyoke Hospital, Inc., Holyoke, Massachusetts.

---

8. Design Monitors To Assess Conformance with the Indicator

Monitors are sets of information gathered over time, compiled and organized in numerical terms, and used to compare performance related to quality indicators with previously collected data. Consistent use of the same relevant data sets, arranged in the same format, facilitates serial measurements of performance and comparison with previous reporting periods. Responsible parties can note performance at a glance and can begin immediately to take remedial steps as necessary.

For each indicator there needs to be a data set for evaluating quality care. In some institutions forms used previously for quality assurance programs may be adapted and used as monitors. For example, forms may be in use for auditing nutrition care or determining patient satisfaction with food service, such as the forms included as Exhibit 10–4. These can often be used as they are or modified to better address newly developed quality indicators. It is not necessary to have a completely separate and unique monitor for each indicator. Monitors or data sets can be developed to measure multiple indicators. Development of forms is addressed more completely in Chapter 11.

9. Complete the Plan for Data Collection

The time period for monitoring (weekly, monthly, quarterly), the person(s) responsible for data collection, and the method of data collection need to be specified. Monitors plus these specifications determine exactly what, when, how, and who collects information to evaluate conformance with the indicator. Exhibit 10–8 shows an actual indicator information work sheet giving procedures for data collection.

Data on some indicators are collected continuously, whereas other indicators require data to be collected during a preset time frame. For high-volume indicators, both the method of sampling and data collection times need to be specified. Sampling strategies are discussed further in Chapter 11.

Data for each indicator may be collected from one or several sources, such as

- patient records
- lab reports
- incident reports
- department logs
- committee meeting minutes or reports
- infection control reports
- patient satisfaction surveys
- sanitation records
- direct observation
- patient interviews
- dietitian, technician, or other employee interviews
- computerized reports
- admission and discharge records
- dietary records such as Kardex cards.

10. Develop a Schedule for Monitoring and Evaluation

Calendars for monitoring quality performance are discussed in Chapter 14. A timetable for evaluation is needed for each indicator as it is developed; the timetable corresponds with the frequency stated for monitoring the respective indicator. As each new indicator is developed, it can be added to the annual monitoring schedule.

SOME PRACTICAL POINTERS

Others who have been through this exercise offer the following suggestions to help ease the burden of the indicator development process (Lehmann 1989; Marder 1989; Podgorny 1991; Skipper 1991; Srp et al. 1991):

- Keep it simple. Develop and use a limited number of indicators and design indicators for which data collection and interpretation are relatively easy. (A compendium of quality indicators developed by the Quality Council of the University Hospital Consortium is provided in Appendix C. Note that these indicators include both generic and diagnosis-specific indicators; a few require cross-functional cooperation.)

Exhibit 10–8 Indicator Information Work Sheet

DIETETIC SERVICE
FY *93*

1. Important function/aspect of care: Assessment of protein/Kcal needs and provision of supplemental nutrition for nutritionally compromised acute care patients in a timely manner.

2. Indicator statement: Assessment of protein needs of nutritionally compromised acute care patients.

3. Type of indicator: Choose one: Sentinel event:____; Rate based __X__.
   Choose one: Process __X__; Outcome____.

4. Threshold: 100%

5. Rationale for selection of indicator and threshold. (Include reference if applicable.): Studies have shown that nutritional status has a definite effect on length of stay and overall recovery time of hospitalized patients. Timely nutritional intervention has been shown to reverse malnutrition and diminish length of stay. A threshold of 100% was chosen because the goal is to provide optimum nutritional care for all compromised patients.

Reference: Messner, RL, et al. (1991). Effect of Admission Nutritional Status on Length of Hospital Stay; *Gastroenterology Nursing*, Spring 1991.

Dietetic Service Policy: 120A-1, May 1992.
Dietetic Service Policy: 120A-5, July 1992.

6. Description of indicator population:

   Nutritionally compromised acute care patients assessed for protein/needs
   ─────────────────────────────────
   Nutritionally compromised acute care patients who have been hospitalized more than 3 working days or more than 5 working days when screened by CDT

7. Indicator data collection procedure: Medical records of randomly selected acute care patients who have been hospitalized for more than 3 working days or more than 5 days when screened by CDT who are classified as nutritionally compromised (Categories III & IV). Documentation of nutritional assessment and plans for supplemental nutrition will be noted.

Developed by:_____ Date:_____

Chief of Service/Program: Chief, Dietetic Service   Date:_____

Courtesy of Donna M. Bashara, Chief, Dietetic Service, Olin E. Teague Veteran's Center, Temple, Texas.

- Be sure everyone knows how the indicators will be used. The quality management process is designed to evaluate and improve structures, patient care processes, and outcomes, not to appraise individual caregivers. Employees who view the quality improvement process as a threat to their security may resist involvement in quality improvement.

- Rely on existing data, when available. Many forms used for quality assurance audits can be incorporated into the quality improvement process. Also, previous experience may be used to derive thresholds; select high-risk, high-

volume, high-cost, or problem-prone procedures; and produce databases or procedures for easy collection of data.
- Standardized forms will facilitate accurate and consistent data collection. When developing forms, for best results work with those who will be using them. Computers can aid both the organization and analysis of data. Request the help of individuals who can assist with designing computer programs for this purpose.
- Indicator development is a time-consuming and tedious process. Avoid the temptation to become frustrated and abandon the entire project. Remember that dietitians at most other hospitals encounter the same difficulties as you, and most are novices going through similar developmental stages. The authors are grateful to those who were willing to share examples of their quality management programs to assist others in their pursuit of improved quality care.

REFERENCES

American Dietetic Association. 1976. *Guidelines for evaluating dietetic practice*. Chicago.

American Dietetic Association. 1993. Learning the language of quality care. *Journal of the American Dietetic Association* 93:531–532.

Angaran, D.M. 1991. Selecting, developing, and evaluating indicators. *American Journal of Hospital Pharmacy* 48:1931–1940.

Batalden, P.B. 1991. Organizationwide quality improvement in health care. *Topics in Health Record Management* 11, no. 3:1–12.

Bernstein, L.H. 1992. Monitoring quality of nutrition support: A chemical marker. *Dietetic Currents* 19, no. 2:5–8.

Brown, D.S. 1992. A conceptual framework for evaluation of nursing service quality. *Journal of Nursing Care Quality* 6, no. 4: 66–74.

Characteristics of clinical indicators. 1989. *Quality Review Bulletin* 15:330-339.

Donabedian, A. 1988. The quality of care: How can it be assessed? *Journal of the American Medical Association* 260:1743–1748.

Dougherty D., and C. Frankmann. 1991. JCAHO—agenda for change: Are you ready? *Clinical Nutrition Management Newsletter*, 10, no. 3:8–9. July.

Dowling, R.A., and C.G. Cotner. 1988. Monitor of tray error rates for quality control. *Journal of the American Dietetic Association* 88:450–453.

Escott-Stump, S. 1988. Quality assurance for dietetic interns during clinical staff relief rotations. *Dietitians in Nutrition Support Newsletter*, 9 no. 6:3–4.

Frankmann, C. 1990. Agenda for change: JCAHO's focus on quality. *Clinical Nutrition Management Newsletter* 9, no. 4:1–4.

Hopkins, J.L. 1992. Teaching ways to measure outcome. *QRC Advisor* 8 no. 9:3–6.

Joint Commission on Accreditation of Healthcare Organizations. 1986. Monitoring and evaluation of the quality and appropriateness of care: A hospital example. *Quality Review Bulletin* 12, no.9:326–330.

Joint Commission on Accreditation of Healthcare Organizations. 1990. *Primer on Clinical Indicator Development and Application*. Oakbrook Terrace, Ill.

Joint Commission on Accreditation of Healthcare Organizations. 1993. *1994 Accreditation Manual for Hospitals. Volume I, Standards*. Oakbrook Terrace, Ill.

Katz, J., and E. Green. 1992. *Managing quality: A guide to monitoring and evaluating nursing services*. St. Louis, Mo: Mosby–Year Book, Inc.

Lehmann, R. 1989. Forum on clinical indicator development: A discussion of the use and development of indicators. *Quality Review Bulletin* 15:223–227.

Lohr, K.N. 1988. Outcome measurement: Concepts and questions. *Inquiry* 25:37–50. Spring.

Marder, R.J. 1989. Joint Commission plans for clinical indicator development for oncology. *Cancer* 64 (suppl):310–313. July 1.

Marder, R.J. 1990. Relationship of clinical indicators and practice guidelines. *Quality Review Bulletin* 16, no. 2:60.

Myles, S. 1989. Monitoring patient outcomes in an oncology unit. *Journal of Nursing Quality Assurance* 4, no. 1:35–39.

New indicators target key cross-department functions. 1989. *Joint Commission Perspectives* 9, no. 11/12:1, 6–7.

Podgorny, K.L. 1991. Developing nursing-focused quality indicators: A professional challenge. *Journal of Nursing Care Quality* 6, no. 1:47–52.

Powers, T., et al. 1991. A nutrition support team quality assurance plan. *Nutrition in Clinical Practice* 6:151–156.

Queen, P.M., et al. 1993. Clinical indicators for oncology, cardiovascular, and surgical patients: Report of the ADA Council on Practice Quality Assurance Committee. *Journal of the American Dietetic Association* 93: 338–344.

Reinertsen, J.L. 1993. Outcomes management and continuous quality improvement: The compass and the rudder. *Quality Review Bulletin* 19, no. 1:5–7.

Schiller, M.R., et al. 1991. *Handbook for clinical nutrition services*. Gaithersburg, Md: Aspen Publishers, Inc.

Seidel, L., and E. Mitchell. 1991. Dietetic service indicators. *QRC Advisor* 7, no. 4:7.

Skipper, A. 1991. Collecting data for clinical indicators. *Nutrition in Clinical Practice* 6:156–158.

Smith, M. K., and D. Fullen. 1985. Quality assurance audit: Sodium restricted menus. *Journal of the American Dietetic Association* 85:1320–1321.

Srp, F., et al. 1991. Quality of care concepts and nutrition support. *Nutrition in Clinical Practice* 6:131–141.

Walters, F.M., and S.J. Crumley. 1978. Patient care audit: A quality assurance procedure manual for dietitians. Chicago: American Dietetic Association.

Williams, A.D. 1991. Development and application of clinical indicators for nursing. *Journal of Nursing Care Quality* 6, no. 1:1–5.

Winkler, M.F. 1992. Quality management for enteral nutrition. *Dietitians in Nutrition Support Newsletter* 14:11–13. April.

Wooldridge, N.H., and N. Spinozzi eds. 1990. *Quality assurance criteria for pediatric nutrition conditions: A model*. Chicago: American Dietetic Association.

Appendix 10-A

COUNCIL ON PRACTICE QUALITY MANAGEMENT TASK FORCE
Current Dietetic Practice Group Projects/Publications

Clinical Nutrition Management
Patient Acuity Clinical Nutrition Staffing Study [In Progress]
Standards of Practice Project for Clinical Nutrition Managers [In Progress]

Consultant Dietitians in Health Care Facilities
Standards of Practice for the Consultant Dietitian [Published]

Consultant Dietitians in Health Care Facilities and Dietitians in Nutrition Support [Joint Project]
Guidelines and Clinical Indicators [In Progress]
- Adults <65 Years of Age with Stage 2, 3, or 4 Pressure Sores in Nursing Facilities or Home Care Settings
- Adults >65 Years of Age with Unintentional Weight Loss or >10% of Actual Body Weight in Six Months or >5% in One Month in Long Term or Home Care Settings

Diabetes Care and Education
Review Criteria [Published]
- Gestational Diabetes Mellitus

Nutrition Practice Guidelines [Proposal Approved]
- Type 1 Diabetes Mellitus
- Gestational Diabetes Mellitus

Dietetics and Developmental and Psychiatric Disorders
Quality Assurance Criteria and Clinical Care Indicators [Published]
- Developmental Disabilities
- Psychiatric Disorders
- Substance Abuse

Dietetics in Physical Medicine and Rehabilitation
Practice Guidelines [In Progress]
- Dysphasia in Rehabilitation

Gerontological Nutrition
Standards of Practice & Practice Guidelines [Proposal Approved]
- Those ages 60–74
- Those ages 75–84
- Those ages >85

Dietitians in General Clinical Practice
Pocket Sized Practitioner Cards [Published versions have not been field-tested]
- Adult Nutrient Requirements
- Sodium Values of Selected Foods
- Potassium Values of Selected Foods
- TPN Guidelines (Adults)
- Tube Feeding Guidelines
- Guidelines for Diabetic Meal Plans
- Guidelines for Nutritional Care of the AIDS Patient
- Guidelines for Nutritional Care of Acute and Chronic Renal Failure
- Guidelines for Nutritional Care of Hyperlipidemia

Weight Management Tool

NOTE: Italics indicates project proposal included a field-test of the guidelines.
*Indicates clinical indicators included.

Courtesy of the Quality Management Task Force, The American Dietetic Association, Chicago, Illinois.

- A collection of camera ready materials designed for the nutrition counselor to photocopy as needed to individualize handouts for patients

Dietitians in Nutrition Support

Standards of Practice for the Nutrition Support Dietitian (Joint Project with The American Society of Parenteral and Enteral Nutrition (A.S.P.E.N.)) [Published]
**(Clinical Indicators included)*
- *Guidelines for Initial Screening*
- *Assessment and Nutrition Management of Patients Receiving Enteral Nutrition Support*
- *Assessment and Nutrition Management of Patients Receiving Parenteral Nutrition Support*
- *Assessment and Nutrition Management of Patients Receiving Home Nutrition Support*
- *Assessment and Nutrition Management of the Older Adult*
- *Assessment and Nutrition Management of the Patient with AIDS*
- *Assessment and Nutrition Management of the Patient with Burns*
- *Assessment and Nutrition Management of the Patient with Compromised Pulmonary Function*
- *Assessment and Nutrition Management of the Patient with Diabetes*
- *Assessment and Nutrition Management of the Patient with Multiple Systems Organ Failure*
- *Assessment and Nutrition Management of Oncology Patients*
- *Assessment and Nutrition Management of the Patient with Renal Disease*
- *Assessment and Nutrition Management of the Patient with Short Gut Syndrome*
- *Assessment and Nutrition Management of Patients Receiving Bone Marrow Transplantation*
- *Assessment and Nutrition Management of Patients Receiving Solid Organ Transplantation*

Pediatric Nutrition

Quality Assurance Criteria for Pediatric Nutrition Conditions [Published]
- *Normal Healthy Infants, Birth to Age 6 Months*
- *Normal Healthy Infants, Age 6 to 12 Months*
- *Normal Healthy Child, Age 1 to 10 Years*
- *Pregnant Adolescents, Age 17 Years or Younger*
- *Normal Healthy Premature Infants, Appropriate for Gestational Age*
- *Pediatric Intensive Care, Age 1 Month to 18 Years*
- *Failure to Thrive (Nonorganic), Birth to Age 3 Years*
- *Short Bowel Syndrome, Age 0 to 18 Years*
- *Patients with Phenylketonuria, Age 0 to 18 Years*
- *Patients with Aminoacidurias, Age 0 to 18 Years*
- *Patients with Insulin-Dependent Mellitus, Age 0 to 18 Years*
- *Patients with Disabilities (Generic), Age 0 to 18 Years*
- *Infants and Young Children with Isolated Cleft Lip and/or Palate, Birth to Age 2 Years*
- *Patients with Myelodysplasia, Age 0 to 18 Years*
- *Oncology Patients, Age 0 to 18 Years*
- *Cystic Fibrosis Patients, Age 0 to 18 Years*
- *Bronchopulmonary Dysplasia, Inpatient and Outpatient, Initial Diagnosis to Age 5 Years*
- *Patients with End-Stage Renal Disease and Dialysis Intervention, Age 0 to 18 Years*
- *Patients with End-Stage Renal Disease and Renal Transplants, Age 0 to 18 Years*

Quality Assurance Criteria for Pediatric Nutrition Conditions Supplement [Published]
- *Home Parenteral Nutrition*
- *Thermal Injury*
- *Inflammatory Bowel Disease*
- *AIDS*
- *Chronic Liver Disease*
- *Obesity*
- *Congenital Heart Disease*
- *Anorexia/Bulimia*

Public Health Nutrition

Quality Assurance Criteria [In Press]
- *Prenatal*
- *1st Postpartum Visit*
- *Breast Feeding*

Renal Nutrition

Suggested Guidelines for Nutrition Care of Renal (Second Edition) [Published]
- Guidelines for Nutrition Care of Adult Dialysis In-Center Patients
- Guidelines for Nutrition Care of Adult Home Dialysis Patients
- Guidelines for Nutrition Care of Adult Hospitalized Chronic Renal Dialysis Patients
- Guidelines for Nutrition Care of Patients Receiving Conservative Treatment for End-Stage Renal Disease
- Guidelines for Nutrition Care of Pediatric Renal Patients
- Guidelines for Nutrition Care of Hospitalized Adult Renal Transplant Patients
- Nutrition Care of Pregnant Dialysis Patients
- Nutrition Care of Pregnant Patients with Renal Insufficiency
- Nutrition Care of Pregnant Patients with Renal Transplant

Chapter 11

Collecting and Organizing Data

When using the ten-step process for quality improvement, steps one through five deal with the planning process. These steps, addressed in Chapters 9 and 10, form the basis for a departmental quality plan. They are essential before an effective monitoring and evaluation system can be implemented. This chapter focuses on the first part of the implementation phase of the quality improvement process: collecting and organizing data. Topics include development of monitoring forms, sampling and data-collection methods, and summary and presentation of the data.

DATA COLLECTION FORMS

The Joint Commission's (1993) 1994 *Accreditation Manual for Hospitals* specifies (PI.3) that "the organization has a systemic process in place to collect data...." As shown in Chapter 10, the design of each quality indicator specifies what data will be collected and how, when, where, and by whom. In many instances new data collection forms will have to be developed or adapted from those in use currently.

General Considerations

The effectiveness and efficiency of data collection can be fostered by using forms that have certain characteristics. Following are some guidelines adapted from Skipper (1991) for use when developing forms:

- Keep the forms as simple as possible.
- Design forms so that each contains only one data set. Do not mix data from two sources (e.g., from dietetic records [Kardexes] and medical records) or those collected by two individuals (e.g., dietitian and dietetic technician) on the same form.
- Standardize forms as much as possible. For example, dietitians from hospitals that are part of a larger health care system can use the same forms, making it possible to compare data between food and nutrition departments within the system.
- When developing forms, get input from those who will use them. Individuals responsible for data collection have practical experience, and their insights can be invaluable in terms of organization, format, spacing, and content of forms.
- Design forms so that specific information is generated. Forms are less accurate and less useful when they include ambiguous terminology or wordy responses, or when data are open to personal interpretations.
- Create forms as checklists or a series of boxes. Such forms are easy to complete, tend to be more accurate, and facilitate tabulation.
- Order forms in such a way that content is consistent with the way data are encountered. If data are collected from several sections of the medical record, put the columns in order to eliminate the need to jump back and forth through the record.
- Pilot new forms and revise them as necessary before printing hundreds of copies. It is best to have several individuals use the form during the test situation; this helps ensure reliable data collection.

- Tabulate several completed forms to test the smoothness of the process and the value of the data before finalizing a data-collection form.
- Adapt forms for use with a computer whenever possible. Computerization facilitates data organization and analysis, makes it easier to handle large volumes of information, and simplifies comparison of information over time.

Creating New Forms

It is generally very easy to develop a new form. Simply follow these nine steps:

1. Decide what data the form is designed to collect. This information flows directly from the indicator to be monitored.
2. Determine how many data are needed, such as numbers of charts, observations, days, or cases that will be reviewed each time the indicator is monitored. This information indicates the appropriate number of rows and columns for data entry. Do not provide room to collect data on the same form for more than one monitoring period. Fresh forms should be used each time data are collected. Used forms need to be submitted for tabulation and analysis and may not be returned in a timely fashion. Also, used forms are easily misplaced, making planned reuse time consuming if forms have been lost.
3. Give the form a title. State the indicator for which the monitor is used. Make space for the date, unit (if appropriate), and name of the individual completing the form.
4. List criteria either on the data-retrieval form or on an accompanying sheet that is easily accessible every time data are collected.
5. Determine how retrieval items will be listed on the form. Will they be enumerated down the side or across the top?
6. Decide how data will be displayed. Will data be entered in boxed columns or blank lines, or will precoded responses be circled?
7. Select an easy method for responding, such as one of the following codes:
   - Yes, No
   - + = Met criteria; – = Met exception; 0 = Variation; J = Justified variation
   - NA = Not applicable; A = Appropriate; Not A = Not appropriate
   - Check mark (X) if criteria are met
8. Give directions for completion of the form. That is, indicate on the form what coding system will be used.
9. At the bottom of the page leave room for summary information such as total number or percentage of observations in compliance and noncompliance, number of cases that met criteria, and total number of cases observed. The summary will correspond to the indicator and information needed to evaluate it.

Several data-collection forms are given in Appendix D. These can be used as examples for developing tools for your own institution.

How Many Different Forms Are Needed?

The number of different forms in use depends on the number of indicators being monitored. Also, the number will differ if all indicators are general or if some are diagnosis specific or care specific. For example, the department of nursing at Loyola University Medical Center has developed 96 monitoring tools corresponding with ten important aspects of care (Podgorny 1991). These monitors were developed by 23 unit-based committees. Some tools are used throughout the institution (patient satisfaction and medication administration), and others are unit specific (pulmonary artery catheter care and care of the patient receiving mechanical ventilation). A similar pattern of monitor usage is typical in food and nutrition departments of large medical centers.

Small hospitals with one or two dietitians may have as few as two or three nutrition care monitors, whereas larger and more complex institutions may use 30 or more tools. At least one indicator should be monitored to assess each important aspect of care selected to receive high priority for the year.

Linking Forms with Indicators

The purpose of developing and using forms is to provide a method for collecting accurate, consistent, useful information. This information, in turn, is used to monitor elements associated with quality indicators, thereby continually assessing performance related to important aspects of care.

When several monitors are used, it is possible to lose sight of each one's purpose. To avoid this problem, Loyola University Medical Center initiated the use of a "face sheet," adapted for use in dietetics and displayed in Exhibit 11–1, to accompany each monitoring tool. The face sheet identifies the important aspect of care and the related indicators, policies, and procedures being monitored. The threshold for evaluation, target population, time frame for data collection, and data-retrieval methods are also identified. This face sheet fosters consistency in use of forms and facilitates the annual review of the quality planning process.

Exhibit 11-1 Face Sheet to Accompany Monitoring Tools

> **Important Aspect of Care:** Nutrition intervention for high-risk patients.
>
> **Indicator:** Nutritional assessment and care are documented in the medical record.
>
> **Criteria:** Evaluation note includes date, problem list, nutrient intake analysis, need for follow-up.
>
> **Standard of Care:** Documentation conforms to institutional guidelines and uses SOAP format.
>
> **Standard of Practice:** Documentation includes objective data such as ideal body weight and weight as a percentage of ideal body weight, evaluation of pertinent clinical laboratory data, assessment of gastrointestinal functioning, nutrient needs, goals of therapy, and evaluation of appropriateness of current diet order.
>
> **Threshold:** 90%
>
> **Target population:** Patients identified during screening as potentially at high nutritional risk.
>
> **Critical Time:** Documentation should occur within 3 days of admission.
>
> **Data Retrieval:** Patient medical record.
>
> **Expected Sample Size:** Minimum of 20 charts on each dietitian's unit
>
> **Sampling Method:** Beds numbered; patients selected by random number. Unoccupied beds skipped and the next number in order chosen for review.
>
> **Data Collection Period:** Once per month on randomly selected days; results reported every three months.
>
> **Date Originally Developed:** July 1992
>
> **Last Date of Review:** July 1992
>
> *Source:* Adapted from Podgorny, K.L., Developing Nursing-Focused Quality Indicators: A Professional Challenge, *Journal of Nursing Care Quality,* vol. 6, No. 1, pp. 47–52, Aspen Publishers, Inc., © 1991.

SELECTING THE SAMPLE

One aspect of quality management that dietitians find troublesome is knowing how many observations to include in the data collection. Indicators related to high-risk, problem-prone situations may be considered as sentinel events; each occurrence will be reviewed and recorded. Schroeder (1991) also suggests that data be continually monitored for all activities/events when indicators have thresholds set at 0 percent and 100 percent. This recommendation evolves from the fact that when performance allows for no deviations (thresholds of 0 percent and 100 percent), the concern is for individual cases, not patterns or trends.

Monitors for high-volume indicators, on the other hand, will target a representative sample. The challenge is to ensure that the selected sample truly represents the total population for which performance is being evaluated.

What Is a Sample?

A sample is a subset of measurements selected from a specific population (Miller and Knapp 1979). It is the sample that is observed and studied. The objective is to select sampling units that are truly representative of all events whether the events be patient trays, nutrition care procedures, patient wards, patients, rooms, or days. Samples may be random or nonrandom.

- Random or probability samples are samples selected according to some chance mechanism (such as a random-number table) that ensures that all cases within a specified population have an equal opportunity to be drawn from the population.
- A nonrandom or nonprobability sample is one in which the sampling units (for example, patients or trays) are not selected according to a random selection scheme. Some examples of nonrandom samples are all patient trays for only one unit on a rotating schedule, subjects who volunteer to participate in a study, or samples of convenience (those who attend a certain class).

Sample Size

When samples are used for data collection, the ultimate objective is to make statistical inferences or draw conclusions about a total population based on information obtained in a sample. Sample size and data-collection frequency are based on three principles (Schroeder 1991):

1. how often the event/activity occurs
2. how important or significant the event/activity is
3. how long the event/activity has been problem free.

These principles give some flexibility, allowing for maximal use of time and human resources.

Sample sizes can be calculated mathematically, depending on the accuracy desired within the estimates. However, this process can become quite complex and requires the use of statistical formulas (Miller and Knapp 1979, 93). To simplify this process, the Joint Commission recommends the use of 20 cases or 5 percent, whichever is greater (Patterson 1989). For example, for the indicator "Quality and accuracy of food items served to patients meet department standards" (Southwest General Hospital, Middleburg Heights, Ohio), the first question is, "Will all meals be surveyed, or only noon and evening meals?" Assuming that trays for all three meals will be evaluated, a hospital with an average census of 120 patients would serve about 335 trays per day, or 2345 trays per week, taking into account NPO patients and those receiving enteric and parenteral feedings. Five percent of this figure would be 117 trays per week or 470 trays per month. This number is greater than 20 observations; therefore 5 percent of trays would be audited.

Consider another monitor, "Patients who are NPO or on clear liquids more than five days." Experience may show that a total of five patients a day from an average census of 400 may meet this criterion. In a week's time 35 patients may represent the population; potentially there would be 140 cases per month. Using the 5 percent guideline, only seven patients per month would be selected—an inadequate number for monitoring. Thus data on 20 patients would be collected in order to do a monthly evaluation.

Katz and Green (1992) recommended that the sample size vary with the type of study. This model, shown in Exhibit 11–2, calls for increasing the sample size when preliminary data suggest that a problem exists. For a *routine review* the rule of 5 percent or 30 observations is used as described above. As the seriousness of the situation increases, sample size is increased:

- A *query review* occurs when data demonstrate a variance outside the established threshold parameters that cannot readily be explained or justified. Here the recommended sample size is 10 percent or 40 cases, whichever is greater.
- An *intensive review* is conducted when unusual occurrences or trends demonstrate a negative impact on patient outcomes. In this case a sample size of 15 percent or 60 cases is recommended.
- *Sentinel events* call for analysis of each case. Sentinel events may include such things as adverse reactions to foods, unidentified malnutrition, food poisoning, death, or nutritional problems that may extend the patient's length of hospital stay. Sentinel events require 100 percent compliance or 0 percent occurrence and monitoring to identify each variance.

Thus when it appears that quality is on the line, effective quality managers step up their efforts to determine whether performance is below expected levels. Increased sample size is one way to increase the potential for uncovering possible deviations.

Exhibit 11–2 Katz–Green Guidelines for Monitoring

Type of Study	Sample Size
Routine review	5% or 30 (whichever is greater)
Query review	10% or 40 (whichever is greater)
Intensive review	15% or 60 (whichever is greater)
Sentinel event	100% (every event)

Source: Reprinted from *Managing Quality: A Guide to Monitoring and Evaluating Nursing Services,* by J. Katz and E. Green, p. 26, with permission of Mosby-Year Book, © 1992.

Selecting a Representative Sample

After determining sample size, the next step is to decide which specific medical records or events will be reviewed to give a random sample. Several methods can be used to select a sample that is truly representative of the total population under review. Remember that the "total population" includes only those events, patients, or activities for whom the monitor applies. The "total population" is the population given in the denominator when indicators are first developed. For instance, if you are monitoring the number of diabetic patients who understand their diets after diet counseling, the total population is the number of diabetic patients who received counseling, not the total number of diabetic patients admitted to the hospital or treated at the clinic.

One of the most popular methods used to obtain a random sample is selection from a table of random numbers such as the one given in Appendix E. Randomness must be ensured. That is, every member of the population must have an equal chance of being selected. To select all cases from a single nursing unit, or to pick every *n*th item in a series would not give each even an equal chance of being audited. The table can be used in several ways.

- In checking tray accuracy, for example, trays might be selected if they correspond to random numbers on the table. First, decide how many trays need to be evaluated at this time. Say the number is to be 25 out of 380 trays assembled. Decide which three digits of the five-digit series will be used; the first three, the middle three, or the last three. For the series 54463 these would be 544, 446, or 463. Using both the first three digits and the last three digits in a series, and reading the table down the trays 074, 283, 174, 366 . . . would be selected. Skip any digits that exceed 380 (the total number of trays) and continue to select numbers until there are 25. Selected numbers are ordered numerically and used for designating which trays are evaluated this time. The use of a random table requires more mental concentration than simply picking every *n*th tray, but the result is truly a representative sample.
- Another way to use the table to achieve randomness is to use the last two digits of the number to select days of the month on which monitoring will occur. Numbers exceeding the number of days in the month are excluded. Consecutive numbers are used until the desired number of days is achieved. If consecutive numbers on the table are 51152, 66953, 60257, 01848, 56157, 82738, 427*25,* 96664, 18195, 531*05*, . . . , data collection would occur on the 25th and 5th days of the month. If weekend days are not used for data collection, those days would be passed over and additional numbers considered to obtain the necessary number of dates for data collection.

- To select patient units randomly for data collection, each unit can be numbered in the overall quality management plan. At the beginning of the year units to be used for weekly, monthly, quarterly, or semiannual data collection can be determined by picking those that correspond to selected positions of digits from consecutive random numbers.
- To select patient medical records randomly, a different strategy can be used. If five charts are needed from a 40-bed ward, for example, the first two and the last two digits of random numbers might be chosen, skipping any numbers over 40. In designing a strategy for numbering beds, such as in two-bed wards, bed A could equal even numbers and bed B could equal odd numbers. If the table of numbers reads *24384*, *32101*, *62318*, *62803*, *79921*, . . . , patients in bed numbers 24, 32, 1, 18, 3, and 21 would be selected for review.
- When reviewing a specific clinic population, use the method shown in Exhibit 11–3 (Verhonick and Seaman 1978). As noted, assigned numbers or social security numbers can be used to identify cases. A table of random numbers is then used to select the desired sample.

As Donabedian (1988) pointed out, those who assess care are not always interested in obtaining a representative sample, or even an illustrative picture of care as a whole. Their purpose is more managerial, namely, to "identify and correct the most serious failures in care and, by doing so, to create an environment of watchful concern that motivates everyone to perform better." In practice, then, samples are often selected from among the most important diagnoses for which known problems exist and where the possibility of improvement is both achievable and significant.

Frequency of Monitoring and Evaluation

Questions often asked are "How often should data be collected?" and "Do data need to be analyzed as they are collected?" Answers to these questions depend on the nature of the indicator, institutional resources (such as personnel, time, and computers), and the number of indicators being monitored routinely. According to Schroeder (1991) the most commonly used schedules are

- weekly monitoring and monthly evaluation
- monthly data collection and review
- monthly data collection and quarterly evaluation
- quarterly data collection and evaluation.

The schedule selected for both monitoring and evaluation should correspond with the nature of the indicator and the value of its contribution to quality care. Most nutrition service departments assess tray accuracy weekly and evaluate results monthly; they may monitor nutritional assessments monthly and evaluate the data on a quarterly basis; and they may both collect and evaluate data related to patient satisfaction quarterly.

Data-collection schedules may deviate from the plan, however. If, for example, monitoring data show an undesirable trend and a program is instituted to solve the problem, sufficient time should be given for the intervention to be completely implemented before monitoring is repeated. In other instances there may be insufficient cases to provide a reliable sample, such as during low census periods, or when key staff are away for vacation or conferences. Also, data may be skewed by particularly high census periods or when patient acuity is unusually high.

Exhibit 11–3 Example of How To Use Random Numbers To Select a Representative Sample of 20 Cases

Table of Random Numbers

The target population contains 80 cases; select a random sample of 20 cases.

Procedure

1. Assign a number to each case from 01 to 80, or use the last two digits of Social Security number, patient number, or arbitrarily assign a number.

25	19	64	82	84
23	02	41	46	04
55	85	66	96	28
68	45	19	69	59
69	31	46	29	85

2. On the table of random numbers, arbitrarily pick a two-digit column.

37	31	61	28	98
66	42	19	24	94
33	65	78	12	35
76	32	06	19	35
43	33	42	02	59

3. With closed eyes, select a random start in that column.

28	31	93	43	94
97	19	21	63	20
82	80	37	14	20
03	68	03	13	60
65	16	58	11	01

4. Beginning with the starting number, continue to select sequentially every 2-digit number in that column and the next if necessary until 20 cases have been selected.

24	65	58	57	04
02	72	64	07	75
79	16	78	63	99
04	75	14	93	39
40	64	64	57	60

5. In the event a random number not included in the sequence 01 to 80 occurs (i.e., 96) skip that number and proceed to the next one listed.

06	27	07	34	26
62	40	03	87	10
00	98	48	18	97
50	64	19	18	91
38	54	52	25	78

Source: Reprinted from *Research Methods for Undergraduate Students in Nursing*, by P.J. Verhonick and C.C. Seaman, p. 75, with permission of Appleton & Lange, © 1978.

Many food and nutrition departments maintain daily or weekly data on volume indicators. Schiller et al. (1991) summarize such volume indicators as

- total number of patients seen
- total number of patients seen by diagnosis
- number of nutrition screenings
- number of basic assessments
- number of comprehensive assessments
- number of follow-up contacts
- number of reassessments
- number of nutrition consultations
- number of brief diet instructions
- number of diet counseling sessions
- number of patients on enteric and parenteral feedings
- number of diet changes recommended
- number of group classes
- number of individuals attending group classes
- average length of stay.

Perhaps only a few pertinent volume indicators would be selected for continuous monitoring, such as the number of patient consultations, number of patients counseled, number attending classes, amount of billed services, and number of incident reports.

DATA COLLECTION

The schedule, procedures, and responsibility for data collection are usually determined by the departmental quality coordinator or a quality committee. Ordinarily several individuals will be involved in data collection, including dietitians, dietetic technicians, supervisors, and other dietary employees. Those involved in data collection should be trained as necessary to ensure that procedures are followed and that data are accurate.

Many hospitals prepare a general plan, such as the one shown in Exhibit 11–4, for data collection at the beginning of the year. Note that data for some monitors are continually collected, such as for sentinel events and continuous monitors; reports are prepared according to the given schedule. In this manner everyone in the department knows ahead of time when audits will be conducted and which monitors will be reported in any given month.

ORGANIZING DATA

Once data are collected, they need to be organized in such a way that trends in care or opportunities for improvement are readily identifiable. This can be accomplished in a variety of ways. Often data are presented in tabular form, comparing desired performance (thresholds) to actual performance. Graphic displays such as bar graphs, pie charts, and line graphs can be used to supplement and clarify a narrative report. Narrative summaries may be used alone, but they can become lengthy. Time constraints and poor writing skills may make it difficult to present data clearly and in an objective and factual manner. Regardless of the method chosen, data need to be clear, explicit, and complete (Schroeder 1991).

Tabular Data Presentation

Tables of one kind or another are probably the most frequently used means of presenting quality management data for analysis. Assume that drug-nutrient interaction counseling is monitored by using the policy, procedure, and form shown in Exhibit 11–5. A summary of hypothetical data can be presented in tabular form, highlighting at least three different dimensions of activity, as illustrated in Exhibit 11–6.

Pooled monitoring results show overall performance of 56 percent, whereas the threshold is 65 percent compliance. When the data are looked at in several different ways, the nature of the problem may be revealed. Summary by unit, as shown in the first part of Exhibit 11–6, reveals that performance is below the threshold in three units. Problem analysis and solution strategies can focus directly on these three areas. In the second example, analysis by drug reveals that performance is below threshold for four drugs; counseling for the others is above threshold. Finally, the third example shows that counseling is provided by mail in over 50 percent of the cases, a reality that bears consideration, since face-to-face counseling is preferred. Further analysis might be conducted to determine whether this level of mailed counseling persists across all units or if the problem is localized on a few units.

As shown, tabular information can be reformatted to provide invaluable insights to further clarify problems and identify trends that may be masked by simply pooling all information. When problems are identified in certain units or for specific activities, more frequent focused reviews may be conducted in those areas to ascertain the benefit of steps taken to resolve certain problems.

Graphic Presentations

The adage "A picture is worth a thousand words" can be applied to monitoring data, as well as to cartoons, artwork, and other illustrations. Data can be presented both simply and easily by using three different kinds of graphs.

Bar Graphs

Bar graphs can be used effectively to illustrate differences between several observations during one monitoring period or

Exhibit 11–4 Calendar for Reporting Quality Monitors

FY 94 Dietetic Service Quality Assurance Reporting Schedule												
MONITORS	OCT	NOV	DEC	JAN	FEB	MAR	APR	MAY	JUN	JUL	AUG	SEP
Sentinel Event: 1. Imminent Danger to Pts/Employees	X			X			X			X		
CONTINUOUS:												
2. NTA			X			X			X		X	
3. Assessment of Pro/CalFluid needs & Provision on Supplement Nut...		X			X			X			X	
4. Pt. Sat. Surveys			X			X			X		X	
FOCUS REVIEWS:												
5. Dept. Minutes			X									
6. HAACP								X				
7. JCAHO "Mock Review"				X								X
RISK MANAGEMENT:												
8. Inventory of food		X										
9. Security					X							
10. Patient Accept. of Meals									X			
11. Starvation in Hospital. Pts.											X	
QUALITY CONTROL MONITORS:												
QC#1/CCP#1: Foods are Served at Accept. Temperatures	X			X			X			X		
QC#2/CCP#2: Employee Infection Control		X			X			X			X	
QC#3/CCP#3: The Service is designed and equipped to facilitate the safe, sanitary, and timely provision of food served to patients.		X			X			X			X	
QC#4/CCP#4: Provision of Subsistences for Recipe Preparation			X			X			X			X

Courtesy of Donna M. Bashara, Olin E. Teague Veterans' Center, Temple, Texas.

between observations for the same activity from one monitoring period to the next. Figure 11–1 shows a bar graph constructed from the tabular data shown in Exhibit 11–6. Note how dramatic the data appear from units whose performance is below threshold.

Bar graphs are particularly useful when threshold levels are incorporated into the basic graph. Notice in Figure 11–1 that noncompliance jumps out when the threshold line is in place. When the target figure is not shown, those analyzing the data must interrupt their thought patterns to picture results in relation to thresholds; the thought process can easily be facilitated by giving the threshold line.

Pareto charts and histograms described in Chapter 8 can be classified as bar graphs. In these specialized quantitative quality tools, bar graphs are used to illustrate relative importance of a problem (Pareto chart) or to illustrate the distribution and frequency of numerical data (histogram). By formatting data in these unique ways, trends and patterns often reflect impor-

Exhibit 11–5 Sample Policy, Procedure, and Form for Drug-Nutrient Interaction Counseling

Department of Dietetics
Division
Nutrition Services

Quality Assurance: Drug-Nutrient Interaction Counseling Assessment

POLICY:

To provide ongoing assessment of drug-nutrient interaction counseling that is provided for patients, a review will be conducted bi-annually by the Coordinator of Nutrition Services or his/her designee.

PROCEDURE:
1. The Pharmacy printout is obtained to determine patients to be reviewed.
2. Form No. 855.01 is completed on all patients* listed on the pharmacy printout.
    a. Record bed number, medical record number, dates admitted to hospital and to floor.
    b. Check drug patient received.
    c. Indicate if patient given instruction or information mailed and date.
    d. Summarize total number of patients *needing* drug-nutrient interaction counseling and total number of patients who *received* counseling.
    e. Prepare list of all patients needing instruction and give to registered dietitian (R.D.) assigned on unit to complete corrective action.
3. A summary report (Form 855.02) is submitted to the Associate Director, Dietetics.
4. Results are compared to established monitoring criteria and included in the quality assurance reports.

*Patients in the intensive care units are excluded from review. Patients hospitalized ≤ 72° (3 business days) are excluded from review.

Date: _____

Auditor: _____

R.D. on Duty: _____ Unit: _____

Total No. Beds Occupied: _____

Policy No: 855.00: WRITTEN: _____
DATES REVISED: _____
FORM NO. 855.01

Drug-Nutrient Instruction Audit

Bed No.	Medical Record Number	Date Admitted to Hospital	Date Admitted to Floor	Drug Patient Received							Patient Presented with		Dates of Corrective Action
				Theophylline	Griseofulvin	Phenelzine Sulfate	Warfarin Sodium	Quinidine	Methenamine	Tetracycline	Mailed Information and Charting (Date)	Discharge Instruction and Charting (Date)	

Total number of patients in need of drug-nutrient interaction instruction: _____

Total number of patients who received drug-nutrient interaction instruction: _____

continues

Exhibit 11–5 Sample Policy, Procedure, and Form for Drug-Nutrient Interaction Counseling (continued)

Drug-Nutrient Instruction Audit Summary

Total number of patients needing FDI counseling _____

    B/F _____: % of total = _____
    H/I _____: % of total = _____

Total number of patients instructed on FDI _____

% of total number needing FDI _____

    B/F _____: % of total needing FDI _____
    H/I _____: % of total needing FDI _____

After corrective action:

Total number instructed on FDI _____

% of total needing FDI _____

    B/F _____: % of total needing FDI _____
    H/I _____: % of total needing FDI _____

Courtesy of Hildreth A. Macy, Associate Director, Food and Nutrition Services, Henry Ford Hospital, Detroit, Michigan.

---

tant relationships missed when data are presented in tables or check sheets.

Pie Charts

Pie charts lend themselves to illustration of data for five or six different variables. Pie charts would be the best way to present some, but not all, results of monitoring for drug counseling. Results of monitoring by drug (Figure 11–2) show at a glance that three drugs account for 78 percent of unmet criteria. Note that this graph reformats information to show what percentage of noncompliance events can be attributed to each of the seven drugs.

Pie charts may be used to reflect differences in performance between units if the number of units is small, e.g., fewer than five or six. Although pie charts could be used for larger units, it is sometimes difficult to grasp significant differences between segments of a pie cut into ten or more portions.

The percent of counseling sessions as discharge counseling or as mailed instructions could be illustrated in a pie chart. Improvements would be clearly evident as the segment representing mailed instructions hopefully shrinks after intervention strategies are instituted.

Line Graphs

Line graphs are often used to show results from one month to the next. Trends are easily grasped from such illustrations. In fact, line graphs are used to construct trend charts or run charts, described in Chapter 8. Figure 11–3 is a line graph of inpatient nutrition consultations for one year. A look at the figures would show an increase in the number of consultations, but visualization enhances the jump in volume from October onward. This improvement may have followed institution of a strategy to increase the number of consultations. In this case, results are instantly apparent to dietitians who put forth extra effort to make the new system work. Such a line graph can be shared with administration to reflect accomplishments and improved productivity.

Line graphs are often constructed to show 13 months. This practice not only shows activity during the previous year, but it also allows comparison with the same month last year. Comparing results from a year ago amplifies differences and can be used to calculate changes in activity from the previous year. For example, the data shown in Figure 11–3 show a 56 percent increase in consultations from April 1991 to April 1992.

USE OF A COMPUTER TO ORGANIZE DATA

Use of a computer may facilitate organization of data in either tabular or graphic format. Walters (1991) recommends asking four questions to decide whether a computer should be used in the quality management process:

1. Exactly what will be measured, studied, or monitored?
2. Will your institution support computerization?
3. Will computerization save time?
4. Will use of a computer save money?

Larger institutions clearly may benefit by using computers because the volume of data often exceeds easy compilation and tracking of results for numerous monitors. In smaller institutions or hospitals using only a small number of monitors, computation of data may be handled just as easily by hand. However, construction of run charts, Pareto charts, bar graphs, histograms, pie charts, or other data analysis tools is significantly easier when computerized data management and graphic software packages are integrated.

Exhibit 11–6 Examples of Tabular Presentation of Data

A. Summary by units
   Threshold = 65%

Unit	No. Meeting Criteria	% Meeting Criteria
1	8 of 9	89%
2	2 of 7	29%
3	2 of 5	40%
4	6 of 11	55%
Total	18 of 32	56%

B. Summary by drug
   Threshold = 65%

Drug	No. Meeting Criteria	% Meeting Criteria
Theophylline	2 of 3	67%
Griseofulvin	0 of 4	0
Phenelzine Sulfate	2 of 5	40%
Warfarin Sodium	6 of 6	100%
Quinidine	1 of 5	20%
Methenamine	1 of 2	50%
Tetracycline	6 of 7	86%
Total	18 of 32	56%

C. Summary by
   type of counseling
   Threshold = 65%

Type of counseling	No.	%
Mailed information	11	59
Discharge instruction	7	41
Total counseled	18	
Total in need of counseling	32	
Percent compliance		56

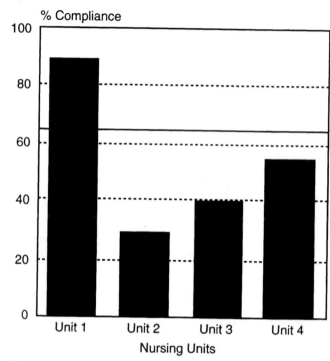

Note: Threshold = 65%.

Figure 11–1 Bar Graph Illustrating Monitoring Results of Food-Drug Interaction Counseling

In any situation, use of a computer for data tabulation, data display, and statistical analysis will facilitate quality management. For example, computerization permits analysis of total population data rather than mere samples, thereby increasing validity of results. Owens et al. (1989) described how a concurrent quality assurance program for a nutrition support service was built around computerized information. Weathers et al. (1986), Ford (1987), and Duffy et al. (1989) showed how the computer is used for clinical dietetic management systems. Leyshock and Tracy (1992) showed how a database is used for monitoring clinical productivity.

Data generated from these and similar computer applications can be integrated easily with quality improvement programs. Some quality applications for computerized data described in the article by Ford (1987) are delineated here to illustrate the types of lists generated as well as their possible uses in a quality management program:

- Menu error report identifies patients at nutritional risk by virtue of multiple diet restrictions or inadequate menu selections.
- Nutrition care plan list helps dietitians identify nutritionally stressed patients.
- Nutritional risk list shows patients on NPO, clear liquid, full liquid, or purée diet for three days or longer, and helps identify and track nutritionally stressed patients.
- Diet lists (diabetic, weight control, protein restricted, modified fat) facilitates referrals for follow-up care and

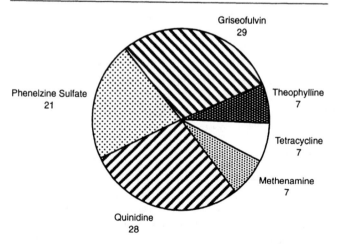

Note: Calculated from data in Exhibit 12–4.

Figure 11–2 Pie Chart Illustrating Percentages of Missed Counseling Sessions by Drug

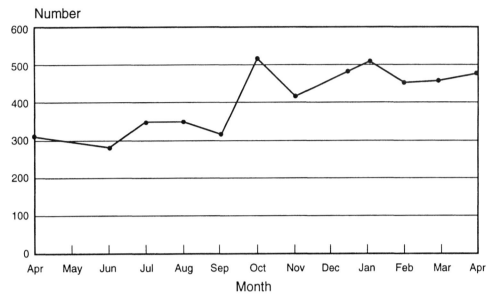

Figure 11–3 Line Graph of Nutrition Consultations at Southwest General Hospital

guides improvements in nutrition education programming.
- Tube feeding list facilitates daily monitoring and generates information to assess clinical indicators.

Recent literature illustrates how computerization could help support ongoing nutrition services activities. For example, Fletcher and Johnson (1991) described procedures for tracking NPO orders and explained how manually recorded data were used in quality assurance activities. Their preliminary results indicated that about 40 patients per month were monitored and that the number of patients NPO or on clear liquid diets decreased and the number of patients receiving nutrition support within 5 days increased. Although the needed information was kept on a handwritten log, use of a computer would facilitate both data collection and analysis.

Stockard (1992) developed a database management system specifically for monitoring the quality of unit-based nursing care at a 600-bed hospital. In this system nursing quality assurance reports are standardized and key quality indicators are used to track quality performance. Data entry is done quarterly by a secretary; it takes only four to five minutes to enter data from each report. Data can be retrieved in several different ways, such as quarterly report by aspect of care; quarterly report by outcome; quarterly list of indicators, thresholds, and compliance rates; and compliance rates by unit. A system such as this could easily be developed for food and nutrition services.

When considering the purchase of computer hardware or software for quality management, it is necessary first to decide how the computer will be used in the following situations:

- Monitoring standard criteria: Are randomization and selection from a large database required? Will data be automatically compared with standards or will this be done manually? In dietetics, monitoring is generally done manually, but a few departments are linked with institutional systems in which standard criteria are monitored automatically.
- Information storage: Do standards and norms need to be kept on-line for reference and comparison? What is the sample size in relation to computer memory? Many nutrition support teams store information and use on-line information for continuous monitoring.
- Case identification and data retrieval: Could a computerized system facilitate case retrieval or sample selection? Are data collection forms set up for easy computer input? Many departments currently use the computer to generate random numbers, identify samples, and summarize frequency data.
- Statistical analysis: Are complex computations or large numbers of repetitive calculations required that would justify computerization? What types of statistical packages will be used? Will the computer be used to generate Pareto charts, histograms, or other analytical tools?
- Reporting of results: Will programs be purchased to allow integration of word processing and data analysis systems? A few departments in large medical centers now have this technology available and are using it to facilitate problem identification, analysis, and reporting.
- Documentation: Computers can enhance the quality of input just as much as they improve the quality of output.

Computerization of quality management activities requires staff members to think logically and conceptually. It increases their ability to see the links between activities performed and quality of care and to report results of services provided (Wilbert 1985).

If the decision is in favor of using computerization, next come decisions regarding hardware and software. An institutional system (Schultz 1990) or the use of a mainframe computer has the advantages of multiple users, high-volume data-entry ability, and applications created for high-level institutional use. Such systems allow integration of dietary data with those from other departments and development of customized software programs to meet the unique needs of the hospital. If a hospitalwide system is in place, or if a department-specific program can be developed, the department's quality coordinator should meet with the computer programmers to determine how data from nutrition services will be integrated into the system.

One dietary department, faced with staffing cuts, found that much of the information used in screening patients for nutritional risk was routinely entered into the hospital computer system; however, the data were not organized for easy retrieval for screening purposes (Langbein 1992). In collaboration with medical information system personnel, specific information was defined, and a computer-assisted screening program was devised. Standards for height and weight, diagnosis, and albumin were defined; the computer flags the name of any patient whose weight falls outside the given parameters, or whose albumin is below normal, or who presents with a high-risk diagnosis. A computer screening printout is generated daily and contains the names of patients "at risk" every three days from admission. By using these computer-generated data as beginning information, dietetic technicians complete the screen by adding weight history and diet order. The entire process takes about five minutes. Quality improvement audits show that 100 percent of eligible patients are screened within 72 hours of admission.

Because they are more "user friendly," it is often easier to use a personal computer (PC) than a mainframe computer if an institutional software package is not available. However, such programs are necessarily limited to data entered by nutrition services personnel, ruling out data generated in other departments such as nursing, pharmacy, and admitting. As hospitals invest in both computer hardware and tools to facilitate quality improvement, single-department computer applications will become obsolete. If personal computers are used, three types of programs are needed.

1. Data Management Program

Data management programs facilitate compilation and organization of data. Some prepackaged software programs such as spreadsheets and database systems are available for quality management at the institutional level, but programs for departmental use are difficult to find. Also, such programs tend to be very costly because the buyer is paying for the programming as well as tests of validity and reliability. Furthermore, prepackaged programs frequently cannot be modified for institutional needs. As a result, most departments that use a PC for quality management also purchase a data-management system and create their own programs for data organization and analysis. Some popular software programs are listed in Exhibit 11–7.

When using these and similar software packages to set up a data-management system for departmental quality initiatives, Walters (1991) offers the following suggestions:

- Learn what systems are currently available at the institution. Plan to integrate administrative, clinical, and patient information whenever possible.

Exhibit 11–7 Popular Software Programs for Quality Assessment and Improvement

Software	Manufacturer
Word processing	
WordPerfect	WordPerfect Corporation
Word	Microsoft
Word Star	Word Star International
Spreadsheets	
Lotus 1-2-3	Lotus Development Corporation
Excel	Microsoft
Quattro Pro	Borland
Database	
D Base	Borland
Fox Pro	Microsoft
Paradox	Borland
Graphics	
Harvard Graphics	Software Publishing Corporation
DrawPerfect	WordPerfect Corporation
Power Point	Microsoft
Statistical	
SPSS	SPSS Incorporated
Minitab	Minitab Incorporated
BMDP	BMDP Statistical Software
Integrative	
Appleworks	Claris Corporation
Microsoft Works	Microsoft Corporation
PFS: First Choice	Software Publishing Corporation
Communications	
Carbon Copy Plus	Meridian Technology, Incorporated
Maclink Plus	Data Viz, Incorporated
Procomm Plus	Datastorm Technologies

Source: Reprinted from Ventura, M.R., Ackerman, M.H., Gugerty, B., Skomra, R., and Crosby, F.E., Selecting and Using Computer Application Software for Quality Assessment and Improvement, *Journal of Nursing Care Quality*, Vol. 7, No. 1, pp. 16–28, Aspen Publishers, Inc., © 1992.

- Specify your goals. Here, be specific. For example, do not say you wish "to monitor dietetic entries in medical records." A more specific way of stating the goal would be to say, "Compare screening and assessment report data for each unit with number of patient admissions per month; compare this month's activity with activity over the past 12 months, compare monthly averages with the year-to-date averages, and give a semiannual and annual average."
- Decide who will enter the data. Keep in mind ease of use as well as training needs of any new hires who may be expected to work with the system.
- Keep each spreadsheet as simple as possible; support personnel can be trained for data input. Thus formats and data points should be understood as easily by nonprofessionals as by dietitians who set up the system.
- Determine what the source document will be for data entry. Set up each application to correspond with the form from which data will be compiled. Logical flow of information will reduce both the time needed for data input and frustration of those who use the system.
- Storage capacity of the system needs to be considered. How many data will be held on-line? This may differ from one monitor to the next, but generally each application should allow for handling data for at least 13 months, or three years for sentinel events, to allow tracking over time and observation of performance trends.
- Keep spreadsheets simple. Some data will be collected daily or weekly; other data will be collected quarterly or semiannually. With a brief refresher, those who routinely enter data should be able to enter data and generate summary reports with little difficulty, even for those data encountered two or three times a year.
- As the spreadsheet is developed, prepare an accompanying manual to guide new users or to use as the basis for training new personnel. In the manual, for each application, illustrate how the program is set up, how data are transferred from forms to the computer, and how summary data are to be generated. Any published references used in developing the applications should be referenced.
- Prepare draft reports produced from the spreadsheet to be sure they are not difficult to generate and that they provide the desired amount of detail. Limit the number of reports to those that will be used within the department for decision making or reporting to others, such as the hospital quality committee. Any needed statistical computations should be built into the system, but they should be kept as simple as possible.
- Train secretarial support, staff dietitians, technical personnel, or volunteer staff to enter data. Training costs can be high, especially if many individuals are trained to use the system or if turnover is high among data processors. Thus only those individuals who will be called upon regularly to enter data need to be trained.
- Find and train a substitute troubleshooter. It should not be assumed that the one who sets up the program will spend time in data entry, although that may be necessary until bugs are worked out of the system. A backup person should be trained in the system to handle problems in the absence of the "computer expert."

The department should consider using a consultant, from either within or outside the institution, to set up departmental computer applications for quality management. As Walters (1991) points out, use of a consultant is likely to speed up the process and ensure that the project will be completed.

2. Graphics Programs

Graphics software programs are used to display information as graphs or charts and to construct tools used in quality management such as bar graphs, Pareto charts, or histograms. Simple tables showing data summaries can be generated from the data-management or word processing system; a graphics program is not needed for tabular presentation of results.

To create a visual display of findings using such programs, data are first gathered and summarized either manually or by computer. Quantitative findings are entered at a computer terminal, and the desired graphic display is generated on command. Most graphics programs first prompt the user to stipulate the desired type of display (bar graph, pie chart, line graph). Directions are then displayed on the computer screen, prompting the user through a series of decisions regarding number of bars needed, shading desired, naming of vertical and horizontal axes of graphs, and so on. Numerical and descriptive data are entered, restricted to the space allowed, and the graphic is printed in the desired format. As described earlier, graphic displays enhance quality reports, highlight trends, and have a strong visual impact when there is wide variation between desired and actual performance.

3. Statistical Programs

Although simple numbers can be added to give summary information, today's quality management programs call for more sophisticated data analysis methods. For example, a study of patient satisfaction with food services is most useful when scores are tabulated showing the percentages of responses in each category, such as "very satisfied," "satisfied," and so on. This type of analysis by computer allows computation of the mean rating and standard deviation and t-tests to compare results of this period with those of the last analysis. Both of these tests are useful in analyzing results, developing targets for improvement, and measuring effectiveness of intervention strategies.

CONCLUSION

Accurate data collection requires use of well-developed forms, selection of appropriate samples, and careful recording of information. Data may be organized in narrative, tabular, or graphic formats. Although data may be analyzed by hand, the use of a computer greatly facilitates the process and offers numerous opportunities for innovation, creativity, and in-depth analysis.

REFERENCES

Donabedian, A. 1988. The quality of care: How can it be assessed? *Journal of the American Medical Association* 260 no. 12:1743–1748.

Duffy, K.M., et al. 1989. A computer-assisted management information system for nutrition services. *Journal of the American Dietetic Association* 89:1296–1300.

Fletcher, J., and E. Johnson. 1991. Tracking NPO diet orders. *Clinical Nutrition Management Newsletter* 10:11. September.

Ford, M.G. 1987. The computer as an aid in clinical management. *Journal of the American Dietetic Association* 87:497–500.

Joint Commission on Accreditation of Healthcare Organizations. 1993. *1994 Accreditation Manual for Hospitals. Volume I, Standards.* Oakbrook Terrace, Ill.

Katz, J., and E. Green. 1992. *Managing quality: A guide to monitoring and evaluating nursing services.* St. Louis, Mo: Mosby–Year Book, Inc.

Langbein, V.E. 1992. Teaming up with computers for timely nutrition screening (Abstract). *Journal of the American Dietetic Association* 92 (suppl), A-34.

Leyshock, P.J., and D.L. Tracy. 1992. Using a computerized system to monitor clinical dietetic productivity. *Topics in Clinical Nutrition* 7 no. 2:69–77.

Miller, M.C., and R.G. Knapp. 1979. *Evaluating quality of care.* Gaithersburg, Md: Aspen Publishers, Inc.

Owens, J.P., et al. 1989. Concurrent quality assurance for a nutrition-support service. *American Journal of Hospital Pharmacy* 46:2469–2476.

Patterson, C. 1989. Standards of patient care: The Joint Commission focus on nursing quality assurance. *Nursing Clinics of North America* 23 no. 3:625–638.

Podgorny, K.L. 1991. Developing nursing-focused quality indicators: A professional challenge. *Journal of Nursing Care Quality* 6:47–52. October.

Schiller, M.R., et al. 1991. *Handbook for managing clinical nutrition services.* Gaithersburg, Md: Aspen Publishers, Inc.

Schroeder, P. 1991. *The encyclopedia of nursing care quality.* Vol. 3. Monitoring and evaluation in nursing. Gaithersburg, Md: Aspen Publishing, Inc.

Schultz, R. 1990. Making computerized quality assurance work. *QRC Advisor* 6:1–8. June.

Skipper, A. 1991. Collecting data for clinical indicators. *Nutrition in Clinical Practice* 6:156–158.

Stockard, R.R. 1992. Using a database management system to manage quality assurance data. *Journal of Nursing Care Quality* 7 no. 1:29–34.

Ventura, M.R., et al. 1992. Selecting and using computer application software for quality assessment and improvement. *Journal of Nursing Care Quality* 7 no. 1:16–28.

Verhonick, P.J., and C.C. Seaman. 1978. *Research methods for undergraduate students in nursing.* New York: Appleton-Century-Crofts.

Walters, J.A. 1991. Using automated systems for quality assurance. In *The encyclopedia of nursing care quality.* Vol. 1. Issues and strategies for nursing care quality, ed. P. Schroeder, 101–125. Gaithersburg, Md: Aspen Publishers, Inc.

Weathers, B.J., et al. 1986. Computerized clinical dietetics management system. *Journal of the American Dietetic Association* 86:1217–1223.

Wilbert, C.C. 1985. Computers: Quality assurance application. In *Quality assurance: A complete guide to effective programs*, ed. C.G. Meisenheimer. Gaithersburg, Md: Aspen Publishers, Inc.

Chapter 12

Data Analysis and Evaluation

After data are collected, summarized, and organized they must be analyzed and evaluated. This chapter includes a discussion of the purpose and expected outcomes of data analysis, desired attributes, statistical methods of data analysis, guidelines for interpreting data, and, finally, some pointers for evaluating services found to deviate from established thresholds or standards.

PURPOSE AND EXPECTED OUTCOMES

The purpose of data analysis is to identify the probable cause(s) of deficient care so that improvement plans can be devised. Analysis and evaluation occur whenever observed results are at a variance with the established threshold or standard for the specific event under review (Joint Commission 1992). This evaluation is intended to be an "in-depth review to examine trends, identify specifics when the cause and scope of a problem are unknown, or to ascertain further information that may help identify ways to improve care" (Elrod 1991). Deviations, even for single important events, need to be analyzed thoroughly to ascertain unexplained or unacceptable variances from established thresholds or standards.

The result of data analysis is the identification of a specific problem. From this, the quality-management coordinator or a designated team will be able to enumerate underlying reasons for the observed deviations and take initiatives to improve performance.

ATTRIBUTES OF EFFECTIVE DATA ANALYSIS

Principles of Quality Improvement stipulates:

> The organization analyzes the impact of its clinical and nonclinical systems on the quality of patient care and services in a timely, ongoing fashion, and provides data for use in quality improvement activities that are clear, concise, and understandable. (Joint Commission Quality-Improvement Task Force 1990)

This principle sets forth several qualities desired in an effective data analysis system. Such systems are

- Timely. Data are collected and organized within defined timelines, e.g., weekly, monthly, quarterly, semiannually. Within seven to ten days after the data-collection period the designated committee should hold a meeting to analyze the data. Long lags between data collection and analysis give the impression that results are unimportant. Also, it is desirable to evaluate data while details are fresh and problems associated with variances can be identified.
- Consistent. Data need to be collected in the same format time after time to facilitate legitimate comparisons with past performance.
- Ongoing. Monitors are set up to provide routine and ongoing assessment of performance. A good example is the

weekly report generated from a computer program to track daily monitoring of metabolic and nutritional data for patients receiving total parenteral nutrition, described by Geibig et al. (1991). Concurrent monitors are preferred to retrospective audits because ongoing data collection allows deviations to be recognized and resolved immediately so that quality standards are maintained.

- Clear. Raw data, summary information, comparative data, and evaluative notations need to be free of ambiguity and confusion. This attribute needs to be taken into account during planning stages; forms and document outlines should foster both conciseness and clarity. Often, numerical data are used to reduce vagueness or unnecessary complexity that may arise with narrative summaries.
- Concise. Summary data should be kept to a minimum. That is, only those monitors needed to assess quality of high-volume, high-risk, or problem-prone indicators ought to be tracked. Reports should contain only pertinent details.
- Understandable. Anyone reviewing raw, summary, or evaluative data should be able to grasp its meaning without further explanation. Ambiguity often can be prevented by careful structuring of quantifiable measures during the planning stages.

Departmental leadership, the quality management coordinator, or other quality team members need to ensure that data are valid and reliable before taking steps to rectify apparent problems. Validity and reliability can be assessed by answering questions such as the following (Elrod 1991):

- Do the monitors truly measure the quality indicator?
- Is the sample size adequate to reflect trends?
- Was the sample drawn at random?
- Can the results be trusted as an accurate reflection of actual performance?
- Was the data-collection methodology appropriate?

If the answer to any of these questions is no, it is important to resolve validity and reliability issues before steps are taken to address problems identified on the basis of potentially inaccurate information. If, for example, a spur-of-the-moment review of six medical records from different units shows absence of nutritional assessment documentation, one should question the small sample size, actual deviation from standards and policy, and whether the selected patients met criteria for conducting in-depth nutritional assessments.

One of the most important things to remember during data analysis is that failure to meet quantitative standards does not necessarily indicate poor nutritional care. It is always possible that the quality assessment was invalid, the criteria for identifying deficient performance were incorrect, or the assessment standard was too high (Williamson 1982).

METHODS OF DATA ANALYSIS

Several different methods can be used for data analysis. Considered here are a few techniques commonly used by dietitians. These include simple tabulation of findings, statistical analysis, use of analytical tools, and extended explorations such as interviews and focus groups.

Simple Tabulations

Raw data or percentages may be studied to determine variations from established norms or thresholds. This is the crudest method of data analysis but the one most often used in nutrition services. Tabular data are often extracted from data-collection forms used in the department, such as the patient food service survey shown in Exhibit 12–1.

Note that departmental data may be biased if dietary personnel whose performance is being evaluated distribute and

Exhibit 12–1 Departmental Patient Food Service Survey Form

THE OHIO STATE UNIVERSITY MEDICAL CENTER

Patient Food Service Survey

Please share your impression of Nutrition and Dietetic services at The Ohio State University Medical Center. Our goal is to improve patient satisfaction.
Please check the appropriate box below. Thank you.

	Satisfied	Not Satisfied	No Opinion
1) The temperature at which the food is served			
2) The daily printed menu and the Special Menu variety (adequate choices to select from?)			
3) The taste of the food			
4) The timeliness at which the food is served			
5) Received food choices ordered			
6) The presentation of the food (attractiveness of the tray)			
7) The personal service provided by Nutrition and Dietetics staff (are they respectful and courteous?)			

Please List Any Additional Comments/Suggestions You May Have
_____
_____

Courtesy of The Ohio State University Medical Center, Columbus, Ohio.

collect forms. For example, a dietary worker who collected forms from that employee's area of responsibility went through the forms before submitting them to the quality committee. The employee extracted and destroyed all forms containing negative comments. This employee feared not meeting high standards of performance and deleted all data that could make the employee look bad. To maintain integrity of departmental opinion surveys it is best to have the forms reviewed for potential biases and to have supervisors or volunteers not connected with the services distribute and collect survey forms.

Results may be presented in different formats. For example, Table 12–1 shows summary data for all units. If a threshold of 85 percent indicates acceptable patient satisfaction scores, overall results suggest that all criteria have been met. However, if many negative comments are written on the forms, a closer analysis of data is warranted.

Table 12–2 contains the same data in greater detail to indicate how patients on each nursing unit rated the food service on all seven criteria. An analysis of these data show that all patients on 7 Doan (maternity) were satisfied with all aspects of both food and service. On the other hand, fewer than the desired 85 percent of patients on 9E Doan were satisfied with either food or service. Fewer than 80 percent of patients on 7E Rhodes were satisfied with the food, but 93 percent or more were satisfied with meal service. Only 75 percent of patients on 9E Rhodes were satisfied with temperature, timeliness, and accuracy, but all these patients were satisfied with food variety, taste, and presentation. This disparity may suggest problems in the galley. Staff members in each unit were rated as very courteous, a basis for positive reinforcement of food service workers.

This example shows the importance of breaking out data to show performance by units or other subgroupings. Summary data are not as useful as unit data when pinpointing problem areas or working with employee groups to improve performance. In the example above, employees on 7 Doan should be rewarded for providing high-quality service, and a plan should be developed to improve the accuracy of employees on 9E Rhodes. Methods of identifying problems and dealing with delinquent performance are discussed elsewhere in this manual.

Statistical Analysis

Tabular data can be enhanced greatly by using a few simple statistics. The types of statistics used, however, are determined by the nature of the data collected or the types of scales used to measure the indicator. Also, data are usually tabulated on a computer if statistics are to be calculated. Commonly used statistical tests include frequency distributions, measures of central tendency, dispersion of data, and inferential statistics (t-tests, analysis of variance, correlations).

Measurement Scales

The type of statistical measure used for data analysis depends on the nature of the data. Data can be grouped into four different categories or measurement scales: nominal, ordinal, interval, and ratio. These types of data are described below. Exhibit 12–2 gives a summary of these measurement scales, the type of information provided, appropriate statistical tests, methods of data presentation, and some examples of each.

Nominal Data. Nominal data are derived from categories that can be "named," thus the title *nominal*. These data require the classification of events or subjects according to specified categories. For example, if the variable "disease types" is used, the assignment of patients to categories such as "diabetes," "coronary heart disease," and "cancer" would provide nominal data. Typical examples of nominal data in dietetics are hospital service units, nutrition care procedures, staff positions, disease or treatment complications, marital status, and gender.

Simple statistical techniques such as trends and central tendencies are used to analyze nominal data. Bar graphs or pie charts may be used to show the percentages of cases that fall into each of the categories studied. The most frequently reported result (mode) is highlighted in representing these data.

A check sheet such as the one shown in Exhibit 11–5 (Drug-Nutrient Counseling) can be used to classify data into categories, such as drugs that may cause food-nutrient interactions. The summary of results may include a tally of instructions by

Table 12–1 Results of Patient Satisfaction Survey

All Units (n = 174)	Satisfied	Not Satisfied	No Opinion
1. The temperature at which the food is served	151 (89%)	16 (9%)	2
2. The daily printed menu and the Special Menu variety (adequate choices to select from?)	146 (85%)	20 (12%)	6
3. The taste of the food	138 (81%)	22 (13%)	11
4. The timeliness at which the food is served	148 (87%)	14 (8%)	9
5. Received food choices ordered	152 (89%)	12 (7%)	7
6. The presentation of the food (attractiveness of the tray)	158 (92%)	5 (3%)	9
7. The attitude of all Dietetics staff (are they respectful and courteous?)	166 (98%)	2 (1%)	2

Please List Any Additional Comments or Suggestions You May Have

Courtesy of The Ohio State University Medical Center, Columbus, Ohio.

Table 12-2 Frequency Table Showing Percentages of Patients Satisfied with Different Aspects of Meal Service by Nursing Unit

Unit	Temperature	Variety	Taste	Timeliness	Accuracy	Presentation	Courtesy
6 Doan	80	87	60	86	80	87	100
7 Doan	100	100	100	100	100	100	100
9E Doan	64	64	50	57	79	77	93
9W Doan	86	93	92	71	93	93	100
10E Doan	85	77	62	62	92	92	100
10W Doan	80	60	60	60	80	100	100
11E Doan	100	87	100	100	89	100	100
7E Rhodes	79	73	67	93	93	93	100
7W Rhodes	92	92	91	92	92	92	100
8E Rhodes	100	90	95	95	86	95	95
8W Rhodes	100	90	90	80	90	90	100
9E Rhodes	75	100	100	75	75	100	100
10E Rhodes	83	94	89	89	88	89	94
11E Rhodes	100	70	80	100	100	90	90

Courtesy of The Ohio State University Medical Center, Columbus, Ohio.

drug category, as illustrated earlier. Dividing data into categories helps to segment summary information so that follow-up evaluations can occur on specific types of cases.

Ordinal Data. Ordinal data are rankings of a qualitative nature, usually along a continuum. These data suggest that some sort of "order" exists within the category. Items in the continuums under study are in ordered relationship to each other, such as measures of illness severity (terminal, critical, serious, fair, good condition), quality (satisfactory/not satisfactory; acceptable/optimal), or opinion (strongly disagree to strongly agree). Little is known about distinctions between measures; it is impossible to measure the distinction between "acceptable" and "optimal," for example. Verhonick and Seaman (1978) give some examples of ordinal scales that might be used in assessing quality of care.

- Graphic rating scales, such as quality of care rated from "highest quality care" to "poor quality care"
- Thurston-type scales, in which the evaluator is presented with a battery of 20 to 25 statements. The respondent checks those items most closely associated with his or her feelings.
- Likert-type scales, such as giving a series of statements to which a respondent can react. For example, an item might read, "I received the best possible dietetic services while in the hospital," and alternative responses might be 1 = strongly disagree, 2 = disagree, and so on.

The patient satisfaction survey data shown in Table 12-1 are ordinal data. They show the percentage of patients satisfied and not satisfied with different aspects of food and food service. The mode can be computed to show whether more patients are satisfied or not satisfied with any aspect of food or food service. The median or "middle" score can be calculated, especially when the response continuum gives three or more categories such as "always courteous" to "never courteous." The median is calculated by listing or ranking all raw scores in order, from the lowest to the highest, and finding the halfway point. The median is the point where 50 percent of the scores are above and 50 percent are below that score. Both the mode and median can give important information for problem solving, especially when thresholds are given in these terms.

Interval Data. Interval data are quantitative in nature. They are based on data initially in the form of continuous measurements such as weight, height, blood pressure, length of stay, service charges, and age. Points on these measurements are equidistant from each other. Likert-type scales in which numerical values are assigned to the continuum of responses are

Exhibit 12-2 Summary of Data Types, Type of Information Provided, and Appropriate Statistical Analyses

Scales	Type of Data	Statistics	Examples of Monitors
Nominal	"Named" Groups	Percentages, mode	Workload distribution, levels of performance, numbers of patients at acuity levels, number of cases meeting criteria
Ordinal	Qualitative	Percentages, median, mode	Patient satisfaction surveys
Interval	Quantitative	All statistical tests	Changes in weight status, cholesterol levels, hospital days, charges
Ratio	Quantitative	Two-dimensional grids	Compare satisfaction with food service for patients on regular and modified diets

often treated as interval data, expanding the possibilities for data analysis. Interval scales are less prone to individual interpretation than ordinal scales, and they are more objective. Thus they provide "hard" data that can be used extensively for statistical analyses.

Both mode and median can be calculated for interval data. In addition, the average or mean score can be determined. Thus interval data provide valuable information for data analysis such as frequency distributions, measures of central tendency, range, and variability, discussed below.

Ratio Data. Ratio scales are similar to interval scales, with the exception that ratio scales have an absolute zero point. In quality management, ratio scales usually measure the frequency of occurrence of one event relative to the occurrence of some other event. Some examples of ratios in dietetics include

- male-to-female ratios for satisfaction with diet counseling
- level of satisfaction with food service in relation to type of prescribed diet
- weight loss in relation to the type of dietary regimen
- percentage of patients with nutritional risk factors relative to major disease categories
- incidence of infection relative to feeding modality.

Ratio data are usually summarized on a two-dimensional grid such as line graphs and scatter plots, or they can be organized by using bivariate and multivariate tables (Jackson 1992, 107). Bivariate tables, for example, provide a method for comparing the percentages of patients on regular and modified diets who rate food as poor, fair, good, or very good. Data-collection forms can be set up to categorize and summarize data readily; figures can be translated to graphic displays to portray meaningful ratios.

Frequency Distributions

A representation of the frequencies with which individual values or groups of values occur within a set of measurements is a frequency distribution. Tables 12–1 and 12–2 show frequency distributions for a departmental patient satisfaction survey and Table 12–3 shows the sample frequency distribution of scores on a hospitalwide computerized patient satisfaction survey showing opinions about meal service. Such distributions may provide helpful information; for example, it can be said from these figures how a majority of patients respond to stated evaluation items.

Percentages or relative frequencies provide a common basis for comparing groups composed of different numbers by using a common base of 100 percent. Percentages are often used as the basis for presenting information in tables or graphs to compare results from one period to another. Such comparisons are most useful when nominal data (instead of ordinal or interval) are used in data collection. For example, Table 12–4 illustrates how computerized data may be used to compare results from one period to the next.

Measures of Central Tendency

The average or mean rating is a measure of central tendency. This statistic is often used to answer the question, "What is the average or mean patient satisfaction score?" An average or the arithmetic mean is the sum of all scores divided by the number of scores in the set. This can easily be calculated by hand, but when hundreds of survey forms are returned, the task can be accomplished more easily on a computer.

A mean score can be looked at as a single "typical" score to get a general understanding of how most patients rate meal service. The average or mean score might be used later as a basis for setting improvement goals. Table 12–4 shows mean scores for patient satisfaction on both the four aspects of meal service (taste and temperature of food and courteousness and responsiveness of staff) and satisfaction with dietetic staff members who provide nutrition counseling and education. Note that questions related to food (taste and temperature) will assess quality of the product, while questions about staff (courteousness and responsiveness) will address quality of service. Questions related to professional staff members were recently added to the survey to ascertain patient satisfaction with professional services.

Cumulative means or averages are frequently used in quality management when data are collected every month. Sometimes monthly data are averaged, giving a year-to-date overall score. A score for the current month can then be compared with year-to-date scores as well as the same month last year.

The mode is another measure of central tendency. It is the most frequently reported score or the response most frequently given on a data-collection form. The mode is easily computed

Table 12–3 Frequency Distributions of Responses on Hospital Patient Satisfaction Survey

My meals were:	Strongly Disagree (%)	Disagree (%)	Neither Agree nor Disagree (%)	Agree	Strongly Agree (%)
Appetizing	2	3	14	63	18
Served at right temperature	4	8	8	67	13
Served by courteous staff	0	1	3	74	22
Served by responsive staff	1	1	4	77	17

Table 12–4 Mean Scores and Standard Deviations on Meals Derived from Hospital Patient Satisfaction Survey

	1st Quarter		2nd Quarter	
	Mean	S.D.	Mean	S.D.
My meals were:				
Appetizing	3.50	1.10	3.57	1.08
Served at the right temperature	3.66	1.03	3.73	0.99
Served by courteous staff	4.09	0.76	4.13	0.78
Served by responsive staff	4.05	0.79	4.07	0.84
I was satisfied with:				
Nutrition and Dietetics (provides nutrition counseling/education)				
Courtesy	4.08	0.79	4.09	0.84
Responsiveness	4.05	0.82	4.03	0.89

Courtesy of The Ohio State University Medical Center, Columbus, Ohio.

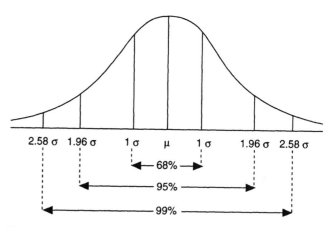

Figure 12–1 Normal Frequency Distribution Curve. *Source:* Reprinted from *Evaluating Quality of Care: Analytic Procedures—Monitoring Techniques*, by M.C. Miller and R.G. Knapp, p. 23, Aspen Publishers, Inc., © 1979.

by listing or ranking all results and identifying the response or item that occurs most frequently. For example, the mode tells us from data presented in Table 12–2 that patients on 7 Doan are the most satisfied with food service at University Hospitals. The mode also indicates that patients overall are most satisfied with courtesy of the food service personnel.

A median score, or the score that falls in the "middle" of a data set, is another measure of central tendency. The median is calculated by listing all raw scores, from lowest to highest, and finding the one in the middle—the point just above which 50 percent of scores lie. By looking at patient satisfaction with temperatures of the food (Table 12–2), we can say that the median of all patient units is 86 percent (seven units have higher percentages of patients satisfied, and seven units have lower percentages of patients who indicated that they were satisfied with the temperature of food).

Dispersion of Scores

Commonly used measures of dispersion for interval data are range and standard deviation.

- *Range*—highest and lowest scores along a continuum. Ranges are often used for interval scales, such as the range in patient variables, e.g., of age, weight, and blood pressure. Ranges also may be used in reporting nominal data or ordinal data over time. For example, a summary report might indicate that over the past eight quarters 79 percent to 83 percent of patients were satisfied with the temperature at which food was served.
- *Standard deviation*—average distance from the mean for a set of scores; one standard deviation includes 68 percent of responses as shown in Figure 12–1. Large values of the standard deviation suggest that scores are widely dispersed around the mean; small values indicate tight clumping of scores around the mean.

Standard deviations provide an indication of how widely scores are dispersed around the mean (Miller and Knapp 1979). The data in Table 12–4, for example, show that patients' opinions about food (standard deviations greater than 1.0) are more spread out along the response scale than are opinions about staff (standard deviations of 0.76 to 0.84). Generally a large standard deviation means that some patients are very satisfied and others are less satisfied, perhaps even dissatisfied. In such a case, analysis of responses by nursing units or other convenient units is helpful to determine whether the disparity of opinion is consistent throughout the hospital or whether good scores on one unit are offset by poor scores on another. The small standard deviation on items regarding staff is consistent with data from the departmental survey (Table 12–2), which showed that more than 93 percent of patients were satisfied with the courtesy of dietary personnel.

Inferential Statistics

When samples are used to draw conclusions about the larger population, inferential statistics are sometimes used. Descriptive statistics such as percentages and means may be used to provide preliminary information for the study. Sophisticated methods might then be used to examine the data further. Inferential statistics are more commonly used in research, but they can also be applied to quality assessment. For example, a dietitian may use a *t*-test to determine whether there is a statistical difference in mean cholesterol levels of patients receiving four sessions of nutrition counseling as opposed to those receiving only one counseling session. Other inferential statistics less commonly used in quality management are analysis of variance (ANOVA) and tests of association such as correlation coefficients.

INTERPRETING THE DATA

The concept of quality management hinges on the systematic evaluation of current practice. In other words, efforts to improve the quality of practice require that structures, processes, and outcomes related to nutritional care be analyzed and that results be interpreted. Interpretation "consists of taking the summarized data, examining these data for patterns, describing these patterns, drawing conclusions, and determining implications for changes to improve patient care" (Driever 1991).

Interpretation is essential to the process of comparing the predetermined standards (thresholds) with actual performance as measured by various monitors. It involves the "search for meaning and implications of the findings" (Driever 1991) obtained during data collection. Generally this activity is carried out by clinical dietitians, departmental leadership, or a quality team with substantial input from those involved in the element of care under scrutiny.

Comparing Results with Standards

After data are collected and summarized, they need to be examined through the use of quantitative measures and analytical tools that suggest how often and why a certain response or a data item occurred. Quantitative measures, usually given in terms of percentages, indicate the level of compliance with the item under study. The level of compliance can then be compared with the preset thresholds for that indicator. Interpretation of the data should not stop at this point. Rather, further discussion and interpretation should occur to elucidate the *meaning* of such findings.

As shown earlier in this chapter, it is useful to look beyond an overall score to assess individual item scores related to a particular indicator. Such an analysis offers insights into which aspects of care exceed thresholds and which areas might be targeted for improvement to increase overall quality of care. Such an interpretation gives meaning to the data and helps elucidate any patterns and trends observed within the data collected.

Interpreting a Change in Results

Trend data should also be interpreted. Figure 12–2 shows results of monitoring patient satisfaction with courtesy of dietary staff. At the beginning of a new fiscal year (July 1) a campaign was launched to improve this aspect of patient satisfaction. Trend data showed little effect in August. Rapid improvements are rarely attained because everyone needs time to adjust to new expectations and to develop new behavioral styles. September trend data showed marked improvement, however.

How big do differences in scores need to be before they are considered "significant"? Is the change in courtesy scores from 3.91 (August 1991) to 4.06 (September 1991) (difference of 0.15) a substantial improvement, or is the change so minuscule that chance alone could account for this apparent increase

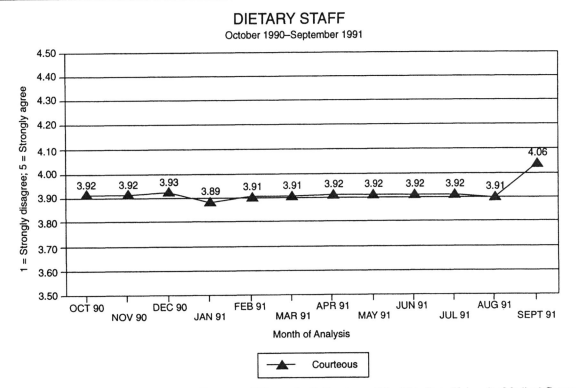

Figure 12–2 Trend Chart (Line Graph) Showing Courtesy of Dietary Staff. Courtesy of The Ohio State University Medical Center, Columbus, Ohio.

in patient satisfaction? To assess accurately the significance of a difference between two or more scores, one must take into account the number of respondents and the variance (standard deviation) in responses.

Assuming a sample size of 125 and standard deviation of 0.85, the following guidelines can be used to determine the significance of differences in results on a five-point patient satisfaction scale (Davis et al. 1992):

- less than 0.15, probably not meaningful
- 0.16 to 0.24, small meaningful difference probable
- 0.25 to 0.35, meaningful difference
- greater than 0.36, stronger difference

Thus, in the example above where the difference in scores was 0.15, what appears to be a major improvement when looking at the line graph is probably not meaningful. However, as shown in Figure 12–3, responsiveness scores from July 1991 (3.78) to September 1991 (4.01) (difference of 0.23) indicate a probable meaningful improvement. If this upward trend in scores continues and the difference in scores lies within the 0.25 to 0.35 range, it is certain that the improvement is meaningful. Such a substantial change in scores verifies the success of an intervention program.

Data can also be compared with those of other institutions if similar data-collection techniques and methods of analysis are used. This is commonly known as "benchmarking," or using data from another institution as the basis for making judgments about quality of food or services at your own hospital. For example, hospitals or extended care facilities in a multi-institutional system may have the same quality management plan and use the same quality indicators and monitors. Nutrition care service units could compare study results and use findings as a basis for setting triggers or goals for improvement.

Even when results are above an established threshold, negative trends may warrant further analysis and interpretation. This may be the case when it appears that problem-solving interventions are beginning to backslide, or when other factors such as staff changes or increased patient census may have a forceful impact on job performance. Figure 12–4 shows a downward trend in timeliness of nutrition screening. This area was targeted for improvement during the year. A comprehensive plan was developed and implemented. After slow progress, and some backsliding in the beginning, results showed marked improvement May through August and then a continuous downward trend September through December. Although the monthly decline was small, it brought performance close to the threshold by year's end. Careful analysis of records showed that the decline in performance could be attributed to scheduling of insufficient personnel on weekends

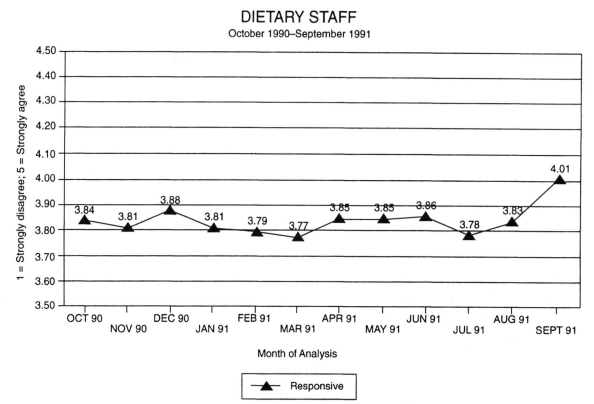

Figure 12–3 Trend Chart (Line Graph) Showing Responsiveness of Dietary Staff. Courtesy of The Ohio State University Medical Center, Columbus, Ohio.

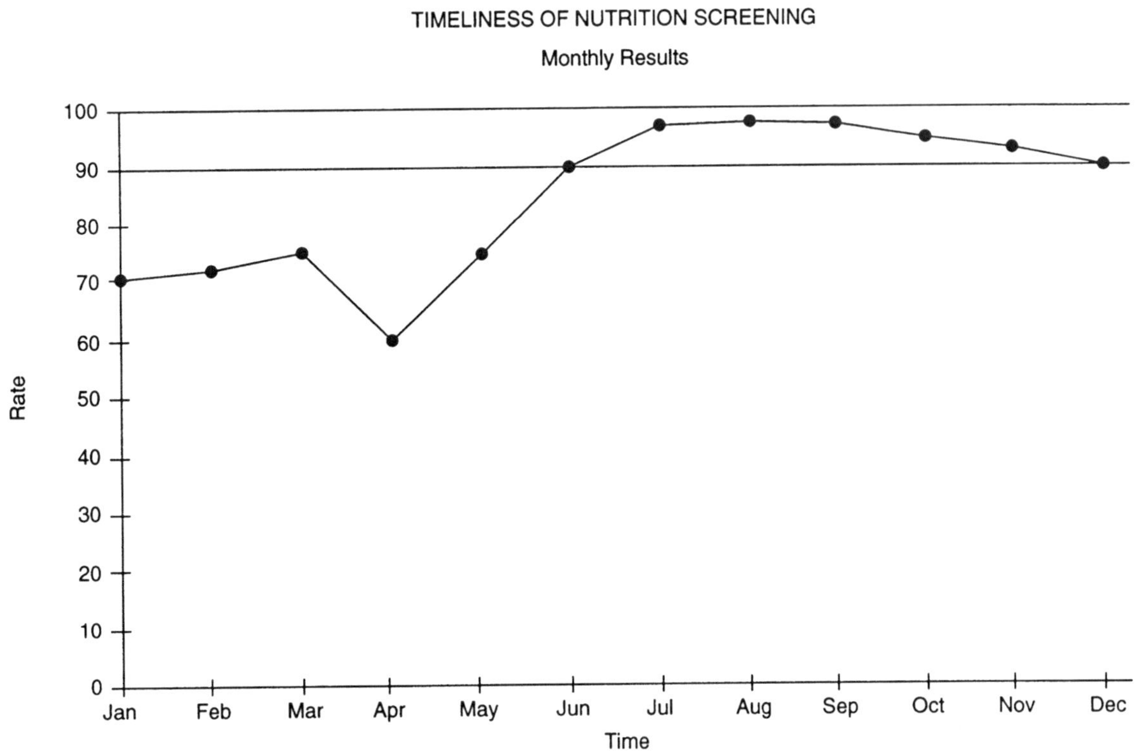

TIMELINESS OF NUTRITION SCREENING
Monthly Results

Figure 12–4 Trend Chart Showing Downcurve in Results. *Source:* Reprinted with permission of American Nutri-Tech, Inc., from *Continuous Quality Improvement for Nutrition Care.* Copyright © 1992 by Rita Jackson.

to complete screenings in a timely manner (Jackson 1992). In this case, analysis and interpretation of trend data, showing a slow deterioration of an important aspect of care, helped identify a problem and correct it before performance slipped below the established threshold.

Narrative Information

When graphics are used in quality reports, narrative interpretative information should be provided. The narrative ought to give conclusions of the study; a brief description of trends, patterns, and themes; and an analysis of what these results mean in terms of quality care. An explanation of why these results occurred can also be included along with a summary of in-depth analysis and rationale for recommended actions.

PROBLEM CLARIFICATION METHODS

An evaluation of care is necessary when findings show that data are at variance with established thresholds. Such an evaluation may include an in-depth review of trends, analysis of the situation to define the nature and scope of a problem, and determination of why the variance occurred (Elrod 1991).

Frequently the analytical tools presented in Chapters 7 and 8 can be used to help clarify a situation, uncover the cause of an underlying deficiency, or pinpoint the root of a problem.

A variation decision tree such as the one developed by Katz and Green (1992) and shown in Figure 12–5 can be used to guide analysis of the variation. The quality committee or staff dietitians first ask whether results show deviation from established thresholds. If not, the data are looked at more carefully to see whether any downward trends are apparent that need further analysis. When no deviations or downward trends are observed, monitoring continues according to the pre-established schedule.

If deviations are present, it is necessary to determine whether the substandard results are a product of patient acuity or characteristics, staff variables, or system variables. Depending on the nature of the problem, different tools and techniques (highlighted later in this chapter) are used to explore factors contributing to an undesirable situation.

Figure 12–6 offers an overview of major quality-improvement tools and how each is used in the problem-identification and problem-solving processes (Berwick et al. 1990). For instance, when downward trends are apparent, staff may use brainstorming to identify possible causes. Additional data may be collected through a time study or a survey to ascertain staff priorities. A Pareto chart can be prepared to illustrate major causes of the problem. These few problem factors are then selected as the target for further analysis and solution.

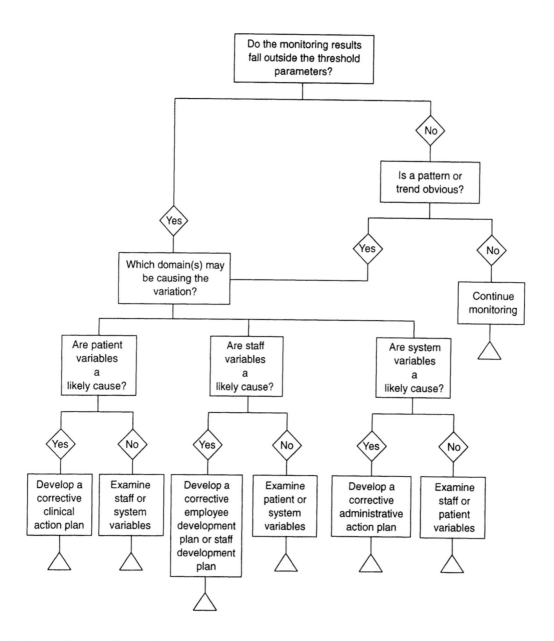

Figure 12–5 Variance Decision-Making Tree. *Source:* Reprinted from *Managing Quality: A Guide to Monitoring and Evaluating Nursing Services,* by J. Katz and E. Green, p. 157, with permission of Mosby-Year Book, Inc., © 1992.

Data from surveys, questionnaires, and structured data-collection sheets sometimes fail to provide sufficient information to understand adequately patient expectations or reasons for poor quality performance. In such cases, focus groups or other creative strategies, such as the ones listed in Exhibit 12–3, can be used to probe the situation (Wilbert 1985).

The focus group is a particularly useful technique for obtaining consumer input (Black 1985). Consumers include patients, nurses, physicians, students, cafeteria or catering guests, patient families, outreach groups, and home-care clients. In this technique, a trained facilitator and a group of six to ten individuals hold discussions of one to two hours to obtain information regarding the group such as concerns, attitudes, expectations, observations, preferences, desired services, and perceptions of service.

Focus groups require the use of a seasoned facilitator and careful selection of participants to ensure true representation of the specified consumer group (Peters 1993). When properly planned and conducted, advantages of using a focus group include the following (Black 1985):

- A focus group is informal, flexible, and easily used in any setting.
- Patients' views are solicited in a nonthreatening environment, and patients recognize that their opinions are valued by the institution.

Exhibit 12-3 Creative Techniques for Obtaining Additional Information Needed for In-Depth Analysis of Problem Areas

- Telephone interviews—query patients discharged from certain units, those treated for specific diseases, or those hospitalized for extended periods. Contact 6-8 days following discharge.
- Exit interviews—query patients at discharge on certain aspects of care such as understanding dietary or medication regimen, awareness of risk factors, or other systems.
- Observer recordings—a "critical incident technique" in which services are delivered while an observer keeps a diary or log of events. This gives a qualitative appraisal of the event in addition to quantitative monitoring data.
- Complaint inquiries—analyze complaints, such as negative and positive comments about food service written on patient survey forms. These can be categories as positive or negative to provide baseline data and to help set improvement goals.
- Patient questionnaires—used widely in food and nutrition departments to obtain patients' opinions about meal service. Satisfaction with both inpatient and outpatient nutrition counseling can be assessed also. For large patient populations, computerized surveys are recommended.
- Patient quizzes—can be used to assess patient knowledge of diet instructions, risk factor recognition, and need for lifestyle modifications. This tool is a useful way to document outcomes of patient diet and nutrition counseling.
- Group techniques—include focus groups and nominal group techniques to provide consumer information from patients and families in a nonthreatening environment.
- Staff surveys—method of obtaining staff attitudes, comfort level, and feelings of anxiety, as well as opinions about the level of staffing, knowledge, and support services needed in a particular area.
- Criteria maps—decision trees, fishbone diagrams, scatter diagrams, Pareto charts, or other graphic illustrations used in quality management to analyze problem areas or to look at data in a new way.
- Computer-assisted studies—use of the computer for data management, statistical analysis, generation of graphs and charts, or similar applications.

Source: Adapted from Wilbert, C.C., Selecting Topics/Methodologies, in *Quality Assurance: A Complete Guide to Effective Programs*, C.G. Meisenheimer, ed., pp. 103-131. Aspen Publishers, Inc., ©1985.

STEPS IN PROBLEM SOLVING		Flow Diagrams	Brainstorming	Cause-Effect Diagrams	Data Collection	Graphs and Charts	Stratification	Pareto Analysis	Histograms	Scatter Diagrams	Control Charts
Defining The Problem	1. List and prioritize problems	○	○		●	○	○	●			
	2. Define project and team		○			○	○				
The Diagnostic Journey	3. Analyze symptoms	●			●	○	○	●	○		○
	4. Formulate theories of causes	○	●	●			○				
	5. Test theories			●		●	●	●	●	●	●
	6. Identify root causes			●		●	●	●	●	●	●
The Remedial Journey	7. Consider alternative solutions	●	●	○		○					
	8. Design solutions and controls	●			●	●	○			○	●
	9. Address resistance to change	○	●	○							
	10. Implement solutions and controls	●				○		○	○	○	
Holding The Gains	11. Check performance	○			●	●	●	●	●	○	●
	12. Monitor control system	○				●	●	●		○	●

Legend: ● Primary or frequent application of tool ○ Secondary, infrequent, or circumstantial ☐ None or very rare

Figure 12-6 Summary of Quality Management Tools and Their Application to the Problem-Solving Process. *Source:* Reprinted from *Curing Health Care: New Strategies for Quality Improvement*, by D.M. Berwick, A.B. Godfrey, and J. Roessner, p. 179, with permission of Jossey-Bass, Inc., © 1990.

- This technique uncovers problems and issues not ordinarily identified through other means.

Also, customers of service (patients, nurses, physicians, and administrators) can be interviewed individually regarding their opinions about the quality of a given service, and employees can be interviewed to help analyze performance in a particular area. Interviews are even more time consuming than focus groups because each is conducted one-on-one. Objectives of the interview process should be defined, questions should be prepared ahead of time, confidentiality should be assured, and individuals selected should be those who are most likely to provide the desired information. For example, if preliminary data show that dietetic technicians complete screenings for only 40 percent (threshold 90 percent) of patients within 48 hours of admission, each technician might be interviewed to help determine reasons for this variance and to get suggestions for what can be done to improve this figure. Although such a discussion can occur in a group setting, individual interviews would

- allow more time for each technician to speak
- pinpoint the discussion toward the units for which each technician is responsible
- avoid embarrassment of any individual who may complete fewer than the average screenings
- demonstrate the seriousness of the problem and genuine interest in working with individuals to improve structures or processes that may inhibit optimal performance.

Deviations from thresholds may be attributable to a deficiency in provider knowledge or performance, lack of patient compliance, or weakness in the system or organizational structure. Any action plans for improvement should be targeted directly toward a resolution of the reason for the deviation. Problem solving and development of solutions are considered in Chapter 13.

REFERENCES

Berwick, D.M., et al. 1990. *Curing health care: New strategies for quality improvement.* San Francisco: Jossey-Bass Inc., Publishers.

Black, M.K. 1985. The consumer: Product of our efforts. In *Quality assurance: A complete guide to effective programs,* ed. C.G. Meisenheimer, 295–314. Gaithersburg, Md: Aspen Publishers, Inc.

Davis, R.M., et al. 1992. *The Ohio State University Hospitals' patient satisfaction measurement system (PSMS).* Columbus: Ohio State University.

Driever, M.J. 1991. Clinical interpretation and analysis of monitoring data. In *Monitoring and evaluation in nursing,* ed. P. Schroeder, 239–252. Gaithersburg, Md: Aspen Publishers, Inc.

Elrod, M.E. 1991. Quality assurance: Challenges and dilemmas in acute care medical-surgical environments. In *Monitoring and evaluation in nursing,* ed. P. Schroeder, 27–56. Gaithersburg, Md: Aspen Publishers, Inc.

Geibig, C.B., et al. 1991. Quality assurance for a nutrition support service. *Nutrition in Clinical Practice* 6:147–150.

Grossman, D., and J. Neubauer. 1992. Basic statistical concepts in quality improvement. *Journal of Nursing Care Quality* 6, no. 4:1–8.

Jackson, R. 1992. *Continuous quality improvement for nutrition care.* Amelia Island, Fla: American Nutri-Tech, Inc.

Joint Commission on Accreditation of Healthcare Organizations. 1992. Quality Assurance. In *Joint Commission 1992 Accreditation Manual for Hospitals.* Oakbrook Terrace, Ill.

Joint Commission Quality Improvement Task Force. 1990. *Principles of quality improvement.* Oakbrook Terrace, Ill: Joint Commission on Accreditation of Healthcare Organizations.

Katz, J., and E. Green. 1992. *Managing quality: A guide to monitoring and evaluating nursing services.* St. Louis, Mo: Mosby–Year Book, Inc.

Miller, M.C., and R.G. Knapp. 1979. *Evaluating quality of care: Analytic procedures-monitoring techniques.* Gaithersburg, Md: Aspen Publishers, Inc.

Peters, D.A. 1993. Improving quality requires consumer input: Using focus groups. *Journal of Nursing Care Quality* 7, no. 2:34–41.

Verhonick, P.J., and C.C. Seaman. 1978. *Research methods for undergraduate students in nursing.* New York: Appleton-Century-Crofts.

Wilbert, C.C. 1985. Selecting topics/methodologies. In *Quality assurance: A complete guide to effective programs,* ed. C.G. Meisenheimer, 103–131. Gaithersburg, Md: Aspen Publishers, Inc.

Williamson, J.W. 1982. Applying clinical problem solving to quality assurance and cost containment: A five-stage approach. In *Teaching quality assurance and cost containment in health care,* ed. J.W. Williamson and Associates, 182–206. San Francisco: Jossey-Bass Inc., Publishers.

Chapter 13

Taking Action To Improve Care

All of the steps taken until now are preparatory for a true, continuous, quality-improvement program. Selecting indicators, determining thresholds, conducting the monitoring process, and collecting and analyzing data should identify many opportunities for change. This chapter presents several models nutrition professionals can use to identify opportunities for maximizing quality improvement. Each step in the problem-solving process is described, accompanied by relevant examples from clinical nutrition practice.

CHOOSING A GAME PLAN

Implementing changes to improve care is definitely time and work intensive. Yet valuable, high-quality improvements can be achieved in a reasonable amount of time if a rational plan of action is utilized. There are several advantages to following a model action plan.

- A model action plan presents a framework for new quality task force participants to follow.
- Each of the steps, beginning with selecting a problem all the way through to implementing a solution, is identified and outlined in chronological fashion.
- Using a model helps groups overcome the tendency to skip steps or jump to "obvious" conclusions when in the problem-solving mode.

There are many such models in existence. One such model is the seven-step IMPROVE approach to problem solving (Ernst & Young 1990). The first letter of each step spells out the goal of the quality-improvement process:

- **I**dentify a problem.
- **M**easure.
- **P**rioritize.
- **R**esearch the causes.
- **O**utline a solution.
- **V**alidate the plan.
- **E**xecute the plan.

This model contains the essential elements of an improvement plan. If a quality team is new to the quality-improvement process, it may require a more detailed explanation of how to go about the business of improving care.

Another approach is the FADE method of effective problem solving (Krasker et al. 1992). FADE is an acronym for *focus, analyze, develop,* and *execute*. The full FADE cycle is pictured in Figure 13–1. Each phase of the FADE model is further divided into a three-step process. This is the problem-solving model suggested by the Joint Commission for use by quality action teams (Krasker et al. 1992).

A third approach is the model of progress developed by Scholtes (1988) and shown in Figure 13–2. This is a top-down flowchart showing the general progression of events for solving a quality problem. It begins with goals, incorporates a quality-improvement loop, and ends with closure activities. Some users, especially those new to the quality-improvement

Models of Effective Problem-Solving

FADE:

Phases of FADE problem-solving:

Focus—choose a problem and describe it
Analyze—learn about the problem by collecting and analyzing pertinent data
Develop—develop a solution and a plan
Execute—implement the plan, monitor the results, adjust as needed

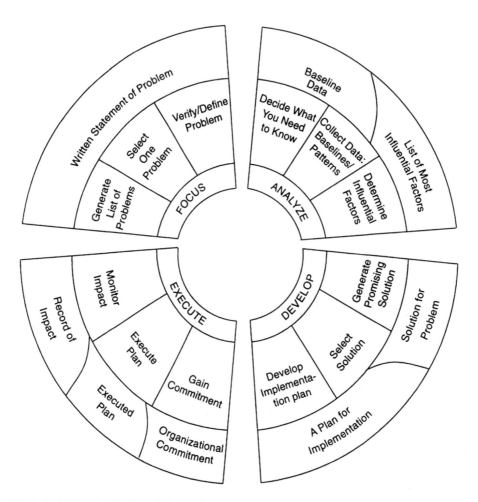

Figure 13–1 FADE Method of Effective Problem Solving. *Source:* Reprinted from *Quality Action Teams: Team Members Workbook,* by Organizational Dynamics, Inc., p. 646, with permission of Organizational Dynamics, Inc., © 1987.

process, may prefer this model because it clearly outlines the actions to take in a written format.

Another popular model is the FOCUS-PDCA cycle (Duncan et al. 1991) shown in Figure 13–3. This model includes the following steps:

- <u>F</u>ind a process to improve.
- <u>O</u>rganize a team that knows the process.
- <u>C</u>larify current knowledge of the process.
- <u>U</u>nderstand causes of process variation.
- <u>S</u>elect the process improvement.
- <u>P</u>lan the improvement.
- <u>D</u>o the data collection, data analysis, and improvement.
- <u>C</u>heck data for process improvement and customer outcome.
- <u>A</u>ct to maintain and continue the improvement.

This model is used to outline the problem-solving process in this chapter. Note how this model incorporates parts of the ten-step process for quality monitoring and evaluation. The "D" or "do" step of the FOCUS-PDCA Model includes step eight: take action to improve care. The last two FOCUS-

Model of Progress

This is a top-down flowchart showing the general progression of events in project teams. In the early part of an improvement project, team members clarify what it means to be on the team: what process they will work on and what kinds of improvements are expected. From these goals and expectations they draft an improvement plan. The first few meetings are typically devoted largely to team building and education. After team members have been exposed to quality and scientific principles, they are ready to begin work in earnest on the process. Usually, they study the process to learn more about how it operates and to identify problems. All theories are checked by collecting data, and appropriate actions determined after analysis. The loop of problem analysis and data collection continues until the team is satisfied that it has identified and eliminated the root causes of problems.

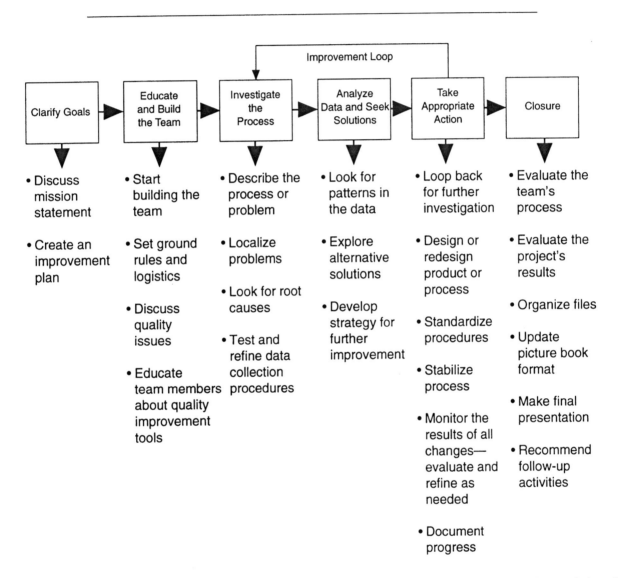

Figure 13-2 Model of Progress. *Source:* Reprinted from *The Team Handbook,* by P.R. Scholtes et al., pp. 4–39, with permission of Joiner Associates, Inc.

PDCA steps, "check" and "act," are actually part of step nine: assess the effectiveness of actions and maintain the gain. Let us examine steps eight and nine (take action to improve care, assess actions and document improvements) of the ten-step quality-improvement process using the FOCUS-PDCA model.

FOCUS-PDCA: FIND A PROCESS TO IMPROVE

Select a Well-Defined Problem

A quality improvement task force needs a clear-cut problem to solve. Initially, the simpler the problem to be solved, the

Focus—PDCA (Hospital Corporation of America)

Expands on the PDCA Cycle by including "preliminary steps"; i.e., FOCUS

- **Find** process improvement opportunity
- **Organize** a team that knows the process
- **Clarify** current knowledge of the process
- **Uncover** root causes or process variation
- **Start** improvement cycle

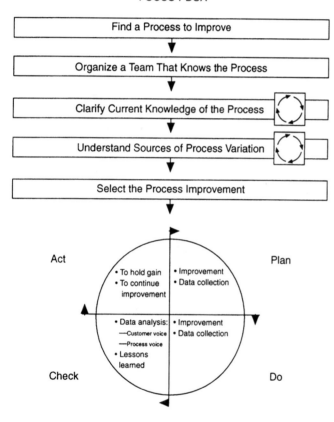

Figure 13–3 FOCUS-PDCA Cycle. *Source:* Executive Briefing: 1993. AMH Standards and Survey Process (p. 4), 1992, Oakbrook Terrace, Illinois. Reprinted by permission.

better. The first time through the FOCUS-PDCA process, the group must learn the problem-solving method while at the same time addressing the problem at hand. An inexperienced but enthusiastic quality-improvement team may want to attack an elephant-sized problem. Rather, the team members should cut up the elephant into sushi-sized pieces that can be analyzed and solved, and for which progress can be demonstrated within 90 to 120 days (Newbold and Mosel 1990). This shortens the group's initial learning curve.

According to the Pareto principle, 80 percent of the trouble comes from 20 percent of the problems. If the team selects, defines, and solves one real quality issue, it will make a substantial contribution to quality improvement (*Quality Force* 1991). Small successes will help build team confidence. As word of success spreads across the organization through formal and informal channels, it will strengthen momentum and support for quality-improvement efforts. The real lesson here is, "Don't go hunting for elephants when you should go fishing for sushi" (Newbold and Mosel 1990, 12).

Devote the Necessary Resources

Another aspect in selecting which problem to address is an assessment of its feasibility, or how "doable" it is. Does the team have what it will take to solve the problem? Will there be adequate resources to implement its proposed action plan? The primary resources to be examined are funding and staffing. For dietitians these issues are practically one and the same, since the single largest line-item expense in the operating budgets of most clinical nutrition departments is staff salaries. Limited resources are a fact of life in today's hospital environments. Clinical nutrition departments have not been exempt from the strain on resources. When managers are faced with making adjustments to accommodate budget reductions, the choices they make may have a negative impact on staffing. Yet the standards of nutrition practice for providing quality patient care and for meeting accreditation requirements are rarely adjusted downward to compensate for the reduction in staff.

There is probably not a dietitian practicing today who has not had to deal with some component of "doing more with less." This is not to suggest that limited resources should be an excuse not to proceed with a quality management program. In fact, quality improvement might be especially relevant in an area undergoing financial challenges. The quality-improvement process helps the department identify crucial practices and key quality indicators, and keeps the staff focused on providing and improving these important aspects of care. Ultimately, evaluating the feasibility of solving a particular problem will increase the odds of achieving success for the quality team.

All resources are flexible commodities that change over time. Although staff and budget dollars are fundamental resources, there are others to consider (*Quality Force* 1991).

Management Support

Managers are more likely to appropriate the necessary resources to solve a problem if they feel that it is central to improving department operations. Managers can be surveyed to assess how they prioritize a group of problem-solving opportunities. This may indicate those efforts most likely to receive managerial support.

Time Commitments

Scheduling constraints fluctuate according to a host of factors such as patient census, seasonal variations, holiday and vacation requests, and the number of staff on leave at any given time. When a quality-management committee is rank ordering a number of problems, they might want to make a list of people best suited to work on each one. Comparing this list of employees with planned work schedules will help define whether the appropriate staff is available to work on issues given top priority. This information may actually be the determining factor in choosing which problem to work on first.

Knowledge

Clinical dietitians are a well-educated group who are very knowledgeable in all aspects of clinical care. The nature of the problem will dictate to what extent dietitians' knowledge of clinical nutrition contributes to the problem-solving process. For example, it would be appropriate for a group of dietitians to be integrally involved in developing a flowchart to monitor enteral feeding practices. On the other hand, dietitians may have a marginal role on a cross-functional team working on a process with which they are less familiar. Similarly, clinical dietitians may have minimal or no part in a quality team working to improve health department inspection scores to implement a hazard analysis critical control point (HACCP) system. A department's knowledge resources should be targeted to the areas where that knowledge can be best applied.

Interest

How interested and motivated the group is in solving the problem will be a major factor in determining its enthusiasm in seeking solutions. The team roster can be modified so that those with the most to gain, or the most time to give, are part of the problem-solving group.

If this is the first time the department is addressing a problem utilizing a formal problem-solving process, it should be certain to select an issue for which adequate resources are available to investigate and resolve the problem.

Determine Cost-Benefit Ratio

To determine whether solving the problem is truly worthwhile, evaluate its cost-benefit ratio. Remember to examine costs not only on the basis of dollars spent but also in terms of time lost. Time can be spent in redoing a task or in "cleaning up the mess" of a project gone awry (*Quality Force* 1991). Cost analysis can be simple or complex. It may be easy to determine the dollars wasted on a preprepared formula that is never used if the cost is based on the price of the product alone. The analysis becomes more complex if labor costs are included. This computation includes not only the hourly wage paid to staff in the nutrition department, but also the wages paid to all the individuals involved in every step of the order and delivery process. A cost-benefit analysis must also include the potential cost of implementing any solutions the group reaches.

Use of a work sheet helps simplify the cost and benefit calculations. A work sheet may contain data on adjustments in salaries, wages, and supply costs (both proposed increases and decreases) and include projected revenues and capital expenditures. An equation for comparing expected returns with required expenses is shown in a sample cost-of-quality work sheet (Exhibit 13–1). At first this six-step outline and summary sheet may appear formidable. But, depending on the problem, not all parts of all sections will apply.

Evaluate the Hassle Factor

Consider the hassles the problem creates. The hassle factor is determined by the amount of discomfort or unpleasantness created for the persons who have to deal with the problem (*Quality Force* 1991). The hassles may be especially relevant when the department is addressing a problem with the delivery of a service and the customer who receives that service is another hospital department. For example, a great deal of distress and poor public relations are generated for the nutrition services department among nursing personnel when an inaccurate tube feeding is sent to the unit for the third day in a row.

Selection Process

The method for choosing which problem to solve often depends on the complexity of the situation and staff priorities. Many different approaches are used in the problem-selection process.

Clear-Cut Issues

The first method is simple: the "just do it" approach. When the problem is obvious, clearly defined, and very limited in scope, this simple approach may be the best route to go. This approach is more appropriate when all the players and functions involved are contained within a single department or area.

Complex Problems

The second approach is a more formalized one. Many quality-management challenges cross departmental boundaries. Because of their scale, investigating complex problems may require hospitalwide coordination, cooperation, and resource investment. One hospital (Medical Center Hospital) utilizes a quality-improvement project requests plan (PRP) to accomplish this. The PRP lists criteria for prioritizing projects such as the following:

- supports the hospital vision statement
- is customer focused

Exhibit 13-1 Cost-of-Quality Work Sheet: Expected Returns

PROJECT NUMBER:

I. SALARY/WAGE REDUCTIONS

Who It Affects (Title/Job Code)	Avg. Hrs. per Activity Decrease	Avg. Hrs. per Year Decrease	Avg. Salary/Wage per Hr.	Savings per Year

Total Salary/Wage Reductions (A) _____

II. SUPPLIES/OTHER REDUCTIONS

What It Affects	Cost of Item	Avg. Quantity per Year	Savings per Year

Total Supply/Other Reductions (B) _____
Total Expense Reductions (C) _____

III. REVENUE INCREASE (DECREASE)

Description/Procedure	Charge per Activity	Avg. Quantity per Year	Revenue per Year

Total Revenue Increase/(Decrease) (D) _____

Other Information
_____
_____

IV. SALARY/WAGE INCREASES

Who It Affects (Title/Job Code)	Avg. Hrs. per Activity Increase	Avg. Hrs. per Year Increase	Avg. Salary/Wage per Hr.	Cost per Year

Total Salary/Wage Increases (E) _____

Courtesy of Medical Center Hospital, Chilicothe, Ohio.

V. SUPPLIES/OTHER INCREASES

What It Affects	Cost of Item	Avg. Quantity per Year	Cost per Year

Total Supply/Other Increases (F) _____
Total Expense Increases (G) _____

VI. CAPITAL EXPENDITURES

Description of Equipment/Renovations	Cost per Item	Quantity Needed	Capital Cost

Total Capital Expenditures (H) _____

Other Information
_____
_____

Contact Person:
Facilitator:

Brief Description of Project
_____
_____

EXPECTED RETURNS
  Expenses
    Salaries   (A) _____
    Supplies/Other   (B) _____
    Total   (C) _____
  Revenues
    Total   (D) _____

REQUIRED EXPENDITURES
  Expenses
    Salaries   (E) _____
    Supplies/Other   (F) _____
    Total   (G) _____
  Capital
    Total   (H) _____

Other Information (If Applicable)
_____
_____
_____

Reviewed by COQ Committee   Date

- has intradepartmental scope
- has anticipated completion time of less than three months
- has no impact on full-time equivalents
- requires expenditures of less than $1,000

The project may be presented to the hospitalwide quality coordinating committee to decide whether it contains particular elements deemed essential to organizational efforts. For example, if the project will have an impact on a major hospital system or policy or requires capital expenditures, it should be directed to the steering committee for authorization.

Many potential nutrition projects have enough impact to require this type of review. For instance, a nutrition-screening mechanism based on admission data may become part of the patient registration process. Also, a clinical ladder to promote professional growth for dietitians might affect wage and salary structures.

There are other advantages in having a PRP in addition to helping staff select problems in an organized fashion. Departments collaborating for the first time can begin to establish a working relationship during the submission and approval process for the PRP. The plan can serve as a valuable communication tool. Project criteria can be modified over time to reflect changing organizational goals or to zero in on particular areas that need attention. Finally, receiving the steering committee's stamp of approval can legitimize the group's efforts in a very visible way.

Prioritize by Checklist

If the formal request approach seems too complex or perhaps the problem lies somewhere in between the "no preparation required," and the "several months of preparation required," there are several formats that offer a shortened methodology yet still address the primary determinants of the selection process.

One format is a checklist used by Henry Ford Hospital (Exhibit 13–2). It contains ten questions related to the problem's definition, and the feasibility and potential impact of its resolution. The questions can be answered easily by checking yes or no. When choosing among projects, the one that receives the most yes votes should probably be the one the group selects to work on first.

Another approach using the yes/no format is illustrated by Exhibit 13–3. The first set of questions shown in Exhibit 13–3 helps determine whether the opportunity is worth working on and whether its quality-improvement potential is favorable when compared with its cost and hassle. Feasibility is assessed by answering yes or no to four key questions related to support, time, knowledge, and interest. Then, if there is still competition for which problem gets attended to first, the steps are outlined to involve the group in a simple priority-ranking process such as the nominal group process or Delphi technique. It

Exhibit 13–2 Henry Ford Hospital Quality-Management Process

PROCESS SELECTION CHECKLIST

When selecting processes to improve through Quality Improvement Teams at Henry Ford Hospital you should be able to check *yes* to most, if not all, of the items listed. If you cannot, you may need to re-evaluate your choice.

	YES	NO
1. The process is related to key business issues.	___	___
2. The process has a direct impact on the hospital's external customers.	___	___
3. The process or work area has a lot of visibility in the hospital.	___	___
4. Managers concerned with this process at all levels of the hospital agree that it is important to study and improve this process.	___	___
5. Managers, supervisors, physicians, and employees in this area will cooperate to make this project a success.	___	___
6. The process is not currently being changed nor is scheduled for major changes in the near future.	___	___
7. The process can be clearly defined with easily identified starting and ending points.	___	___
8. The process is not being studied by any other group.	___	___
9. At least one cycle of the process is completed each day or two in order for the team to collect enough data over a reasonably short period of time.	___	___
10. The opportunity statement for this team describes a process to be studied and improvement opportunities, *not* a proposed solution.	___	___

Courtesy of Henry Ford Hospital, Detroit, Michigan.

is preferable for the group to attempt to reach a consensus so that each member will perceive the problem selected as the one he or she chose. This promotes continued enthusiasm as the group moves on to resolution (*Quality Force* 1991).

Which way is best? If after reviewing the criteria in place, you are still unsure of whether to pursue a formal or informal approach, review the problem at hand from a "make work" perspective. The planned approach may be preferable if the benefits exceed the amount of work involved in obtaining the committee's blessing in the approval process. If not, go full speed ahead via the "just do it" approach.

Initially, establishing a workable framework is probably just as important as solving the problem itself. Quality teams can always move on to tackling more difficult problems by using the same framework. Laying the initial foundations and developing the team's confidence in the process at the beginning will determine whether they ever "get past go."

Exhibit 13-3 Yes/No Priority-Ranking Tool

THE QUALITY LEADER SEMINAR

1. Worthwhile
   Is the opportunity worth working on?
   - Quality—Does the opportunity involve significant quality problems for patients and/or internal customers?
   - Cost—Does the opportunity involve significant cost to the hospital in rework, penalties, lost time, or otherwise "cleaning up the mess"?
   - Hassle—Does the opportunity create significant discomfort or unpleasantness for the people who do the work?

2. Doable
   Can the team make progress on the situation, given its resources?
   - Support—Can we get enough support from management and others to get the job done?
   - Time—Does this team have enough time to see the work through to completion?
   - Knowledge—Do we know enough about this topic to do good work on it?
   - Interest—Do we care enough about this problem to work hard at it?

Choosing Criteria

To keep this problem from becoming needlessly complex, a team will usually consider just a few of these criteria (or others) when making a decision and will use a very simple scoring system (perhaps "yes" or "no" for each criterion).

Priority Ranking

Another way to narrow your opportunity list is through priority ranking. This technique allows each member an opportunity to rank order their votes.

Step 1—Review the list of opportunities under consideration.

Step 2—Have every team member rank order their top three according to how much they want that problem to be selected.

Step 3—Have people read their choices aloud while results are recorded on a flipchart.

Step 4—Add up the scores, examine the results as a team, and decide whether another round of rankings is needed. If so, first eliminate the lowest ranked problems from Round 1, focusing on a second round of polling on the remaining problems.

Step 5—The opportunity with the highest number of votes becomes the focus for problem-solving.

Special Considerations

In quality improvement teams, it's better to reach decisions by reaching a consensus than through just a vote. Voting indicates which ideas are most popular, but it can leave the "losers" unwilling to pitch in. Even if a vote is taken, strong differences should still be dealt with afterward. Everybody will then be able to abide by the decision, even if they don't support it 100 percent.

If you do use voting, many procedures are possible. Since the circumstances will vary, design your own system using these guidelines:

- Everyone's vote carries equal weight.
- All members have the same number of votes.
- No discussion is allowed during voting.

Source: Reprinted from *The Quality F.O.R.C.E.*, by E.C. Murphy, Ltd., pp. 252–254, with permission of E.C. Murphy, Ltd., © 1991.

What If No Needs Are Identified?

The quality-improvement process is designed to seek out opportunities to improve services. If a department has come this far and opportunities are not apparent, it is time to rethink the processes in place. In the traditional quality-assurance approach, as long as a threshold is not crossed, no action is needed. When this outcome is achieved, often the quality-assurance coordinator and the quality committee breathe a collective sigh of relief. In this scenario, there will be no need to identify what went wrong or, more typically, who is at fault. Also, if all monitors in a traditional audit are within the standard, it signifies (perhaps falsely) that all is well within the department and certainly simplifies the reporting process.

In the quality-improvement process, no news is not good news. Dougherty (1992c) suggests that quality managers "should operate under the theory that if it isn't broken, break it" (Dougherty 1992c, 1). A good way to encourage creativity about new ways to provide a particular service is to discontinue it. If no one misses the service, it was probably not essential and can be discontinued permanently. If the service is missed, customers will articulate why they need it. This information can be used when developing new procedures and objectives when the service is re-established to ensure that customers' needs are met.

If no problems are identified, several types of investigations can be conducted to uncover potential sources for quality improvement.

Review the Monitoring Process

Monitoring is one of the keys to determining where to focus the department's resources. A quick review of monitoring tools and processes, based on the following questions, may assist in unearthing some area for change if absolutely no opportunities are identified. What philosophical approach was used in selecting monitors? Were the monitors truly revised to reflect quality-improvement parameters? Are at least some of the selected monitors outcome oriented? Are the outcomes related to patient expectations or are they reflections of dietitian-imposed standards?

Evaluate Threshold Levels

Many dietitians may be using this manual as a guide for converting a traditional quality-assurance process to a system that focuses on improvement. It may be the first time that outcome criteria or indicators related to patient satisfaction with clinical services are monitored. It is difficult to set realistic yet sensitive thresholds for new indicators. If there is a concern that the threshold level for a particular indicator might be off base, review the criteria used in selecting its range. The following hypothetical example examines how threshold limits are set. A clinical manager was receiving secondhand complaints from referring physicians regarding patients' appointments in the outpatient nutrition clinic. In looking into the matter, her first step was to review the quality reports for the two most recent quarters. The clinic staff were monitoring two important aspects of care for providing nutrition counseling in the ambulatory setting. A data review showed the following:

1. Aspect of care: patient satisfaction with the professional services provided by the dietitian
   - Indicator: services meet or exceed patients' expectations
   - Threshold: 90 percent of patients indicate "satisfied" on follow-up questionnaire
   - Results: satisfaction ratings of 93 percent and 92 percent, respectively, for the previous two quarters
2. Aspect of care: patient satisfaction with the scheduling and appointment process
   - Indicator: time spent waiting to see the dietitian
   - Threshold: average wait time of less than 30 minutes based on appointment sign-in log
   - Results: average wait time of 25 minutes consistent for the previous two quarters

The data indicated that there were no problems with nutrition clinic appointments. The aspects of care were "customer determined" and appropriate for this service. The monitoring schedule was followed. The report summaries were concise. Yet the complaints continued. Upon further investigation, the true fault in the process was identified: It was the 30-minute threshold level set for the second aspect of care.

The research conducted when establishing the quality-improvement process correctly identified that time issues were as important as service issues for outpatients who came to the clinic for their medical care. But, just as quality is customer determined, the delivery of quality service must be measured against the customers' expectations (Lehr and Strosberg 1991). In this case, the dietitian had selected the 30-minute threshold based on her reasonable judgment and the desire to document adequately each patient's progress immediately after the counseling session.

Information gained from a patient/client focus group revealed that the majority of clinic clients were attempting to schedule appointments over their lunch breaks. Clients became anxious when their wait time extended to half the time they had allotted for their appointment. They were preoccupied with getting back to work on time and tended not to ask questions in order to cut the session short. Most clients felt that a ten-minute wait for a prescheduled appointment was reasonable. Based on this client feedback, the dietitian reset the threshold for wait time to ten minutes. She staggered her appointment schedules and altered her charting practices to accommodate the change. Follow-up formal quality reports correlated with informal physician feedback. The new threshold levels were met, and customer complaints declined.

At first, setting thresholds will probably be more of an educated guess than a scientific method. It should be expected that it might take a few years to refine the process. The use of market research data to set thresholds based on customer expectations can help to determine initial limits now and prevent problems later.

Search Records and Forms

If you follow the steps of reviewing philosophy and threshold levels and are still unable to produce any opportunities for change, do some additional searching. Department records and forms contain a gold mine of information that may indicate potential opportunities to improve care. For example, review comments written on patient satisfaction survey forms, review records of employee complaints, and review the nature of calls to patient relations or the customer hot line. Find out the top causes of customer dissatisfaction leading to a call or complaint. Because these contacts are usually patient initiated, they contain valuable information on customer needs and wants.

Review Requests for Capital Expenditures

Capital expenditures represent the department's plans to invest the largest single amounts of its budget allocation. Do capital requests reflect the true quality-related problems with which the department is dealing? For example, the department may be considering capital budget proposals for purchasing a new dish machine and for renovating the cafeteria dining area to provide a much-needed increase in seating capacity. Most likely, funds for only one of these projects will be approved in the next fiscal year. Are there data to support which purchase would have the most impact on quality-related issues and most improve satisfaction? Or are there other, more pressing, quality issues that need to be addressed, such as revamping the tray rethermalization system to improve patient food service? Once the hundreds of thousands of dollars are spent, will the investment pay off in terms of quality issues? These and similar issues need to be addressed when delineating important aspects of care. Are you able to monitor them? If not, the department may need to develop new indicators to include them.

Look through Memo Files

Informal department memos can be used to track issues that require communication across department areas. One department uses an informal message/reply system (Exhibit 13–4) to communicate between dietitians and food service personnel. A review of the memo file indicated that a disproportionate

Exhibit 13–4 Informal Message/Reply System

OSU Reply Memo

Instructions: Sender completes "Message" section and keeps pink copy. Recipient completes "Reply" section, keeps yellow copy and returns white copy to sender.

Message

Subject Snack 11 WD   Date 3/1/94
From   M.O. RD   Dept _____
Address _____
To   _____

Patient, B.F., did not receive ordered snacks on 2-27 at 3:00pm and on 2-28 at 10:00 and 3:00. A snack card was in the galley.

Signed M.O., RD

Reply

To M.O., RD   Date 3/1/94

Upon checking this incident, I found out that the nurse up on 11 WD had requested from the Diet Clerk that the patient's 10:00 and 3:00 snack be held in the refrigerator. The Diet Clerk also mentioned that on several occasions, the patient, B.F., had refused his snack that he had been receiving. Thank you for communicating this matter to me. Please let me know of any other concerns that you might have.

Signed IB Patient Food Service Supervisor

Courtesy of the Ohio State University Medical Center, Columbus, Ohio.

number of memos related to some aspect of nourishments: their delivery, agreement with diet order, and/or patient acceptance. Further investigation yielded several ideas appropriate for quality improvement follow-through such as the following:

- Records of donations to the food bank: The number and type of unused or leftover items being sent to the food bank may indicate a problem in the food-ordering or production process or denote a quality problem with particular recipes.
- Notes on menus: Patients often jot down comments on the menu regarding their opinion of the food and service received. These are often unsolicited comments, yet they provide genuine, unfiltered information regarding patient satisfaction. This practice can be promoted by the department by encouraging patients to respond in this manner. A note to this effect can be included in the menu-completion guidelines.
- Accident, incident, and safety reports: What are the recurrent events that show up on these reports? Are they serious enough to merit this report? Evaluating the systems associated with these high-risk issues may be a top priority for intervention.
- Disciplinary files: Are there particular practices, such as poor attendance or unsafe food-handling techniques, that are addressed time and again in the disciplinary process yet never seem to dissipate? Analysis may identify how best to approach these problems.

Dietitians hold high ideals for their areas of clinical practice. Many have no difficulty identifying opportunities for improvement. Once a list is formulated, selecting the problem to attack first and deciding how to go about the process of taking action to improve care are the next crucial issues. But first, a quality team needs to be designated to work on the problem.

FOCUS-PDCA: ORGANIZING YOUR TEAM

The nature of the problem influences the membership of the team chosen to address it. Team members should have in common a working knowledge of the problem at hand. Members are selected so that each views the situation from a different vantage point. For example, the following participants make up a team evaluating the delivery of enteral feedings. The team is analyzing the process that occurs when the physician's order for a tube-feeding product changes during the middle of the day. The team developed a brief outline that listed each participant, his or her department, and the role each played in the tube-feeding change process.

- Physician/medicine: writes the order for the formula change.
- Ward clerk/clerical assistance: transcribes the order to the diet sheets.
- Diet clerk/nutrition and dietetics: picks up and transports the sheets to the diet office.
- Dietitian/nutrition and dietetics: evaluates appropriateness of the product change.
- Dietetic technician/nutrition and dietetics: transports approved changes to formula room.
- Food service worker/nutrition and dietetics: obtains product from inventory.
- Transporter/transportation: delivers product to nursing unit.
- Nursing assistant/nursing: verifies product against physician order.
- Nurse/nursing: hangs new formula and disposes of former product.

Cross-Functional Teams

A team composed of players from multiple departments is commonly referred to as a cross-functional problem-solving team or task force. This team is composed of nine employees representing five different departments. Each participates in a process that occurs multiple times every day in the hospital. Before becoming part of the team, each player knew only the adjacent persons in the work-flow chain. Each was very familiar with his or her own task and vaguely aware of the next person's function. Initially, team members had no notion of the number of people involved in the process. Nor did they have any conception of the roles of other employees who were more than one link away from their position in the chain. As team members begin to communicate, "they discover that each one of them understands the problem from their own individual perspective according to the part they play in it, but few have considered the problems of other players" (Dougherty 1992a, 3). Representatives from all the departments that affect the ability to deliver a service must understand the skills and constraints of the other team players (Dougherty 1992a). A major advantage of participating on a cross-functional team analyzing a multidisciplinary problem is that the experience broadens the scope of each member and provides the opportunity for learning to occur.

Departmental Teams

The processes for some quality problems may be limited to a single work area and managed by a homogeneous work force. Perhaps the department is experiencing problems with late menus. In one department dietetic technicians are primarily responsible for distributing, collecting, and adjusting menus. The quality-improvement team might most appropriately consist entirely of technician volunteers. This group is an example of a departmental team. Another example of a departmental team is a group made up solely of dietitians working to simplify the data entry for the nutrition services billing process.

Team Differences

There are differences between cross-functional and departmental teams other than the composition of their membership. Cross-functional teams evaluate processes that span two or more departments. Their projects typically are selected by a hospitalwide steering committee to which they are accountable. Cross-functional teams are usually ad hoc. Like a task force, they disband once problem resolution is achieved.

Conversely, departmental teams, made up of department employees, analyze intradepartmental problems. They can exist indefinitely and move on to work on other problems as long as they are productive.

Exhibit 13–5 gives a comparison of cross-functional and departmental teams. Both kinds of teams can be equally effective in achieving success if the group is in synchrony with the nature and scope of the problem.

Team Selection

All team members must be willing and active participants (Duncan et al. 1991). Selection of team members via a volunteer process can promote enthusiasm. If a large number of employees are equally proficient in certain functions, the self-selection process works well. If not, an alternate means of determining team members is by assignment. The person making the assignments can be the quality committee leader, the department head, or—for interdepartmental teams—an institutional quality-coordinating committee or an administrator.

Determining the number of team members is another point where the scope of the challenge dictates the decision. Two or three people may be adequate to address a narrow, well-defined issue such as rotating the floor stock of infant formula. All the steps in this process are performed by two food service supervisors, usually those working the evening shift. They shelve the formula, stock the nursery, and dispose of expired products. This team of two supervisors was effectively able to improve the formula delivery process by adjusting inventory levels to prevent run-outs.

Much larger groups of up to ten employees are appropriate for complex issues (Duncan et al. 1991). Restructuring the tray distribution and collection process, for example, might involve a cross-functional cadre of representatives from the departments of dietetics, nursing, and housekeeping.

Determining Who Does What

Team Leaders and Members

Team leadership need not follow officially sanctioned organizational lines. Often, in fact, it is preferable to select a front-

Exhibit 13–5 Comparison of Cross-Functional and Departmental Teams

---

QUALITY IMPROVEMENT TEAMS
TYPES OF TEAMS

CROSS-FUNCTIONAL TEAMS

Purpose: Cross-functional quality improvement teams evaluate processes that span two or more departments or functions.

Project Selection: Typically selected in accordance with the priority list of improvement projects determined by the Quality Steering Committee.

Accountable To: Quality Steering Committee.

Membership: Team members should represent each functional area affected by the process, and should be selected based on their understanding and ownership of the process being studied.

Composition: Generally composed of department directors, managers and supervisors, and have up to eight members, including the leader.

Life Cycle: Frequency and duration of team meetings will be a function of the process being studied; a typical team life cycle might be one to two hours per week for twelve to sixteen weeks. Cross-functional teams disband once their recommendations have been successfully implemented.

Implementation: Recommendations implemented by responsible managers with the assistance of team members upon acceptance by the Quality Steering Committee.

DEPARTMENTAL TEAMS

Purpose: Departmental quality improvement teams evaluate problems that occur within a department or between two departments or functions.

Project Selection: Typically selected by team members as this broadens employee participation and increases ownership.

Accountable To: Department Manager.

Membership: Team members should represent each activity area within the department and should be selected based on their understanding and ownership of the process being studied. Members may be rotated to ensure that all department employees have an opportunity to participate on the team.

Composition: Generally composed of employees and their supervisors, and have up to ten members including the leader.

Life Cycle: Meetings may be conducted in lieu of or as an adjunct to regular departmental meetings, and may require about one hour per week. Departmental teams can exist indefinitely, as long as members desire and they continue to produce positive results.

Implementation: Recommendations implemented by department managers unless they involve multiple departments or exceed approved budgets, in which case recommendations should be presented to the Quality Steering Committee.

Source: © 1990 Ernst & Young.

line employee who is most familiar with the task at hand to lead the group. A leader is selected based on the job description, a person who by virtue of his or her daily work is most proficient in the systems to be examined. Although it may be tempting to name a manager as group leader, the manager may actually be more valuable in providing insight as one of several team members whose contributions are all considered equally.

The leader's primary responsibilities are to influence what decisions are made and to ensure productive use of team members' time (Ernst & Young 1990). The leader must create and maintain the channels that enable members to do their work (Scholtes 1988). Much of the leader's work may be accomplished outside of the actual meeting time. A leader, for example, may need to intervene in behalf of a member who is encountering difficulty in being released from the daily job routine to attend meetings or complete assignments. The leader is also the contact point for communication between the rest of the organization and the team (Scholtes 1988).

In the most successful teams, members are chosen according to their knowledge and involvement and, ideally, their ownership of the problem at hand. The members selected for a quality team "should consider their participation as a priority responsibility, not an intrusion on their real jobs. The project is now part of the members' real jobs" (Scholtes 1988, 3-14). Members also have responsibilities to participate, to share their knowledge, to ask questions, to make sure they understand everything they are being asked to evaluate, and to carry out delegated assignments.

Facilitators

Some quality teams, especially inexperienced ones, may benefit from having a designated facilitator as well as a leader. Choosing a facilitator may be more challenging, because an individual's personal characteristics come into play. The facilitator must mediate without taking over and must be organized without stifling the group's activity. The facilitator may be the one group member who is not familiar with the problem under study. This lack of expertise of the issues may actually assist in the facilitation process. The facilitator, as a disinterested third party, can focus on keeping the group on track. Her or his role is to manage how decisions are made rather than becoming involved in the content under discussion.

Having a facilitator requires devoting an additional employee's time to the project, which is no small investment. Yet having someone fulfill this role promotes team productivity, and therefore it pays off in the long run. The facilitator can attend to all the ground rules and housekeeping details related to the meeting process. These tasks may be very simple for a small departmental team well versed in solving problems. They may be monumental for a large cross-functional team inexperienced in complex problem solving. The facilitator takes responsibility for the agenda, scheduling, and documentation. She or he can monitor key functions such as attendance and time keeping. It is the facilitator's responsibility to ensure that communication guidelines are followed and that all members have equal opportunity to participate. Guidelines for developing meeting ground rules are found in Exhibit 13–6. A summary of team roles and responsibilities, comparing the

Exhibit 13–6 Developing Meeting Ground Rules

Attendance	Who will schedule meeting, arrange for room, and notify members?
	How will absences be handled?
	Can team members be replaced for absenteeism?
Time Management	How does team define "on time"? Are starting and ending times enforced?
	How will time allotted to agenda items be monitored?
	What is the role of the timekeeper? Who will serve as timekeeper?
Participation	What advance preparation is expected?
	How will participation be monitored to ensure equal contributions?
	How will activities be monitored to ensure productive meetings?
	How are assignments made? What are the expectations for their completion?
Communication	How candid can members be? Is information confidential to team?
	How will discussions be started? What if discussions get off track?
	How will interruptions or side conversations be handled?
	What listening skills are expected?
	What forms of criticism are acceptable?
	How will creativity be encouraged and negative thinking discouraged?
Decision making	How will differences of opinion be expressed and resolved?
	How will conflicts among members be handled?
	What process will be used to reach consensus? To guard against "group think"?
	How will decisions be made?
Documentation	What process will be used to set meeting agendas and allocate time?
	How will agendas and minutes be distributed?
	Who will serve as recorder of minutes?
	Where will documentation be kept?
Other	What meeting interruptions are acceptable and non-acceptable?
	How will breaks be handled?
	Will refreshments or smoking be allowed?
	Who is responsible for meeting room set-up and clean-up?
	What support services are available? Who will coordinate them?

Source: © 1990 Ernst & Young.

purpose and selection criteria for each type of team member, is outlined in Exhibit 13-7. Specific roles of a facilitator guiding group processes are described in Chapter 7.

Team Charters

With all the steps involved in team formation, some organizations have opted to formalize this process by utilizing a team charter process. A charter briefly previews the team's mission, boundaries, resources, and representatives. The charter provides direction by listing expected improvements. It assists in keeping the team on track by defining which systems (or parts of a system) that will and will not be studied by the group. By listing resources available to the team, a charter can promote both realistic goal setting and success. Guidelines for drafting a team charter are listed in Exhibit 13-8.

The ability to initiate the charter process should be at the discretion of each and every employee. Ideas should be reviewed within the department first. If the process to be studied is clearly interdepartmental in nature and best handled by a cross-functional team, the department manager may opt to submit it to the hospitalwide steering committee via the administrator or vice-president.

The steering committee reviews all charter requests. Then, according to its standard review procedure, it elects to approve or disapprove the project. If approved, the committee may endorse the submitted roster of proposed team members or recommend others to participate. The steering committee usually has a mechanism to track projects in process. In this manner, a charter helps prevent unintentional duplication or overlap by ensuring that two teams are not independently working to solve the same problem.

There are many variations of the chartering process. The one described here is practiced at Henry Ford Hospital. The policy and procedure for the charter process and the charter request form are found in Exhibits 13-9 and 13-10. A flowchart of the process beginning with the initial problem identification stage and following through to team formation is tracked in Figure 13-4.

Does it sound as if there is more work involved in assembling a team than the work that the team will actually do? Remember that the steps are extensively described so that groups will have a guideline, one that can be adjusted according to the group's function. The most effective quality team is one that harnesses the creativity of its members and that can reach agreement without endless, agonizing discussion (Leebov 1991).

As an organization matures in its quality initiatives, team building will become a more natural, less rigorous phenomenon. One chief executive officer observed this firsthand. The hospital he governed was well on its way in the transition from quality assurance to continuous quality improvement. The administrator described being stunned when he received a progress report listing 70 active quality teams, only 25 of which he was aware. Many were natural work group teams formed at their own initiative. They had convened extemporaneously, not because of any top-down directive. This insight led the organization to halt the formalized team-chartering

---

Exhibit 13-7 Team Roles and Responsibilities

	Team Facilitator	Team Leader	Team Member
Purpose	To promote effective group dynamics	To guide teams to achieve successful outcomes	To share knowledge and expertise
Major Concern	*How* decisions are made	*What* decisions are made	*What* decisions are made
Principal Responsibilities	☐ Ensure equal participation by team members ☐ Mediate and resolve conflicts ☐ Provide feedback and support team leaders ☐ Suggest problem solving tools and techniques ☐ Provide TQM training	☐ Conduct team meetings ☐ Provide direction and focus to team activities ☐ Ensure productive use of team members' time ☐ Represent team to management and QSC ☐ Document team activities and outcomes	☐ Offer perspective and ideas ☐ Participate actively in team meetings ☐ Adhere to meeting ground rules ☐ Perform assignments on time ☐ Support implementation of recommendations
Position Type	Organization-wide	Team-specific	Team-specific
Selection Criteria	Personal characteristics	Job title and description	Ownership of process

Source: © 1990 Ernst & Young.

Exhibit 13-8 Drafting a Team Charter

Team Mission	What parts of the process or system should be studied?
	What led to selection of this issue?
	What data exists or is required to study the issue?
Expected Improvements	What are the goals or expected outcomes of this study?
	What magnitude of improvement is expected?
	Who will approve and implement recommendations?
Boundaries and Constraints	What parts of the process or system should *not* be studied?
	What time or budgetary constraints are applicable?
	What decision making authority does the team have?
Resources Available	What internal or external experts should be consulted?
	Who may be called upon to assist the team?
	What support services are available such as computers, graphics, presentation materials, etc.?
	Who will cover for members during meetings?
Team Representation	What functional areas will be represented?
	What job titles will be represented by members?

Source: © 1990 Ernst & Young.

Exhibit 13-9 Policy and Procedure for the Charter Process

HENRY FORD HOSPITAL
QUALITY MANAGEMENT STEERING COMMITTEE
QUALITY IMPROVEMENT TEAM CHARTER PROCESS

Purpose:
1. To conduct a formal on-going process for the senior management of HFH to review and recognize QITs that are formed within HFH.
2. To identify key cross-functional processes that impact our external customers and identify opportunities for improvement.
3. To assure resources are provided to the QITs to accomplish their mission.
4. To track QIT activity within HFH.

Identification of QITs:
1. Any HFH employee may suggest a QIT be formed and submit a *QIT Charter Request Form* to the department head.
    - Departments will develop a process to select QITs they wish to submit.
    - Department heads will then present the *QIT Charter Request* to their VP/QMSC sponsor and management team for review and discussion.
    - Each VP/QMSC sponsor's management team will develop a process to select QITs they wish the VP/sponsor to present to the HFH QMSC.
    - At designated QMSC meetings the QMSC will review those QITs presented by VP/sponsors.
2. The QMSC may develop a QIT charter based on the strategic importance of the process to our internal and external customers. The QMSC will assign a leader, facilitator/advisor and suggest members for participation on the team based on its knowledge of customers and suppliers within the process.

Decisions of QMSC:

Potential decisions that a QIT charter request may result in:
- Charter approved and QIT recognized as a "departmental demonstration" QIT.
- Charter approved and a QMP facilitator/advisor assigned.
- Charter not approved and reasons identified.

Approved HFH:/QMSC 11/90
mv/a:QITCP

Courtesy of Henry Ford Hospital, Detroit, Michigan.

process. It was felt that team formation had become second nature in this hospital's quality quest. Teams were considered just another tool, like a flowchart or Pareto diagram, that helped get the job done (Kennedy et al. 1992). This type of spontaneity is a good indication that quality values are being internalized and that the organization's cultural transformation to quality improvement is in full swing.

The director of a nutrition department related how these small but critical incidents were the most telling early signs of progress in her department's efforts to move toward quality improvement. The hospital was still in the education phase and had not yet embraced an organizationwide quality improvement initiative. Individual departments were going ahead with internal endeavors to prepare staff members in anticipation of the metamorphosis. Late one afternoon, a staff dietitian was notified about a couple's 30th wedding anniversary. The wife was a patient with terminal cancer. The dietitian called the chef, who in turn called the purchasing manager for samples of a new entree he had requested for testing. A dinner meal was quickly assembled, prepared with some convenience foods, but with flair. It was served maitre d' style by the evening food service supervisor, on borrowed "VIP" china service.

"The best thing about this story," said the director, "was that I didn't even hear about it until the following day. A nurse manager mentioned how touched the patient was by the gesture and how impressed the staff was with our culinary expertise." At first, nurses had assumed that the dietitian had ordered the meal from a local restaurant (the usual practice) because it had been pulled together so quickly. This was truly

Exhibit 13–10 Charter Request Form

HENRY FORD HOSPITAL
QUALITY MANAGEMENT STEERING COMMITTEE
QUALITY IMPROVEMENT TEAM
CHARTER REQUEST FORM

Henry Ford Hospital QITs should be chartered by the HFH QMSC. If you have a suggested QIT you would like chartered by the HFH:QMSC please complete this form (*type*) and return it to: Mary Vidaurri, Ph.D., A-2 Admin.

PROCESS FOR IMPROVEMENT: _____
OPPORTUNITY STATEMENT: _____
_____
_____
_____
_____

PROPOSED TEAM MEMBERS:

Team Members (leader)	Department/Position
_____	_____
_____	_____
_____	_____
_____	_____
_____	_____
_____	_____

QIT Sponsor: _____
Date Reviewed by HFH QMSC: _____
Charter of Team: ____ Approved as departmental demonstration QIT
____ Approved and QMP Facilitator/Advisor assigned: _____
____ Not Approved MV/c:charter (Rev. 11/90)

Courtesy of Henry Ford Hospital, Detroit, Michigan.

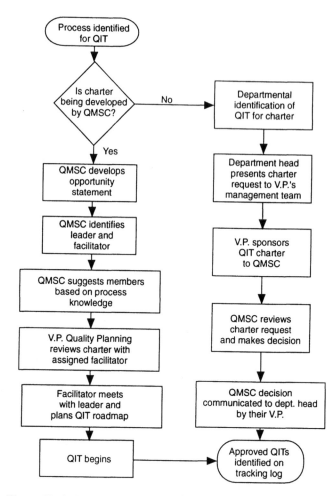

Figure 13–4 Quality Improvement Team Charter Process. Courtesy of Henry Ford Hospital, Detroit, Michigan.

---

an example of an empowered team of employees who coalesced spontaneously to take advantage of a quality opportunity. They were not afraid to make decisions, pull staff temporarily from their normal work roles, and act according to their own initiative. This event may not sound like much of an accomplishment for many full-service nutrition departments. But the hospital had eliminated all catering operations years ago and had no special-order meal service available for hospitalized patients. It demonstrated how far the department had come in working together.

FOCUS-PDCA: CLARIFY KNOWLEDGE AND UNDERSTAND REASONS FOR VARIATION

It is important that the team clearly and concisely define the problem before taking any steps to solve it. This is accomplished by first developing a problem statement, then analyzing the problem, using only factual data.

Developing a Problem Statement

The problem (or opportunity) statement serves as the focal point throughout the clarification process. The problem should be described in terms of factual events or measurable conditions (Quality Force 1991). If the problem is vague, it can perhaps be better defined by listing its symptoms. Symptoms are any facts, conditions, events, or results that provide clues to the underlying cause(s) of the problem. Symptoms can be identified by answering a series of questions, such as the following:

- Who is involved?
- What has happened?
- Where is it occurring?
- When does it happen?

Once an exhaustive list of symptoms is identified, the most pressing item(s) can be incorporated into an opportunity statement. Defining the issue in this way enables the group to deal with the problem in a factual, objective, and measurable manner. A sample format for writing an apropos opportunity statement is presented in Exhibit 13–11. If a team can fill in the blanks it will have stated which part of the process contributes most to the problem, how the process can be improved, and what the expected outcome is.

Facts vs. Factoids

When clarifying problems, avoid using personal opinions, emotions, and factoids. Factoids are pieces of information that appear to be factual but are actually opinions that have grown in acceptance and considered as fact (Kennedy et al. 1992). Dietitians find it frustrating to deal with the many factoids they encounter in daily practice. Factoids are a hindrance to a problem-solving team because they may divert the group on a proverbial wild-goose chase when the group should be following the scent of the hunt. Factoids may be popularly referred to as old wives' tales.

A group of clinical dietitians had no difficulty generating a list of factoids related to the practice of nutrition. Most dietitians will be able to relate to having to deal with some of these beliefs commonly held by other health professionals regarding nutrition services.

- A patient will automatically develop diarrhea once an enteral feeding is started.

Exhibit 13–11 Format for Opportunity Statement

HENRY FORD HOSPITAL
QUALITY IMPROVEMENT TEAM (QIT)
OPPORTUNITY STATEMENT

Definition:
A concise description of the process to be improved, its boundaries, and why it should be improved.

Purpose:
To clearly and specifically identify and communicate the mission of the Quality Improvement Team.

Sample Format:
An improvement opportunity exists in the portion of the _____ process beginning with the _____ and ending with the_____. This portion of the process currently causes _____. Improvement of this process should result in _____ for the (customers) and (suppliers) of this process.

Courtesy of Henry Ford Hospital, Detroit, Michigan.

- The entire amount of parenteral solution that is ordered by the physician is actually infused into the patient.
- All of the food served in the hospital cafeteria is "healthy."
- Dietitians know what foods each patient receives for lunch because they check each tray.
- Patients on renal diets may never eat spaghetti because of the potassium content of tomato sauce.
- Vitamins are a good energy source.
- Patients on modified diets believe they can have all the foods that appear on their menus in any combination or amount. (After all, if it wasn't acceptable, it wouldn't be on the menu, would it?)
- Every dietitian is an expert on weight control.

The team can avoid having its efforts ambushed by factoids posing as data by relying only on statistics and relevant facts rather than gut feelings. Again, a new team might benefit from the expertise of a facilitator or strong leader so that there are fewer false starts during the problem clarification phase.

Key Quality Characteristics and Variables

The key to success in selecting the appropriate action plan is to identify precisely the key quality characteristic, the singular item that plays the biggest part in the desirable quality outcome. Once the one fundamental quality characteristic is known, the group can examine all data and determine the key process variable(s), that is, the one or two steps most related or most critical to achieving the quality result (Duncan et al. 1991). Identification of key quality indicators and variables is the clarification step having greatest impact on the outcome. Quality teams must be very careful to not skip over problem clarification or jump to obvious conclusions. By doing so they may gloss over the key process variable and end up implementing a whole new procedure that does not solve the problem at hand.

Using the Tools

Here is where all the quality tools come out of the box. Let us examine how some of them could be applied to clarify and better understand one of the systems mentioned previously in this chapter: the tube-feeding delivery process.

Assume that a cross-functional quality improvement team has already been chartered to study the process. The team is made up of the nine employees, each of whom has some role in enabling a tube-feeding product to get to a patient when there has been an order for a product change. The dietitian has been selected as the team leader because of her knowledge of enteral feedings and her understanding of department processes. Each person has his or her own set of beliefs of what the problem is and how best to fix it. Each brings these personal agendas with them to the first meeting:

- *Physician:* assigned to the team by the hospital steering committee but attends few meetings due to scheduling priorities.
- *Ward clerk:* thinks all tube-feeding products are the same. Believes the solution is to limit the types of products to two or three at most; that way there wouldn't be all this switching.
- *Diet clerk:* believes there is a Murphy's law for tube feedings: there will always be at least one late tube-feeding change on the days she is covering extra floors. She's thinking of asking for an assignment change to the maternity unit where she won't have to deal with this extra work.
- *Dietitian:* thinks the physicians on service usually make errors in recommending a formula change when acting independently. She believes changes should be made only during morning rounds, when all team members are available to provide input.
- *Dietetic technician:* can't believe how casually everyone treats this issue. She knows this is the patient's only source of nutrition. Tube feedings should be available on a "stat." basis, just like emergency labs.
- *Food service worker:* believes this is a conspiracy to make her job more difficult. She's tired of filling orders in the afternoon for the same patients who were on the morning list. She believes there should be a moratorium on tube-feeding changes beginning one hour prior to when she goes off duty.
- *Transporter:* has more important runs to take care of. He has patients waiting to be picked up in radiology. Those patients are usually tired and cranky after their tests, and he'll have to hear about it as he accompanies them all the way back to their rooms. He believes the product should be stocked on the floor.
- *Nursing assistant:* checks off the product, only to find that one of the bottles is incorrect—again. Dietary always makes these mistakes. Now she'll have to make another call. She believes the employee who made the error should be disciplined. That would put a stop to this carelessness.
- *Nurse:* knows that this diet order was written at 2:00 P.M. today. She is incredulous that it took five hours to obtain the new formula. This reinforces her belief that the hospital is being far too conservative on staffing issues. She thinks this proves more personnel are needed.

Common Errors

If every individual was attempting to improve this process autonomously, each might have been guilty of committing some of the most common errors in the quality-improvement process related to clarification and understanding:

- failing to define the problem
- making decisions based on incomplete or inaccurate information
- skipping steps by jumping to the "obvious" conclusion
- "selecting a desired solution, instead of a process" (Scholtes 1988, 2–3).

Fortunately, a facilitator has been appointed to the team. His job is to make sure the model plan is followed so that the group is not ensnared in these traps. He begins by having the group answer the diagnostic who, what, where, and when questions.

To answer these questions the team used a variety of tools. First, they created a flowchart depicting the nine steps in the process (Figure 13–5). Several team members were concerned not only with the number of formula changes ordered, but also with how long it took the players to accomplish the change. Therefore, the flowchart includes a timeline to address the "when" issues.

Many of the assumptions held by individual group members were revealed in a brainstorming session. The facilitator guided the group in developing two check sheets to gather facts related to the process. One check sheet was used to determine which products on the enteral formulary were used most often by each nursing unit (Exhibit 13–12). The dietetic technician was assigned to collect these data. They were readily obtained by tallying the unit requests on the dietetic distribution form. The second check sheet recorded the number of times that the new formula order seemed questionable based on the dietitian's recommendations and required some follow-up on her part before being processed (Exhibit 13–13). The dietitian volunteered to collect this information.

The group also decided to track the number of tube-feeding change orders received after the 3:00 P.M. cutoff time. They wanted to determine whether there was a pattern. The food service worker agreed to collect this information. The results, over a three-week period, were plotted on a scatter diagram (Figure 13–6).

When the group reconvened they were able to accurately answer the who, what, where, and when diagnostic questions.

- *Who?* All the players and their respective roles were depicted on the flowchart.
- *What?* Tube-feeding orders were regularly changed midday. The products involved in the majority of changes were the 1 calorie/cc formula, 1.5 calorie/cc formula, and the fiber-containing formula.
- *Where?* The bulk of the changes occurred on two nursing units.
- *When?* All but one late change occurred within a one-and-a-half-hour time frame, from 3:30 P.M. to 5:00 P.M.

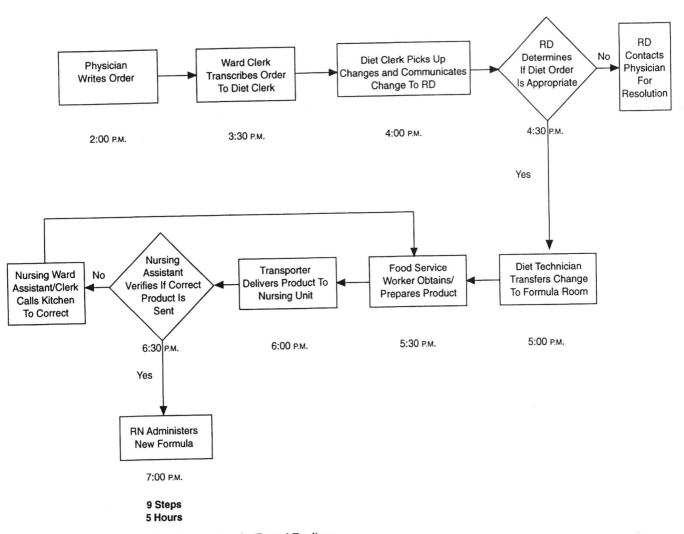

Figure 13–5 Flowchart for After-Hours Orders for Enteral Feedings

The group was able to use this information to define the problem statement by filling in the blanks on the opportunity statement form (Exhibit 13–11).

> An improvement opportunity exists in the portion of the <u>enteral feeding order and delivery</u> process beginning with the <u>change order that is received after the formulas have been prepared and delivered</u> and ending with the <u>delivery of the new product to the floor</u>. This portion of the process currently causes <u>a delay in appropriate patient intervention and frustration on the part of numerous staff members.</u> Improvement of this process should result in <u>improved nutritional intake and adequacy</u> for the <u>patient (by decreasing the downtime between formula changes)</u> and <u>labor-time savings for the suppliers</u> of this process.

The team members learned a lot about the tube-feeding delivery process when they analyzed the information provided by these simple tools. They were able to accept knowledgeably that some of their initial assumptions were not supported by the data. The ward clerk learned that formula products were as different from one another as diabetic diets are from renal diets. The dietitian learned that formula changes were appropriate the vast majority of the time. Even in the single instance in which the original formula was clearly a better choice for the patient, the alternative product that was ordered presented no risk over the short time it was actually infused. All of the members were surprised that a simple change order initiated a nine-step process that took five hours from start to finish. The nurse realized that the real issue was the number of steps and the number of employees involved in carrying them out, not the total number of hospital employees. Critical analysis of the data had effectively expunged the factoids.

By using this information, the group hypothesized that the key quality characteristic was the availability (or lack of availability) of alternative feeding products on the nursing unit. The variables were the number of steps involved in obtaining

Exhibit 13-12 Check Sheet Formula Requisition by Nursing Unit for a One-Week Period

Formula Description	Nursing Unit (number of containers requisitioned)											
	A	B	C	D	E	F	G	H	I	J	K	Total
1 cal/cc		6		28					76	49		159
1.5 cal/cc		14			20	24						58
2 cal/cc		11							5			16
elemental		9							22			31
fiber	35		11			45			17	3	4	115
high cal/pro	2											2
high nitrogen					29			21				50
Total	37	40	11	28	29	65	24	21	120	52	4	431

the needed products when the timing of the requests fell outside routine daily operating procedures for the tube-feeding preparation and delivery system. Now that the group had more precisely identified which parts of the process were key, they were able to zero in and develop specific plans to improve the process.

FOCUS-PDCA: SELECT AND PLAN

The action plan should focus on modifying the key process variable identified earlier in the problem-solving process. A number of alternative solutions may appear feasible. In most cases, it is best to select the most viable solution and implement it in a small-scale pilot first. If the problem has been clarified and analyzed appropriately, planning the best action to take will be simplified.

Exhibit 13-13 Check Sheet for Evaluation of Tube-Feeding Order Changes

Changes in Tube-Feeding Orders	Week	Yes	No
Was the tube-feeding ordered an appropriate selection for the patient?	1	ℍℍℍ ℍℍ	1*
	2	ℍℍ ℍℍℍ ℍ	0
	3	ℍℍ ℍℍ ℍ	0
Total		37(97%)	1(3%)

*Comments: Formula with fiber would have been a better choice.

Plans can be developed to revise and improve key factors: systems, knowledge, and/or behavior. Improving the after-hours tube-feeding delivery process was basically a systems issue, although knowledge of the products and process were contributing factors as well.

In the previous example, the group drafted a revised flowchart that eliminated six of the steps (Figure 13–7). This was accomplished by stocking nursing units with a supply of the two most commonly used formula products. The team's next step was to conduct a trial of the revised procedure on the surgery unit, which had the highest number of patients on enteral feedings.

Systems Issues

In hospitals the vast majority of problems are systems related. Improving a system frequently requires changing more than one procedure by one person. If the team chosen is a good fit for the problem under study, it should be aptly suited to improve the process. There is no shortage of opportunities to improve systems operations that affect clinical nutrition practice. Interviews with dietitians and a review of departmental quality plans and reports reveal a number of system issues that could be improved successfully by quality management. Some examples are systems that relate to

- mailing diet instructions
- responding to requests for nutrition consultations within 24 hours
- obtaining accurate and timely calorie counts
- billing for nutrition services
- recording amounts of infused enteral and parenteral formulae in intake/output sheets
- providing continuity of care when inpatients become outpatients
- addition of medications to enteral formulas
- transferring a patient's tray when his or her room changes
- obtaining special foods for patients on extremely restrictive diets

Knowledge Issues

Plans that revolve around enhancing employees' knowledge are usually more clear-cut and easier to implement than those that involve entire systems. Some appropriate actions that can be taken to provide opportunities for learning include the following:

- developing nutrition classes on new procedures (example: hands-on miniworkshop when introducing a new enteral feeding pump)

Taking Action To Improve Care   171

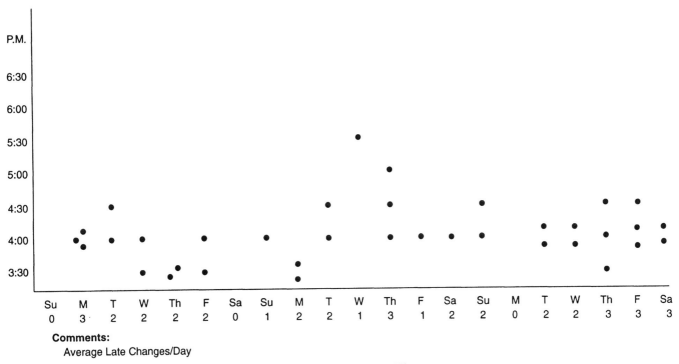

Figure 13–6 Scatter Diagram: Late Tube-Feeding Orders Received in Kitchen

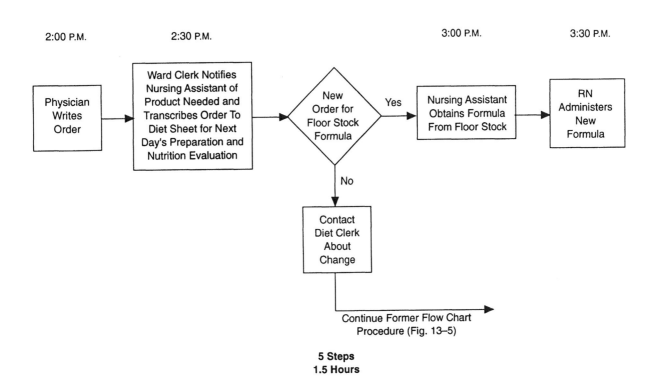

Figure 13–7 Revised Flowchart for After-Hours Orders for Enteral Feedings

- restructuring educational formats (example: adding an article review/self-test option as an alternative to attending an infection-control in-service on preventing foodborne illness)
- providing resources and references (example: purchasing a department subscription to a relevant clinical journal and circulating it among the professional staff)

Orientation, training, and retraining are at the core of any plan that relies on improving the knowledge base of the staff. However, employee training is expensive, making it one of the least-used methods of quality improvement. Is the cost of training justified? It has been estimated that the return for every training dollar spent per employee is $30 worth of productivity (Harris 1989).

For registered dietitians and technicians, continuing education is a fact of professional life. It is one of the factors that keep dietetic professionals in the forefront and ensure visibility as experts in nutritional care. Offering educational programs to all employees, both professional and support staff, targeted at improving a specific aspect of care, can promote the department's efforts to be a leader in the quality quest. The practice of dietetics is a combination of art and science. As the results of nutrition research are incorporated into standards of practice there will continue to be many opportunities for improving the knowledge base of practitioners.

Behavioral Issues

A benefit of using the FOCUS-PDCA model approach is its focus on systems rather than on individuals. The aim of a quality-improvement process is to correct a problem by improving a process. In traditional quality-assurance programs the purpose was often to identify the person at fault who caused the problem. In health care there is no question that accountability is important. But constantly asking "who's accountable?" when a problem is being analyzed really means "who's to blame?" (Kennedy et al. 1992). This negative focus was often punitive in nature and it was at the heart of outmoded peer review programs that had little or no effect on quality care. Focusing on who is at fault tends to foster defensive responses in employees, such as denial and evasiveness.

Obviously, strategies that encourage employees to react negatively do not foster problem ownership and resolution. In fact, the opposite often occurs, and the substandard care is reinforced. This negative cycle flows from poor care to punishment to game playing back to poor care. It does not show a path out, one that will break the cycle. This cycle, typical of traditional quality-assurance methods, conveys the message that there are two types of practitioners: poor practitioners and potentially poor practitioners (Merry 1991). After a peer review monitor is completed, even one in which no performance issues are identified, this method might prompt the following response by the committee chairman: "The numbers look OK this month, but what will the next audit show?"

This type of practice is based on McGregor's Theory X: that people are lazy, irresponsible, and dependent. Theory X and quality-improvement programs are wholly incompatible (Merry 1991). The Theory X approach places all responsibility on the manager to achieve quality improvement by inspection. It conflicts with everything we know about the quality-improvement process. Quality cannot be "inspected" into the workplace through external review. True quality originates within work and by workers. The manager's role is to be the "good cop" who supports and guides the process, not the bad one who rounds up and jails offenders.

Ideally, quality-improvement programs promote a blame-free environment. In the past, a group reviewing a blip on a graph might respond: "Who's this person here, who's number two?" (Figure 13–8). The question needs to change to "Why do we have this kind of pattern?" (Kennedy et al. 1992). Such a question is more effective in that it addresses a system that can be improved rather than an individual problem performance issue that needs to be corrected.

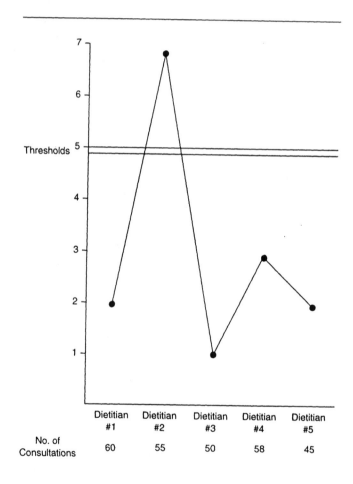

Figure 13–8 Percentage of Late Billings per Dietitian

Organizations can attain dramatic improvement in systems operations without any person's or department's perceiving the quality process as a threat or a reflection on poor performance (Duncan et al. 1991). This is not to suggest that dietitians take a Pollyanna approach to peer review (Merry 1991). Peer review is still a valid method of data collection. Nevertheless, a manager must address truly bad care in the relatively small number of cases where it exists. When doing so, it is important for the manager to keep the performance issue in perspective and handle it confidentially and within the normal chain of command. When sharing negative feedback with a staff member, the manager should ask the employee to "describe how this feedback can be used," instead of admonishing the caregiver to "not let this happen again" (Petersen 1989).

The manager must also ensure that addressing poor performance is not the emphasis of the improvement process. Instead, the peer review process, from the point of data collection to the corrective action stage, should encompass "the entire spectrum of practice, from potential negligence to genuine excellence" (Merry 1991, 317).

An example of a peer review practice that promotes excellent performance across the board is a lecture rehearsal and review policy. Dietitians are frequently invited to share their nutrition expertise by providing in-service programs and presentations. Since other forms of professional communication such as patient education materials and journal submissions are subject to internal review, the staff decided to adopt a similar mechanism to preview their colleagues' presentations during practice sessions. The sessions provided the opportunity to polish the presentation skills of the speaker and improve the mechanics of the talk by editing out typographical errors on audiovisuals or clarifying vague issues in advance. They also provided an opportunity to increase the knowledge base of all participants, those who lectured and those who provided constructive feedback. This practice was initiated to advance the skill levels of the entire staff, not to "punish" a poor performer.

In summary, the keys to an effective behavior-related review process are as follows:

- Take a positive view of health professionals.
- Provide educational feedback of relevant and useful information to everyone under review.
- Focus on systems, not individuals.
- Address collegial issues through established, local mechanisms and protect confidentiality (Merry 1991).

Describe Change Expectations

There is one more step before moving into the action phase. All action plans need to have built-in expectations for change. The team needs to determine in advance what is expected to change, who is responsible for implementation, and when the change is expected to occur.

For the systems-related problem of tube-feeding delivery, the questions might be answered this way:

- *What will change?* (1) The incidence of after-hours tube feedings that need to be sent from the formula preparation area to the ward will change, along with (2) the length of downtime of formula infusion between product changes.
- *Who will make the change?* The dietitian will act as a liaison with food service supervisory staff and nursing personnel to establish a simple inventory system for stocking formulas on units.
- *When will the change occur?* The new procedure will begin the first of next month.

The same series of questions are applicable for knowledge- and behavior-related issues. Two examples follow. The first relates to enhancing the knowledge of dietetic technicians to improve their accuracy when checking renal menus.

- *What will change?* Accuracy of menu checking for renal diets.
- *Who will make the change?* Dietetic technicians.
- *When will the change occur?* After the renal dietitian revises the menu-checking guidelines to reflect current menu items available and provides an in-service program.

Appropriate outcomes for improving the billing process (a behavior issue) may include the following:

- *What will change?* The number of late billings.
- *Who will make the change?* The dietitian.
- *When will the change occur?* After the dietitian completes a refresher course in billing procedures.

FOCUS-PDCA: DO, CHECK, ACT

Once the action plan is implemented, how will you know if you are making progress toward the goal? First, monitor the effect on the key quality indicator or characteristic and ask whether what you did made a difference. The need to validate effectiveness of the actions emphasizes the importance of having selected measurable criteria. Otherwise, you will not be able to determine clearly whether you are succeeding in your quality-improvement efforts.

One department group was working on improving patients' opinions of the meal trays they received during their hospitalization. After discussing the various reasons that patients might not like their trays, the group made the assumption that tray appearance was the primary culprit responsible for patient dissatisfaction. They implemented several changes to increase visual attractiveness. These included designing new tray covers and purchasing new color-matched dome lids. Patients' comments on food and nutrition service did not improve.

The group reconvened and elected to perform a more detailed analysis of the data available from the patient questionnaire to learn why their efforts had not achieved the desired results. The team leader enlisted the help of a student. The student manually categorized all the written comments for the past year on a check sheet and then transferred the information to a Pareto chart (Figure 13–9). This was a tedious process, but well worth the effort.

In this example, the assumption that tray appearance was important to patient satisfaction proved to be a factoid. Data analysis indicated that the key quality characteristics influencing patient satisfaction were the temperature of the food, accuracy of the tray, timing of meal delivery, and the responsiveness of the server. These were all rated as more important than tray appearance, which ranked near the bottom of the list in the number of times it was mentioned.

The department pursued a variety of strategies to address the top four dissatisfiers. They began more in-depth customer service training for new employees and implemented a manager/patient visitation program to nursing units on a rotating, random basis. They reviewed reheating and tray-checking procedures. They also began sharing the survey results with all of the employees by discussing them at staff meetings and posting the graphs in the work area. They met their desired improvement goal within the next two quarters. The moral is that data analysis can be a time-consuming, monotonous task. But it requires the same amount of time investment whether it is done before or after a trial process. Overall time, energy, and resources are better utilized when data analysis is completed upfront. The quality-improvement motto of "do it right the first time" is aptly applied to the process of collecting data.

The tray satisfaction indicator proved to be an easy one for the department to track. Its assessment was part of the evaluation form provided to every patient after discharge. Most de-

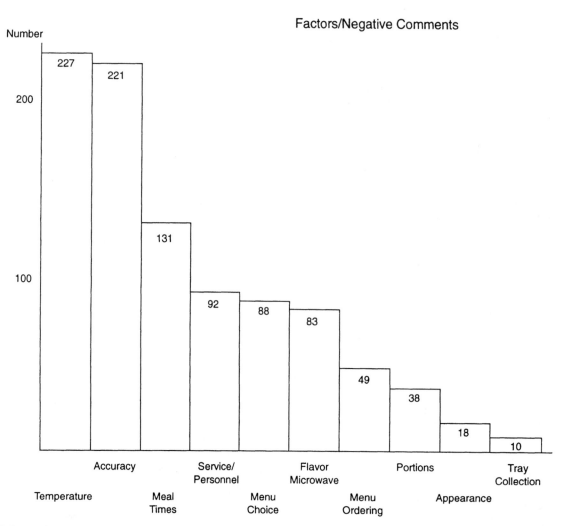

Figure 13–9 Pareto Chart: Patients' Qualitative Comments on Food Service

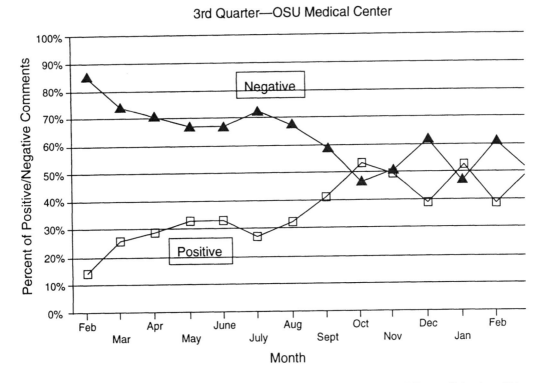

Figure 13-10 Qualitative Trends for Food Service. Courtesy of the Ohio State University Medical Center, Columbus, Ohio.

partments will not have an automatic monitoring system in place and will have to develop one. Figure 13-10 shows the steady progress the changes in tray delivery had on improving patient satisfaction by reducing the number of negative comments.

Check Progress

The team dealing with patient satisfaction with trays committed a common error by basing their first action plan on an assumption rather than a fact. However, they were not guilty of another common mistake: not having measurement criteria and a system to assess the impact of their change. They were readily able to answer the following questions:

- How will the effectiveness be evaluated? By tracking patient comments on the discharge survey.
- Who decides when and how the evaluation will occur? The director of patient relations, who compiles and distributes survey results.
- How often will you evaluate the effectiveness? Results calculated monthly and distributed on a quarterly basis.

Another way in which departments can check their progress is via the benchmarking process. An advantage of using benchmarking as a method of checking for sustained improvement is that it allows the opportunity to compare results against internal and external standards concurrently. First, the change can be assessed according to the predetermined measurement criteria established before implementation of the change. At the same time, benchmarking enables the department to track how it is performing this function in comparison to other departments, including those that are leaders in the field in the provision of this particular service. Exhibit 13-14 provides an example of a survey used to assess outpatient satisfaction at the Cleveland Clinic Foundation. This survey is distributed to all outpatients, including those who receive nutrition counseling. By reviewing results of the survey the nutrition services department can compare how well they are doing in satisfying patients who receive their service over a period of time. It also allows them to assess how well they are delivering these services in comparison to other outpatient areas.

Act To Maintain the Gain

Once the trial is completed and the positive change effected, it is time to institutionalize the improvement. The team will probably be the best group to teach others about the change. The fact that the team is usually composed of those most integral to the system to be changed will greatly facilitate and shorten the education process.

Exhibit 13–14 Example of a Survey Used To Assess Outpatient Satisfaction

The Cleveland Clinic Cares!

Please tell us about today's visit to The Cleveland Clinic Foundation in _____.

Answer the following questions only as they relate to your experience in the above area.

1. Is this your first visit in this area at The Cleveland Clinic Foundation? ☐ Yes ☐ No

2. Please indicate how satisfied you were with each of the following as they relate to this particular area.

		Very Satisfied	Satisfied	Dissatisfied	Very Dissatisfied	Not Applicable
a.	Your overall experience in this area?	☐	☐	☐	☐	☐
b.	The competence of the clerical/reception personnel?	☐	☐	☐	☐	☐
c.	The competence of the technical personnel?	☐	☐	☐	☐	☐
d.	The respect and courtesy of clerical/reception personnel?	☐	☐	☐	☐	☐
e.	The respect and courtesy of the technical personnel?	☐	☐	☐	☐	☐
f.	The privacy of the area in which you were seen?	☐	☐	☐	☐	☐
g.	The clarity of verbal instructions?	☐	☐	☐	☐	☐
h.	The clarity of written instructions?	☐	☐	☐	☐	☐
i.	The ability of personnel to explain the reason why you were being seen and to answer your questions?	☐	☐	☐	☐	☐

3. Were instructions on how to prepare for this visit clear and easily understood?
   ☐ Yes ☐ No ☐ Did not receive any instructions

4. Were you able to arrive on time for this appointment?
   ☐ Yes ☐ No

5. If you were delayed for this appointment was it because of: [Check all that apply]
   ☐ The registration process   ☐ Personal reasons
   ☐ An earlier CCF appointment   ☐ Other
   ☐ Was not delayed

6. How many minutes after your appointment time did you wait to be seen by someone in this area?
   ☐ 0   ☐ 11–20   ☐ 31 or more
   ☐ 1–10   ☐ 21–30

7. If you had to wait more than 20 minutes during any part of your visit, did you understand the reason why?
   ☐ Yes ☐ No ☐ No wait

8. Do you have a good understanding of what follow-up steps to take when you go home?
   ☐ Yes ☐ No ☐ Did not receive follow-up instructions

9. If you are in need of this specific medical service again, will you return to The Cleveland Clinic Foundation?
   ☐ Yes ☐ No ☐ Not sure

10. What can we improve in this area to serve you better?
    _____
    _____
    _____

Name (optional) _____

Return your completed questionnaire to a parking attendant and receive $1.00 off of your parking fee. Please return it to an information desk if you did not use our parking facilities.
Thank you!

Courtesy of The Cleveland Clinic Foundation, Cleveland, Ohio.

The team can do a lot of the legwork involved in the implementation phase. They can even take the final step of writing up the revised process in the organization's accepted format for policy and procedures, but utilizing a fresh approach: all post–continuous quality-improvement implementation policies and procedures should be written with a customer orientation rather than a hospital orientation (American Hospital Association and National Society for Patient Representation and Consumer Affairs 1991).

One way this can be accomplished is by changing the philosophy of a particular policy. For example, a staff of clinical dietitians changed their approach to monitoring the provision of "quality" nutrition consultations. Prior to continuous quality improvement, the focus of the staff's practice plan was on how well and how soon and how many consultations were documented in the medical record. After quality-improvement intervention, the practice plan was revised. Now the primary quality determinants driving the practice plan relate to patient acuity and patient satisfaction.

Existing policies, procedures, and employee practices can be reviewed for phrases and comments that indicate a lack of customer service orientation. Dougherty (1992b) describes some common catch phrases to watch for and adjust:

- "No substitutions allowed." Eliminate policies that get in the way of meeting customer needs.
- "Employees are not authorized to . . . ." All employees must be authorized to meet customer needs.
- "The following items are not allowed on a particular diet." Take extra care so that patients understand the fine line between diet orders and patient requests. Focus on what dietary modifications can be incorporated into a particular regimen.

A more specific approach can be taken in the actual procedural steps for implementing a new and improved process. In the following example, the philosophy has not changed, but the instructions on how it can be achieved are stated in a manner that converts vague policy to action:

- Before: Dietetics' employees will respect the patients' rights to privacy.
- After: If a patient's door is closed, employees should knock and ask for permission before entering the room.

Now is when the manager's support is imperative, or all the team's efforts will have been in vain. The manager must provide the group with the authority to implement the change. The group must also have access to necessary resources and staff. There is nothing more demoralizing to the team or damaging to the quality process than a good idea, with proven results, that is never put into practice. This "thanks, but no thanks" approach will probably ensure that this is the last project that any of the team's members will pursue.

Periodically checking to see whether the gain in quality care is maintained becomes part of future monitoring efforts. This does not mean that every change must be permanently monitored. If this were the case, the more successful the quality program is in making improvements, the more cumbersome the monitoring process becomes. If repeated monitoring of a process indicates that it is consistently within identified thresholds, there are a variety of choices the quality committee can make. The first is to upgrade the standard. For example, members of a nutrition support team developed and monitored outcome-based standards for patients receiving total parenteral nutrition, such as metabolic complication rate, incidence of catheter sepsis, and positive nitrogen balance. If the percentage of compliance during a six-month period was consistently higher than the initial standard, the standard was amended upward (Owens et al. 1989).

A second alternative is to "retire" the standard temporarily and substitute another one. One director described the approach used in the department he manages:

> If an indicator continues to show positive improvement over a six month period, it will be replaced with another one. We will continue to check the original indicator twice for the next six months to ensure that its quality has not suffered because it is no longer checked on a regular basis. (Harris 1989)

Once the tracking process shows that the change has been internalized, it is time to move on to another problem. Would this course of action meet Joint Commission requirements? It should, as long as the department can show documentation to support the decisions made.

This concept of continuous improvement is the hallmark of quality management: There is always the potential to do better. Even once a work process is stabilized, "it is still everyone's responsibility to seek new and better ways to align the process with customer needs and expectations" (Labovitz 1991, 16). In traditional quality assurance, the tendency was to monitor the same indicators over and over again. "How can we continue to expect things to get better when we're doing the same things every year? Doing the same thing over and over again but expecting different results is one of the definitions of insanity" (Kennedy et al. 1991, 87). In ten years do we still want to be wrestling with the same set of issues? If not, we need to look at a different approach to problem solving. We need to solve the problem and move on.

Ten years from now nutrition professionals should be solving problems of the 21st century, ones that we cannot even conceive of today. In a more global sense, dietetic professionals will not be able to achieve the agenda set forth for the nutrition-related initiatives of the year 2000 if they are still working on the problems of the 1990s. Quality management can be a powerful tool to get there from here.

REFERENCES

American Hospital Association and National Society for Patient Representation and Consumer Affairs. 1991. *Defining, measuring and meeting customer requirements: The meaning of quality in a CQI environment.* Teleconference. Chicago: American Hospital Association.

Dougherty, D.A. 1992a. The focus of TQM. *Hospital Food and Nutrition Focus* 8, no. 7:1,3–5.

Dougherty, D.A. 1992b. Quality programs: Satisfying the customer. *Hospital Food and Nutrition Focus* 9, no. 3:5–6.

Dougherty, D.A. 1992c. Total quality management—fad or future? *Hospital Food and Nutrition Focus* 9, no. 1:1,3.

Duncan, R.P., et al. 1991. Implementing a continuous quality improvement program in a community hospital. *Quality Review Bulletin* 17:106–112.

Ernst & Young. 1990. New York.

Harris, R.D. 1989. Returning to the basics: The bottom line to quality improvement. *Hospital Topics* 67, no. 2:19–23.

Kennedy, M., et al. 1992. A roundtable discussion: Hospital leaders discuss QI implementation issues. *Quality Review Bulletin* 18, no. 3:78–96.

Krasker, G.D., et al., presenters. 1992. Executive briefing at the Ohio State University Hospitals: 1993 AMH standards and survey process. Inservice. Oakbrook Terrace, Ill: Joint Commission on Accreditation of Healthcare Organizations. October.

Labovitz, G.H. 1991. Beyond the total quality management mystique. *Healthcare Executive* 62:15–17.

Leebov, W. 1991. *The quality quest: A briefing for health care professions.* Chicago: American Hospital Publishing, Inc.

Lehr, H., and M. Strosberg. 1991. Quality improvement in health care: Is the patient still left out? *Quality Review Bulletin* 17, no. 10:326–329.

Merry, M.D. 1991. Can an external quality review system avoid the inspection model? *Quality Review Bulletin* 7:315–319.

Newbold, P.A., and D. Mosel. 1990. Quality through people: Integrating budgeting, strategic planning, and quality management. *Quality Letter*, 8–15.

Owens, J.P., et al. 1989. Concurrent quality assurance for a nutrition support service. *American Journal of Hospital Pharmacy* 46:2469–2476.

Petersen, M.B.H. 1989. Using patient satisfaction data: An ongoing dialogue to solicit feedback. *Quality Review Bulletin* 15, no. 6:168–170.

The quality F.O.R.C.E. 1991. London, England: EC Murphy, Ltd.

Scholtes, P.R. 1988. *The team handbook.* Madison, Wis: Joiner Associates, Inc.

Chapter 14

Communication: The Vital Link in Continuous Quality Improvement

Every member of the clinical nutrition staff, including the department director, clinical managers and specialists, staff, dietitians, dietetic technicians, and food service workers, can participate in efforts to improve quality. In fact, their participation is crucial in developing and promoting a quality-oriented environment in the workplace. However, it is not possible to have every employee participate in every project or be a member of every team. Therefore, most departments have a committee or some other type of governing structure that is responsible for coordinating quality-improvement activities. This chapter reviews the functions related to the flow of communication surrounding quality management and the forms, records, and documents that enable the work to be done and the results to be implemented.

QUALITY STEERING COMMITTEE

Most hospitals have some sort of organizationwide body to coordinate quality-related activities. This body may take the form of a quality steering committee, or quality council as it is sometimes known. It may even be a separate department designated to coordinate quality endeavors. The discussion of communication of quality-improvement activities should begin at the organizational level because it is at this level that the cultural change is usually initiated. "Before quality improvement can occur, an environment must be created that demonstrates a decisive commitment to change" (Tackett 1991, 30). The hospitalwide steering committee ideally will be the group to set the tone and establish the policies that both promote an institutional climate and determine institutional reporting channels for continuous quality improvement.

Another role of the quality steering committee is to facilitate program implementation. The steering committee may oversee the chartering process that prioritizes and selects from among the many quality-improvement opportunities, those that will be pursued by the organization. The steering committee is responsible for removing barriers and providing the resources that quality teams need to achieve success. Once a quality climate is established within the organization, the number and complexity of the types of problems being addressed will increase greatly. It is the role of the quality steering committee to track the projects, assess their progress, prevent overlap, and reward successes. Exhibit 14–1 summarizes the roles and major responsibilities of the quality steering committee.

The steering committee must have significant leverage in order to perform these broad-ranging functions. This is accomplished by appointing members from the senior executive, administrative, and medical staff, as well as from the ranks of department heads. The committee is frequently chaired by the chief executive officer and coordinated by a quality manager or director of the department of quality management (if one exists).

DEPARTMENT INTERACTIONS WITH THE QUALITY STEERING COMMITTEE

Dietitians must be familiar with the role of the quality steering committee because it has a major impact on the organization as a whole, as well as of quality practices within the nutrition services department. For example, the organization of the department's quality program may be modeled after that of the

Exhibit 14–1 Roles and Responsibilities of a Quality Steering Committee

ESTABLISH POLICY

☐ Create a vision describing the ideal future state of the organization—how the organization will look, act and compete differently—five years in the future. Link the vision to the organization's mission statement.

☐ Determine the motivation for the organization to evolve from its current state to its future state, and outline how TQM can be used successfully as a change strategy.

☐ Develop a policy deployment plan that establishes key breakthrough areas and assigns management annual objectives for enabling the organization to achieve its vision.

☐ Create a definition of quality that is understood by and applicable to the organization's stakeholders.

☐ Formulate an implementation plan that establishes priorities and sets forth internal actions, responsible parties, time frames, milestones and resources required to deploy TQM throughout the organization.

FACILITATE IMPLEMENTATION

☐ Realign management systems—including governance, management reporting, human resources, finance and accounting, information systems, planning and marketing, and quality assurance—to create incentives and support behaviors that are consistent with TQM principles, tools and techniques.

☐ Identify and eliminate systemic barriers that prevent the organization from implementing TQM and/or realizing its vision.

☐ Establish a process which governs how quality improvement teams are formed and disbanded, sets priorities for project selection, determines the boundaries for team charters, and establishes the documentation standards for team activities and results.

☐ Ensure that solutions proposed by quality improvement teams are approved and implemented promptly.

☐ Work with medical staff leadership to develop and execute an implementation plan for involving physicians and their office staffs in both clinical and operational quality improvement initiatives.

PROVIDE RESOURCES

☐ Develop guidelines for TQM curriculum including courses to be offered, course content, frequency of course offerings, and recommended and required curriculum for board members, medical staff members, managers, employees, facilitators and team leaders.

☐ Oversee development of recognition and reward programs that recognize employees for both TQM efforts (e.g., participation on a team) and results (e.g., 50 percent reduction of waiting time).

☐ Establish financial guidelines for TQM including recommended annual budget for the Quality Coordinator, consulting services, investment in training, travel expenses, budget impact of employee participation in training and on quality improvement teams, and overtime policy.

☐ Develop and implement a communication strategy for outlining TQM objectives and for publicizing TQM successes at the individual, quality improvement team, departmental and organizational level.

☐ Ensure that managers, employees, physicians and others are provided with the time and other resources required to fulfill their TQM commitments.

ASSESS PROGRESS

☐ Monitor implementation progress in accordance with the TQM implementation plan, and update the plan as required to meet organizational needs.

☐ Devise measurements and oversee development of a reporting system to track progress towards the organization's vision. Include both outcome measures (e.g., employee turnover) and process measures (e.g., number of quality improvement teams).

☐ Develop and monitor critical indicators of quality based on a systematic process for understanding customer expectations.

☐ Coordinate development of a performance management system to reward behaviors that support TQM principles and modify or eliminate behaviors that contradict TQM principles.

☐ Implement a benchmarking process for assessing organizational performance against industry best practices.

Source: © 1990 Ernst & Young.

hospital's quality-management plan. The structure of the department committee is usually outlined in the department's quality-improvement plan. A variety of structures are effective. Some departments even operate successfully without a formalized structure. Monitoring activities are incorporated into day-to-day operations as part of routine job responsibilities. Committees or teams are formed to address quality-improvement opportunities as they arise. Nutrition services departments should strive to have their efforts to improve quality complement the organization's quality philosophy, no matter which format it chooses to implement to accomplish its continuous quality-improvement endeavors.

It is certainly appropriate for the department to incorporate the four basic quality management responsibilities described by Ernst & Young (1990) into the framework for developing an individual department plan:

1. Establish policy.
2. Facilitate implementation.
3. Provide resources.
4. Assess progress.

There are several other benefits to be gained from having a working knowledge of how the hospital quality steering committee operates. Since there are ample opportunities for improving the care of patients and customers on the receiving end of nutrition services, it is likely that one or more of these projects will be chartered by the hospital. Dietitians may recommend projects for selection, serve on steering committee–sponsored teams, or act as liaisons to the quality committee. If dietitians are familiar with the players and modus operandi of the hospital committee, they will feel more comfortable and confident and be more effective in their interactions with them. The quality steering committee can serve as a valuable resource to individual department efforts to improve quality. This committee controls the physical resources that obviously are precious commodities. Members of the institutional quality committee are a great font of knowledge on continuous quality-improvement processes. It is definitely in the best interest for nutrition services departments to keep current with and maintain a good channel of communication with the quality steering committee. This access can be accomplished in a number of ways.

- Director of dietetics may serve as a quality steering committee member by virtue of being the director of a clinical department.
- Clinical manager or dietitian may serve as the department's quality committee chair and hence be the liaison to the institutional quality steering committee.
- Quality steering committee coordinator may be a member of the quality management department or a designated institutional quality committee member assigned to be the communication link with the dietary department.

Virtually all of these communication routes can be effective if they are used. Dietitians should maximize their ability to gain access to the quality steering committee and its resources to enhance the effectiveness of their own department's plans.

DEPARTMENT COMMUNICATION RESPONSIBILITIES

The key element, actually the final step in the ten-step process, relates to communication responsibilities:

> Communicate results of the monitoring and evaluation process to relevant individuals, departments, or service and to organizations' quality assurance program.

Communicate via Committee and Members

Maintaining and managing the communication flow related to quality-improvement activities is often a primary role of the department quality committee. In fact, committee members may be selected according to the job they hold within the department. If a department is targeting increasing the availability of nutrition services for outpatients in the ambulatory care center, the staff dietitian or manager of this service would be a natural choice to serve on the department committee for that year: first, of course, because of her or his expertise in the area; second, because she or he would serve as a valuable channel for information flowing to and from the committee. The dietitian would be the most logical person to

- be the point of contact with other health care professionals both within the hospital and in the ambulatory center, home care services, and other outreach initiatives
- interact with patients to gain their perspective on how nutrition services can best meet their needs
- bring salient information back to the committee
- communicate with colleagues within and outside of the department on what changes will take place and how they affect current operations.

Report to the Quality Steering Committee

The departmental quality committee is responsible for formally sharing the results of continuous quality-improvement activities with all appropriate parties. The communication plan must be flexible enough to accommodate adjustments, since the information flow will be influenced by the nature of the projects the committee is coordinating at any given time. A summary of quality-improvement activities is communicated to the hospital quality steering committee, usually via the department liaison. How this will be accomplished needs to be delineated in the overall plan. For example, the following section relating to the communication process was excerpted from the nutrition services department plan at Medical Center Hospital:

- Reports will be submitted to the Director, Hospital Quality Evaluation, on a quarterly basis. Reports are due on the 15th of the assigned month.
- The Director, Hospital Quality Evaluation, will send a feedback memo to directors within the seven days of the Quality Evaluation Committee meeting.
- If a response to that memo is necessary, the Director will provide that response by the date listed on that memo.

The committee should have similar guidelines for disseminating information to employees within the department and to those in other departments that either were involved in the quality-improvement process or will be affected by the implementation of its results.

DOCUMENTATION: DOING THE PAPERWORK

Record-Keeping Systems

A thorough record-keeping system is a crucial element in tracking quality-improvement processes. Good records are essential for a number of reasons (Scholtes 1988).

- Quality-management projects take time. They most commonly last six to eighteen months. Members may come and go. "Good records help new members catch up and keep old members informed of developments" (Scholtes 1988, 4–9).
- Clear, illustrated records help to educate and win the support of people who have neither the time nor the inclination to review lengthy, tedious, or detailed papers.
- Frequently, presentations about quality-improvement results, summaries of activities, or in-services about changes initiated as a result of quality-oriented processes are required. Having up-to-date, accurate records makes it easier to prepare for presentations.
- Good records make it easier to retrace steps and track down problems or errors.
- Records facilitate the evaluation process.
- Records of all aspects of quality initiatives are necessary to meet accreditation requirements. These documents provide the necessary evidence to assure the public and regulatory agencies that nutrition practice within the hospital is consistent with community and national standards.

Departments may actually elect to document continuous quality-improvement procedures in a manner that follows the guidelines outlined in the *Accreditation Manual for Hospitals* (Joint Commission 1992, QA 3.1.7). If so, the committee needs to document its

- findings
- conclusions
- recommendations
- actions taken
- results of actions taken

The committee must also be able to show how all of these were reported through established channels.

Since the Joint Commission (Staff, *Trustee* 1992) prohibits the rewriting of records, it is essential that records be completed correctly the first time.

Use of Forms

The great number of forms included in this manual indicate that documentation is a central element of continuous quality improvement. Because forms are widely used and distributed, it is appropriate to review briefly some suggestions regarding their use.

Forms that request specific information leave the user little opportunity for "interpretation." These types of forms for data collection, such as check-off sheets or those that incorporate a yes/no format, are among the easiest to use and tend to generate the most accurate data.

Printing forms can be expensive. It is wise to plan ahead and test a new form on a trial basis before ordering a large number of copies. During the trial period, have as many different people as possible fill out the form. Once the trial is complete, collect the forms and tabulate or summarize the information recorded. If this is easy to do, the form will most likely be a valid tool and can proceed on to be printed (Skipper 1991a).

The role played by the many forms used in carrying out quality-improvement functions is demonstrated in a simple illustration taken from the quality plan of the Department of Nutrition Service, University of Chicago Hospitals (see Appendix A-1). In the section on monitoring, 17 different forms are listed for data-collection and evaluation activities (Exhibit 14–2). Data from a multitude of sources can be compiled into a single document to simplify the analysis process. This can be accomplished by using a form such as the sample quality of care service report (Exhibit 14–3). This form can be used to tabulate the results of the information gathered on clinical indicators over a one-year period. Its use would reduce the amount of work involved in filing and summarizing data sources when preparing periodic reports and the annual summary and evaluation.

Exhibit 14–2 Methods of Monitoring Data-Collection and Evaluation Activities

---

Methods of Monitoring

Data sources include patient medical records, diet kardexes, patient menus and snack labels, patient interviews, and sanitation check lists. All supervisory and professional staff are involved in data collection. Results of data collected are compared with established thresholds, evaluated, and action taken to resolve problems. Monitoring is ongoing in order to assess improvement.

Forms used for data collection and evaluation are as follows:

- Monthly Chart Review Check List
- Height/Weight Monitoring Sheet
- Tube Feeding Monitoring Sheet
- Diet Kardex Card Monitoring sheet
- Chart Audit Worksheet (Dialysis Unit)
- Chart Audit Worksheet (Nutrition Support Service)
- RD Peer Review Monthly Assignments
- Supervisors QA Monthly Assignments
- Menu Accuracy Audit Worksheet
- Snack/Supplement Audit Worksheet
- Calorie Count Audit Worksheet
- Late Tray Audit Worksheet
- Patient Interview Form (Satisfaction Survey)
- Nutrition Counseling Review Form
- Diet/Drug Interaction Monitoring Form
- Bedside Tables Monitoring Form
- Tray Delivery Log Monitoring Form

Courtesy of University of Chicago Hospitals, Chicago, Illinois.

Exhibit 14–3 Nutrition Services Department Patient Food and Clinical Services Quality of Care/Service Report

Indicators	Jan.	Feb.	Mar.	Apr.	May	Jun.	Jul.	Aug.	Sep.	Oct.	Nov.	Dec.	Total
I. Volume Indicators													
1. Total number of meals served to patients (daily average)													
2. Percent of patients on modified diets (Average completed, 1 × Week inc. maternity)													
3. Number of nutrition consults													
A. Diabetic diets													
B. Cardiac rehabilitation patients													
C. Weight loss diets													
D. Cardiac rehab/diabetic and/or weight loss													
E. Other diet instruction													
F. Nutrition/medication interaction													
4. Number of nutrition consults involving multiple dietary modifications													
II. Quality Indicators													
1. Outcome of phone contact with patients who returned Patient Services Questionnaire with complaint													
2. Number of direct complaints from patients													
3. Number of diets ordered in nonspecific terms													
4. Number of diet orders on dietary Kardex at variance with order in medical record													
5. Number of inappropriate diets identified													
6. Additional appropriate dietary modifications identified													
7. Number of patients on below status for longer than 3 days													
A. NPO (not on supplemental tube feeding)													
B. Clear liquid diet													
C. Full liquid diet													
8. % error frequency of menu preparation													
9. Patient Satisfaction Questionnaire results (%)													
A. Nutrition Services Survey													
B. Hospital Patient Services Questionnaire													
10. Trayline food temperature below standard at time of use (%)													

Courtesy of Riverside Methodist Hospitals, Columbus, Ohio.

Meeting Agenda

Agendas for quality committee meetings need not be sophisticated to be effective. All agendas must include the topics to be covered during the meeting. An agenda might also include the following elements (Scholtes 1988):

- the presenters: either the person most familiar with or the one who originated each topic
- time guidelines: amount of time, estimated in minutes, allotted for discussion of each item
- Topic description: whether the item needs to be announced, discussed, or decided
- objectives: to help direct discussion toward achieving results (Dougherty 1992)

A sample form from Henry Ford Hospital (Exhibit 14–4) demonstrates how some of these elements can be appropriately incorporated into the agenda for a quality-improvement team meeting. Ideally, the agenda should be drafted and distributed in advance. If this is not possible, it is appropriate to outline the agenda briefly at the beginning of the meeting.

Meeting Minutes

Meeting minutes are among the most basic and essential records of data in the entire quality-management process be-

Exhibit 14–4 Quality-Improvement Team—Culinary Team Meeting #1

```
                    AGENDA

    I.   Team purpose                    (10 minutes)
    II.  Improvement process—FOCUS       (10 minutes)
    III. The team method                 (30 minutes)
         A. Brainstorming
         B. Multiple group technique
         C. Multiple voting
         D. Group consensus
         E. Flow charting
         F. Data collection
         G. Scientific method
    IV.  Role of team member             (15 minutes)
         A. Contribute ideas
         B. Accept & carry out assignments
         C. Work to improve process
         D. Stick to the agenda
         E. Help set the agenda
    V.   Assignments                     (10 minutes)
    VI.  Next meeting agenda             (5 minutes)
    VII. Meeting evaluation              (10 minutes)

           Next Meeting Monday, 2:00 P.M.
```

Courtesy of Henry Ford Hospital, Detroit, Michigan.

cause they serve a wide variety of important functions. Minutes are a tool for committee and team members because they do the following (Scholtes 1988; Winters et al. 1990):

- list points under discussion
- assist in understanding the history of a long-standing problem
- provide a record of assignments
- remind members of tasks to be performed between meetings
- educate new team or committee members
- record decisions made

Managers, committee chairs, and others who receive and review meeting minutes from quality teams (of which they may not be a member) value the information contained in minutes for other reasons (Scholtes 1988; Winters et al. 1990). For these individuals, minutes are helpful because they

- aid in understanding the issues and challenges facing the team
- are useful for tracking a team's progress
- are a source of data for performance appraisals
- serve as a historical record of achievements
- may be used in developing reports
- are valuable tools for evaluating the quality-management program

In order for minutes to fulfill all these obligations they must contain several pieces of information.

- date
- meeting times (open and close)
- attendees and guests
- topics
- main points of discussion
- assignments
- decisions made
- actions taken
- attachments (tools, forms, etc.)
- future agenda items

Use of a standard format to document meetings offers several advantages. Primarily it simplifies the work of the person assigned to record the minutes and greatly cuts back on the paperwork associated with record keeping. A standard format can be a driving force in committee operations because it provides structure and promotes consistency (Winters et al. 1990). Additionally, a format that revolves around the pertinent quality concerns helps keep the issues in the forefront and the committee focused on action plans.

A sample of a standard format is provided in Exhibit 14–5. The format is geared toward quality team meetings. It incorporates many of the essential elements described in this chapter, so that the minutes recorded on them can be utilized with maximum effectiveness.

One of the weaknesses associated with traditional quality-assurance processes is that their activities tended to be separate and distinct from the day-to-day operations of the department. In the past the person assigned responsibility for quality assurance activities may have been the only person aware of how the department fared in its quality endeavors. If the results of continuous quality-improvement monitoring activities never make it out of committee, the need for improvement may be recognized, but true advancements in quality practice will not occur.

Reports

Reports used for quality-improvement activities may be highly structured forms or free-ranging documents written in a narrative style (Lieske 1985). Regardless of the format used, the basic principles of written communication should be considered when preparing a report. Lieske (1985) lists the common elements found in most quality reports:

- statement of the topics, problems, and/or objectives
- description of relevant data
- conclusions and recommendations
- actions to be taken and responsible parties
- dates, deadlines

Exhibit 14–5 Format Geared toward Quality Team Meetings

Project Team Meeting Record, Part 1

Instructions: Use this page as a blueprint for creating a meeting record tailored to your team. You can then make copies of this form and have the scribe take notes on each topic discussed. He or she would then complete the summary items on this side at the end of the meeting.

Meeting Number _____ Date _____ Location _____

1. Project name _____

2. Mission statement:

   _____

3. ✔ To indicate "present"
   - Member _____
   - Member _____
   - Member _____
   - Member _____
   - Member _____
   - Member _____
   - Member _____
   - Member _____

   Others attending:

4. Agenda: Enter key words indicating the agenda topics. Check off an item when it is completed. Items you do not complete should be carried over to the next meeting.
   - ( ) 1. Warm-up
   - ( ) 2. Agenda review
   - ( ) 3.
   - ( ) 4.
   - ( ) 5.
   - ( ) 6. Set agenda for next meeting
   - ( ) 7. Meeting review

5. Brief summary of topics, decisions, or conclusions and next steps (on reverse side).

6. Futures file: items for future consideration but not for the next meeting.

7. Meeting Review
   "+"   "–"

Next meeting:
Date _____ Time _____ Location _____

Recorder: _____

Signature of recorder: _____

Source: Reprinted from *The Team Handbook*, by P.R. Scholtes et al., pp. 4–11, with permission of Joiner Associates, Inc., © 1988.

The information in reports is best "conveyed in a straightforward, concise and coherent style" (Lieske 1985, 134). Visuals are appropriate if they add clarity to what is being communicated and do not detract from the report's purpose. Creative use of graphics enhances interest and readability and may convert an otherwise dull and boring report into a captivating document. Since reports summarize events that have already occurred, it is appropriate to use past tense. Last, it is preferred but not essential that reports be typed. Any format may be used, however, as long as the report is neat and legible.

Three sample reports are included. The first report (Exhibit 14–6) summarizes the efforts and recommendations for provision of post-ileostomy diets. This narrative-style report is easy to read because of its subheadings and brevity. This is one type of report that might be generated by a group working on improving a process; the report is sent to the department quality committee.

The second report (see Appendix A-5) is an example of a quarterly report from Yale–New Haven Hospital. This narrative summary contains a good deal of information regarding the department's quality-related activities, yet it is easy to read because of its column divisions and outline format. Some of the information in this report is gathered from the Patient Nutrition Care Monitors Question Sheet found in Appendix D-6.

Exhibit 14–6 Recommendations for Diet Provision after Bowel Surgery

---

POST-ILEOSTOMY DIET PROVISION

For the current system to work:

1. The physician must order a "soft ileostomy" diet.
2. The unit secretary must transcribe "soft ileostomy" on the diet changes and, if calling the pantry, relay that this is the diet order.
3. The dietetic technician must enter "softileo" into the clinical nutrition system and write specific instructions on the diet changes so that the food service supervisor correctly adjusts a standard tray.
4. If the pantry has been called for a tray, the pantry worker must know what adjustments to make to an exiting tray so that it complies with the diet order.

Efforts taken to-date have included:

1. Programming the clinical nutrition system to automatically adjust trays for patients on a "soft ileostomy" diet. All vegetables, both raw and cooked, have been excluded from the diet except: cooked carrots, green beans, mashed squash and peeled potatoes.
2. Inservices provided to unit secretaries, pantry personnel and dietetic technicians.
3. Instructing patients not to eat the foods that are not allowed if they receive them on a tray.

Based on subjective feedback, the efforts taken to-date have been marginally successful in correcting the problem.

Assessment:

1. The current system is too complicated. There are too many people involved who have neither the knowledge nor motivation to correct the problem in the long-term.
2. The efforts taken to-date have been made in areas that are probably not the major source of the problem. They have focused on: adjusting the clinical nutrition system (which takes several meals to "catch up" with the patient and presumes that the correct diet order is in effect); putting the onus of responsibility on the pantry person and the patient; inservicing the primary personnel involved (turnover is high and secondary crews are the norm in these areas).

Recommendations:

1. Short-term—remove green beans and carrots from the diet parameters. Simplify initial adjustment (dietetic technician and pantry worker) to "NO VEGETABLES."
2. Track patients to determine where and how frequently the current system breaks down. Based on findings, fix the system.

Courtesy of Cleveland Clinic Foundation, Cleveland, Ohio.

This is the type of report the department quality committee might send to the hospital quality steering committee. The attachments to the report list the results for each quarter and a column for calculating annual totals. This format makes the report especially useful when completing the annual program evaluation.

The third sample (see Appendix A-6) is a year-end report from the Olin E. Teague Veterans' Center, Temple, Texas. As

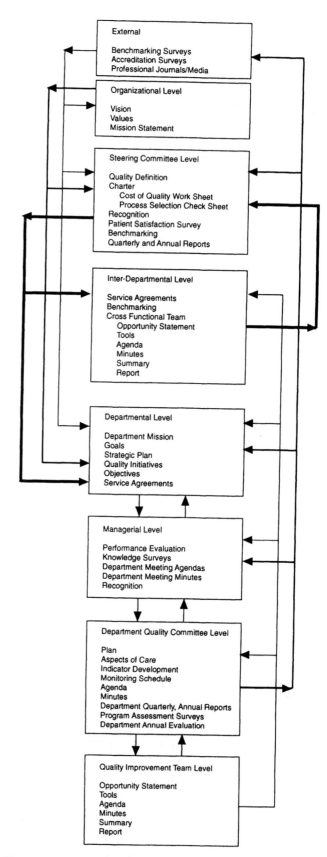

Figure 14–1 Flowchart of Quality Communications

illustrated, this report summarizes key functions, findings of quality-monitoring activities, and outcomes of steps taken to improve quality. The report also contains a brief discussion of overall quality-management effectiveness. The quality plan for the next year is based on results of the current year, illustrating continuous efforts toward quality improvement.

Quality data may also be incorporated into the departmental or service annual report (Skipper 1991b). These annual reports help document clinical care and communicate nutrition services activities to those unfamiliar with these aspects of patient care. Such reports, in addition to giving patient census data, can offer quality improvement data, analysis of findings, accomplishments, and goals for the future. In this format, annual reports can become a marketing tool by increasing visibility of clinical nutrition services, a statistical tool for documenting involvement of dietitians, and an administrative tool for tracking goals and accomplishments.

PAPER TRAIL

Recorded information is meaningless unless it gets into the hands of the people who need it. It is important not to isolate the work of project teams or shield the reports of the quality committee. Those who need to be in the quality communication loop include

- the people directly involved in the quality-improvement activities
- the people affected by its results, the department quality committee and management group
- the department's administrator or vice president
- the hospital quality steering committee

The need for quality-improvement activities to reach this number of areas and people emphasizes the important role of record keeping and communication in the quality-management process. Figure 14–1 provides a flowchart for the types of documents and their pathways in the quality-improvement process. Information can be shared outside of the organization as well. Publishing quality-improvement activities and results in professional journals and other media adds to the larger body of knowledge and serves as a resource for peers.

In traditional quality assurance, data were gathered but frequently held "close to the chest." There were many reasons for doing so. Perhaps results were disappointing and people were uncomfortable sharing them, fearful that doing so would reflect poorly on the manager's performance. People may have felt that controlling information allowed them more control overall (Leebov and Ersoz 1991). Or perhaps disseminating information was such a cumbersome process and created such a volume of paperwork that information sharing simply seemed an overwhelming or Herculean task. Considering the number of people involved in the communication loop, it is easy to understand how this could be so. Limiting the number of people with access to the data would reduce the manager's work. It was possible for quality-assurance activities to succeed (in a traditional sense) with only limited access by the staff involved in them.

Conversely, it is impossible for quality improvement to occur unless there is almost universal access to information. One effective strategy for publicizing quality-improvement activities is to incorporate a quality component into all existing department meetings. Doing so may increase the flow of information while at the same time decrease the need for generating multiple copies of multiple minutes, reports, and so on. Current quality issues can become a routine agenda item, keeping the spotlight on quality. After all, if department meetings do not focus on how well the department is doing in providing its basic services, one may question the necessity of holding the meeting in the first place.

REFERENCES

Dougherty, D.A. 1992. Ten time management tips for managers. *Hospital Food and Nutrition Focus* 8, no. 7:7–8.

Ernst & Young. 1990. New York.

Leebov, W., and C.J. Ersoz. 1991. *The health care manager's guide to continuous quality improvement.* Chicago: American Hospital Publishing, Inc.

Lieske, A.M. 1985. Reporting mechanisms. In *Quality assurance: A complete guide to effective programs,* ed. C.G. Meisenheimer, 133–155. Gaithersburg, Md: Aspen Publishers, Inc.

Scholtes, P.R. 1988. *The team handbook.* Madison, Wis: Joiner Associates, Inc.

Skipper, A. 1991a. Collecting data for clinical indicators. *Nutrition in Clinical Practice* 6:156–158.

Skipper, A. 1991b. How to prepare an annual report: A guide for nutrition support dietitians. *Supportline: A newsletter of Dietitians in Nutrition Support* 13, no. 6:5–7.

Staff. 1992. JCAHO prohibits rewriting records. *Trustee* 45, no. 7: 22.

Tackett, S.A. 1991. The quality council: A catalyst for improvement. *Journal of Quality Assurance* 13:30–36.

Winters, B., et al. 1990. *Tips, tools, and techniques. Documenting medical staff meetings.* Chicago: Care Communications, Inc.

Part IV

Managing a Nutrition Services Quality Program

Chapter 15

Involving Clinical Staff in the Quest for Excellence

From a continuous quality-improvement perspective the hospital is a theater. The actions that take place in front of the audience are a primary influence of their perception of the quality of performance. A customer perceives "an experience," not a department-by-department evaluation. Employees need to recognize the significance of their behaviors, since they are always on stage and always under scrutiny (American Hospital Association and National Society for Patient Representation and Consumer Affairs 1991). This "on-stage/off-stage" concept, utilized by the highly successful Disney theme parks, has been borrowed by a variety of companies to enhance their quality management efforts (Betts and Baum 1992).

Many roles support this production, suited to dietitians, dietetic technicians, and every other member of the clinical nutrition staff. Every employee action can influence the critic's (or in this case the patient's or client's) thumbs-up or thumbs-down rating. The theatrical analogy is not meant to imply that practicing quality requires putting on a false face or that dietitians be anything other than what they are—the nutrition experts—but rather that the clinicians recognize the significance of their roles in the spotlight, practice their professional craft, polish their performance, and understand their audience.

This chapter highlights the roles each member of the clinical nutrition staff can undertake to become a quality champion by promoting the continuous quality-improvement efforts of the department, the organization, and in personal practice.

THE CLINICAL MANAGER

A manager's primary function in the quality-management process is gaining employee commitment to quality. In this role the manager assumes the role of director, whose job it is to coax each and every performer—the stars, those in supporting roles, even the bit players—to execute their roles to the best of their abilities. Gordon (1985) outlined the steps a manager can take to promote employee commitment to quality.

Respect Employees' Needs

Although continuous quality improvement is focused on the customer, it has been said that the effective quality manager places customer needs second, because employees come first. Employees are internal customers. Their satisfaction is an essential element and resource in quality-improvement initiatives. This not only is good management practice in the customary sense, it is also central to achieving quality success. The golden rule of service is that people treat customers with the same level of attention and courtesy they get from their managers (Krupp and Klein 1991).

Organizations have designed programs to promote this caring and collegial atmosphere. For example, the "Intensive Caring" program developed by the Ohio State University Medical Center (Columbus, Ohio) identifies behaviors employees should demonstrate to each other and to patients, families, and visitors. The intensive caring value statements relate to issues of respect, dignity, diversity, and creativity (Exhibit 15–1). The statements reflect the basic ideals of the hospital's mission statement and emphasize the role of the individual as part of a caring team. Since employee commitment to quality is critical for the quality improvement program to succeed, a manager must recognize the importance of this commitment and promote an atmosphere within the depart-

Exhibit 15–1 The Ohio State University Medical Center's Intensive Caring Value Statements

---

I am part of a tradition of Intensive Caring at The Ohio State University Medical Center.

This Intensive Caring tradition means:

- I respect the dignity and diversity of the people we serve.
- I am respectful of the dignity and diversity of my colleagues.
- I value my role in the healing process.
- I strive to develop myself for myself and those we serve.
- I care about others and respect their feelings.
- I listen and respond to others.
- I am creative.
- I am committed to the tradition of Intensive Caring.
- I am part of the Intensive Caring team at The Ohio State University Medical Center.

This tradition embraces all three elements of The Ohio State University Hospitals mission: patient care, education and research.

Courtesy of The Ohio State University Medical Center, Columbus, Ohio.

---

ment that fosters the individual staff member's development in the quality process.

According to Gordon (1985), five levels characterize the hierarchy of employee commitment.

1. *Alienated*: This is the lowest level and describes the grudging employee, one who exudes a sense of resentment and reluctance. This employee's responses on a quality culture readiness survey would probably raise a red flag indicating that the individual is not nearly ready to assume ownership in quality-improvement strategies. Unfortunately, the grudging employee can be found at every level of the organization, including the ranks of the professional staff.

2. *Instrumental*: At this level an employee equates work as the sum of pay plus benefits. These employees continuously evaluate whether to stay with or leave their current positions depending on what other opportunities are available at any given time. They may be too preoccupied with this dilemma to see the big quality picture. Since nutrition departments frequently have many positions that are on the lowest end of the hospital's pay scale, these employees may well perceive their jobs as paycheck-to-paycheck propositions. Yet the employees in these positions, be they tray line workers, cooks, clerks, or tray delivery personnel may have as much, if not more, of an impact as the professional clinical staff on how patients' needs are met. A manager must make sure that all employees' needs are considered and plan a quality-improvement training program and reward system that is applicable and valued by everyone.

3. *Normative*: At this level employees consider "average" as the gold standard for performance. These employees produce no less, but certainly no more, than what is routinely expected. Quality-improvement initiatives may provide many opportunities to empower and challenge these employees to become valuable contributors in the improvement process. One manager, describing this phenomenon, mentioned that for the first time in years she had seen a particular employee excited about something, and it was a quality-oriented project. Motivating the group of normative employees has the most potential for turning the tide in a positive direction.

4. *Affective*: At this level, commitment is based on affection, respect, like, or dislike. Soldiers are much more likely to throw themselves on a grenade to protect their buddies' lives than to protect their concept of their nation. Likewise, dietitians may be much more motivated to work patients' special food requests into a modified diet to fulfill their role on the team and to earn the patients' gratitude rather than to satisfy the hospital's written mission statement. A manager should never underestimate the strength for personal gratification the motivational force has for this type of employee.

5. *Moral*: At this level, the employee identifies with the organization's mission, and this relationship enhances his or her self-respect. If the result of quality-improvement programs is that more employees reach this level, it demonstrates internalization of the values inherent in the quality quest.

Create a Quality Culture

The manager's true challenge is to make all types of employees feel that they are valuable and respected, that each is helping the department and organization reach its goals. This means developing those employees who are functioning at the alienated, instrumental, and normative levels; expanding the motivational factors for those at the affective level; and maintaining the efforts of those who operate from the moral level. Clinical managers must create a culture that encourages contributions from all employees (Staff, Harris Trust 1991). Obviously this requires implementing changes in the work environment. It is commonly believed that people will resist such a change. "People don't resist change; they resist being changed" (Scholtes 1988, 1–21). Managers can employ a variety of strategies to involve staff in the department's improvement efforts to accomplish the necessary culture change.

Drive Fear Away

Driving fear away is a pivotal point. Driving out fear may be a real challenge to clinical managers if they have been schooled over the years in a directing or controlling approach to management (Leebov and Ersoz 1991). Staff who work in fear, who are afraid to speak their minds, or who hide behind meeting the "letter of the law" in their jobs simply will not cooperate or take risks in a quality-oriented environment. "Quality managers recognize the paralyzing power of fear and take steps to build trust, open communication, and adopt a cooperative, experimental spirit with staff and coworkers" (Leebov and Ersoz 1991, 16).

Overcommunicate

To describe the role of the manager in quality improvement as a "communicator" is insufficient. Leebov and Ersoz (1991) state that for managers to be effective quality-improvement communicators they must do the following:

- *Communicate* "any and all information that affects people's motivation, performance, quickness to act, involvement, commitment, and ability to contribute value to the organization" (Leebov and Ersoz 1991, 18).
- *Listen* to their staff, suppliers, and customers.
- *Share* information regarding mission, vision, values, goals, progress, problems, trends, competition, finances, successes, and failures.

Provide Adequate Resources

There are other critical measures a manager can use to promote attention to quality. A manager must consider quality explicitly when making decisions about which services to provide. Then, the manager must ensure the availability of resources and the time needed to effect the change (Frankmann 1990; Hopkins 1990). The manager is obligated to procure equipment, software, and other assets within current budget capabilities. In this role, the manager acts as a producer by providing the backing the program needs to succeed.

Talk and Reward Quality

Talk excellence. Develop a slogan that employees can relate to. Something as simple as Ford's "Quality is job no. 1" is memorable, understandable, and effective. Supervisors who praise good performance and service "are urging a course of action that produces quality" (Gordon 1985, 31). Managers need to support the achievements made in quality service by ensuring that individual and group achievements are publicized and recognized.

One way this can be done at the individual level is to build quality expectations into each employee's performance evaluation. There are several advantages to doing this. First, this approach reinforces that quality is everyone's responsibility. Second, it provides employees with clear, written guidelines for how they can contribute to the department's quality improvement efforts. Finally, it allows the manager an automatic opportunity personally to review and discuss each individual's achievements in this area when conducting performance evaluations. Examples of quality-related performance criteria for a clinical dietitian and associate director are described in Exhibits 15–2 and 15–3.

Salary increases and bonuses are popular reward mechanisms in business and industry. But in the health care environment, these are not always feasible or readily available. They are often difficult to tie, with any immediacy, to the deserving action.

Health care managers may therefore need to be more creative in order to recognize quality achievements appropriately. The winner of an Academy Award receives a statue, yet the Oscar symbolizes the pinnacle of achievement in the motion-picture profession and is awarded with such fanfare that it bestows enormous prestige to its recipient. The clinical manager has a variety of options for recognizing quality achievements that can be accomplished with existing department resources (Exhibit 15–4). Some types of recognition, such as verbal praise, cost absolutely nothing. Yet employees often mention that a personal thank you or other appreciative comment is preferred over more formal recognition methods.

Managers can also nominate individual staff members or teams that department employees participated in to receive organizationwide recognition. Some examples of nomination forms used to recommend a person or a team are found in Exhibits 15–5 and 15–6. Note how these forms incorporate quality-improvement values into the nomination process. They ask for information on how an individual or group has met customers' needs, acted as role model(s), worked as part of a team, and reacted to unusual circumstances. Since the focus of quality improvement today is on its continuous nature, a reward system should recognize contributions to team efforts as well as outcomes. If a true quality-improvement program is in effect, objectives will change regularly. As initial goals are reached, new ones will almost immediately be set, stretching both the hospital and its employees to new levels of excellence (McLaughlin and Kaluzny 1990).

Each employee will differ on what he or she feels is important in terms of recognition. Participating in quality-monitoring and evaluation activities "may satisfy some professional staff's need for responsibility, achievement and recognition" (Bevsek and Walters 1990, 32). Managers should take the time to find out what types of rewards are most meaningful to each staff member and follow up accordingly.

The care and services offered by a department of nutrition have the potential to interface with virtually every patient and staff member. This provides dietetic professionals with the unique opportunity to make significant contributions to the quality-improvement process at every level, as individuals or as part of a team.

Exhibit 15–2 Example of Quality-Related Performance Criteria for a Clinical Dietitian

THE OHIO STATE UNIVERSITY MEDICAL CENTER

Employee __STAFF DIETITIAN__   Department __Nutrition & Dietetics__

Supervisor _____   Date _____ _____
                                      interim   final

SECTION A—CRITERIA/STANDARDS

CRITERION: __40__ % __Clinical Nutrition Practice__
             weight

RATING		SPECIFIC PERFORMANCE STANDARDS: State each standard separately.	COMMENTS: Specific examples to support rating must be given for all standards.
Interim	Final		
		1. Adheres to policy and procedures for nutrition clinical practice at 90% compliance.	
		2. Provides quality nutrition care to patients with an overall patient satisfaction rate of 90%, as measured by patient satisfaction surveys.	
		3. Completes and documents consults for nutrition evaluation/assessment, education and ongoing nutrition intervention according to established standards of practice. Appropriateness of care evaluated quarterly by senior dietitian according to standards of practice.	
		4. Effectively instructs patient and family on nutrition therapy. Evaluated by senior dietitian using patient interviews, direct observation and validated feedback from patients, family, nursing, etc.	

PERFORMANCE RATING SUBTOTAL: _____ × _____ % = _____
                             avg. rating weight   subtotal

Page ____ of ____

Courtesy of The Ohio State University Medical Center, Columbus, Ohio.

Exhibit 15–3 Example of Quality-Related Performance Criteria for an Associate Director

THE OHIO STATE UNIVERSITY MEDICAL CENTER

Employee __ASSOCIATE DIRECTOR__   Department __Nutrition & Dietetics__

Supervisor _____   Date _____ _____
                                                                              interim   final

SECTION A—CRITERIA/STANDARDS

CRITERION: __25__ % __Quality of Service__
                weight

RATING		SPECIFIC PERFORMANCE STANDARDS: State each standard separately.	COMMENTS: Specific examples to support rating must be given for all standards.
Interim	Final		
		1. Assures problem resolution within cost centers to improve customer satisfaction.	
		2. Assures that operations meet the quality, safety, and organizational standard defined by accrediting and licensing bodies as indicated by inspection findings.	
		3. Optimizes the provision of products and services to promote customer satisfaction. Participates in quality improvement efforts at the managerial level and incorporates quality improvement activities in daily operations.	
		4. Assures effective coordination of functions between and within assigned cost centers.	
		5. Promotes the Intensive Caring values in all aspects of job performance.	

PERFORMANCE RATING SUBTOTAL: _____ × _____ % = _____
                                                        avg. rating   weight   subtotal

Page ____ of ____

Courtesy of The Ohio State University Medical Center, Columbus, Ohio.

Exhibit 15-4 Department-Level Resources for Recognition

Verbal thank you

"Congratulations to us" notice

Banners recognizing staff accomplishments

Certificates of accomplishments

Submit summary to hospital employee publications

Announce achievement at department meeting

Send memo to administrator or CEO describing accomplishment

Hold informal recognition event with refreshments

Nominate staff for hospitalwide awards

Reinforce how the achievements contributes to the department's and hospital's goals and successes in conversations with employees

Encourage staff to appreciate one another

Courtesy of Medical Center Hospital, Chillicothe, Ohio.

Exhibit 15-5 Recognition Nomination Form for a Person

Nominee's Name _____ Department _____
Nominee's Position _____
Quarter For Which The Nominee Is Being Nominated 1 2 3 4
　　　　　　　　　　　　　　　　　　　　　　Please Circle
Number Of Years Employed By Medical Center Hospital _____
The Nominee Must Be An Employee For At Least One Year Before Being Eligible For This Award.

1. Addressing each of the characteristics below, please explain why the nominee should receive the award of excellence. Please include examples where possible. Please use the back of this form if more space is needed.
   - Consistently Meets or Exceeds Customer Expectations
   - Works Well As a Team Member and Promotes Teamwork
   - Creates a Positive Image for Medical Center Hospital
   - Is a Role Model for Others by Demonstrating a Belief in CQI Philosophies 100% of the time

2. In addition to what you have said above, what else makes this person deserving of the award of excellence? Please use the back of this page if more space is needed.

_____　　　_____
Nominator　　　　　　　Date

Courtesy of Medical Center Hospital, Chillicothe, Ohio.

Many dietitians have taken advantage of the myriad of improvement opportunities available to them and have earned recognition for their leadership in the quality arena. A brief sampling of areas in which dietitians have excelled follows. Individually, many dietitians have been recognized for their excellence in providing quality patient education. Another dietitian participated on a hospitalwide, cross-functional team that earned a Team Spirit Award. This team created and serviced a postprocedure area to meet the immediate recovery needs of patients undergoing outpatient medical procedures.

The entire nutrition department at MetroHealth Medical Center received its institution's first organizationwide People First Award (Jurdi-Haldeman 1992). Department director Dalal Jurdi-Haldeman described the major quality improvements that earned her and her staff this recognition. First, the department leaders addressed employee needs at three levels. They met personal needs voiced by the staff by providing lockers and an employee lounge. They met employees' training needs by providing an in-depth customer-service training module. They met recognition needs with a "bright ideas" appreciation program that rewarded team, customer service, and cost-cutting efforts.

Professional development for the clinical staff was achieved by challenging dietitians in their areas of expertise. Every dietitian has participated in and coauthored abstracts related to his or her research.

The department took care of its internal customers by revamping cafeteria and patient menus and adding a new coffee and guest meal service. It met the needs of its external customers in the community by obtaining grant funding for geriatric and pediatric nutrition programs.

This example demonstrates how quality-improvement programs are truly a "win-win" opportunity. Patient and customer satisfaction are up significantly. Staff development is an ongoing process. Revenues, across the board, in cafeteria income and professional billings have increased as well.

Present the Whole Picture

Because of the complexity of hospitalwide systems and the breadth of the scope of functions even within departments, managers must take care to set clear expectations. Employees need to know what their job responsibilities are, how their actions affect quality, and how their performance will be evaluated (McCabe 1992).

In the hospital environment, food and quality of service characteristics are positively correlated to the level of satisfaction a patient has with the hospital as a whole (DeLuco and Cremer 1990). The provision of these nutrition services is accomplished by completing a series of tasks, one after another, within the department. These tasks intermingle, cross over, and end around with those of other departments until the flow-

Exhibit 15-6 Recognition Nomination Form for a Team

Intensive Caring Team Spirit Award

This award will be presented to a group of employees (not necessarily from the same department) who have shown outstanding, unusual team achievement under special or unique circumstances or for a special project, task force, committee, or program.

Please list the names of all the team nominees

Name	Department	Phone Number

What did this group of employees do under what circumstances to be considered for this award?

_____
_____
_____
_____
_____
_____
_____
_____
_____
_____

Nominator's Name   Phone Number

Courtesy of The Ohio State University Medical Center, Columbus, Ohio.

---

charts that track their courses resemble mazes. It is often difficult for employees to view their functions even within the context of their own department, let alone within the hospital as a whole. Yet it is the creative manager's job to help them do just this. In this sense the manager acts in part as a casting director. She or he needs to find players best suited to the roles available, then take one step beyond by making sure they "read the script" to understand what scene they are in and its sequence in the plot.

One way to enable dietitians and dietetic technicians to enhance their appreciation for the organization as a whole is to allow them to lead department teams or to participate on hospitalwide committees. A clinical dietitian may chair the department quality committee. She or he may be a member of or have reporting responsibilities to the hospital quality steering committee. In performing these duties the dietitian will have ample opportunity to see how this role affects the department's functions and how the services the department provides relate to the entire hospital.

Finding a similar mechanism to broaden the scope for other department employees can be more challenging. The quality-improvement process provides many opportunities for staff at every level to participate as team members. If employees are given the opportunity to be part of a cross-functional team, it may enhance their perspective by exposing them to the issues and people who are part of departments other than their own. However, encouraging interaction among staff members across department lines need not be limited to the processes related to quality monitoring and evaluation alone. Joint department meetings and training sessions are other ways to foster interdepartment communication and familiarity. Matching employees from two different areas and having them take turns demonstrating their jobs to each other is another way. No matter which route is taken, the manager must ensure that all employees are provided opportunities to see the bigger picture, not just their scene in the play.

Be a Role Model

The manager must be a quality role model. This does not mean that the manager must be the first to arrive and the last to leave each day. Rather, if managers themselves identify strongly with their institution's mission, they will convey this intensity of commitment for their staffs to emulate (Gordon 1985).

A manager should be knowledgeable about quality-related concepts and methods and fluent in quality management jargon. Learning the language is just scratching the surface. Managers must keep in touch, both with what the customer expects and what employees encounter in trying to meet these expectations. Many companies require managers to do a stint every month on the phone lines or the front lines, just so they never lose touch (Flower 1990). Managers can do this on their own, even if no formal policy requirement is in place.

In order to grow and change, managers must have the right combination of attitude and abilities, skills, and mind-set. Jossey-Bass (Leebov and Ersoz 1991) developed a matrix of four kinds of managers and their relation to what is involved in quality management (Figure 15-1). Managers depicted in each of the four quadrants are described as follows:

> *Willing and able managers* continuously develop their skills. They have a mind-set that motivates them to sharpen and use those skills in the service of solving real problems and making continuous improvements that result in increased customer satisfaction.
>
> *Willing but unable managers* lack the ability or persistence to acquire needed skills despite consider-

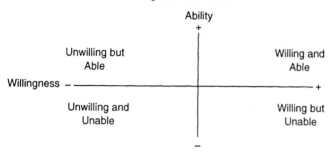

Figure 15–1 Matrix of Four Kinds of Managers. *Source:* Reprinted from *The Health Care Manager's Guide to Continuous Quality Improvement,* by W. Leebov and C.J. Ersoz, p. 19, with permission of American Hospital Publishing, © 1991.

able drive to contribute to the organization's quality improvement.

Unwilling but able managers have the skills or the potential to learn the skills they need but lack the openness to incorporate these key skills into their everyday actions. Or the culture of the organization deflates their motivation and inclination to use the skills that they have.

Unwilling and unable managers do not take steps to acquire skills they need to improve quality. Also, they resist incorporating continuous improvement as an inherent part of their management responsibilities. (Leebov and Ersoz 1991, 19)

A manager's mind-set will drive her or his actions. There are ten essential mind-set shifts that must occur for a manager to make the transition from the traditional quality assurance management style to one that embraces a continuous quality improvement philosophy:

1. From provider orientation to customer orientation
2. From tolerance and getting by to customer orientation
3. From director to coach and empowerer
4. From employee as resource to employee as customer
5. From reactive to proactive
6. From tradition and safety to experimentation and risk
7. From "busy-ness" to results
8. From turf protection to teamwork across lines
9. From we-they thinking to organizational perspective
10. From cynicism to optimism (Leebov and Ersoz 1991, 21)

Managers can evaluate their own mind-sets for quality by completing the self-assessment exercise in Exhibit 15–7. The check-off format and the simple scoring guidelines provide an

Exhibit 15–7 Self-Assessment Exercise

Self-Assessment: Do You Have the Quality Management Mind-Set?
Instructions: Next to each item, place a check in the column reflecting your answer. (Disregard the circles: they will be used for scoring your answers later.)

	Yes	No
1. Do you believe that customers have unrealistic expectations and that you can never really satisfy them?		○
2. Are you and your staff clear about how your internal and external customers define quality?	○	
3. When faced with different ways to do things, do you decide on the basis of what's best for your customers?	○	
4. Do you devote time and energy to collecting feedback from your customers?	○	
5. Are you satisfied with how your department is working as long as you don't hear complaints?		○
6. Do you often communicate a vision of excellence that constitutes a stretch from the way things are currently working?	○	
7. Do you avoid confronting staff even when their performance warrants it?		○
8. Do you communicate ambitious performance expectations to all employees?	○	
9. Do you feel more gratified when you solve problems for your employees than when they solve them without your help?		○
10. Do you feel proud of your employees when they bend the rules creatively to satisfy a customer's need?	○	
11. Do you feel insecure at the thought that your staff can function very well without you?		○
12. Do you see your role as that of coaching, providing tools, and running interference, thereby enabling your employees to serve their customers?	○	
13. Do you balk at employees' requests because you think employees should be grateful for what they already get from work?		○
14. Do you consider your employees as customers whose satisfaction is key to the success of your department?	○	
15. Do you begrudge the time it takes to nurture, recognize, and support employees?		○
16. Do you devote more quality time and energy to the retention of good staff than to the recruitment of new people?	○	
17. When problems arise, do you feel relieved after you've put out the fire and can move on to business as usual?		○
18. Do you usually take on a new project or set a new goal only in response to a request from your boss?		○
19. Are you a person known for making good things happen?	○	
20. Do you feel too busy to do anything more than handle one crisis after another?		○
21. Do you habitually search for new and better ways within your span of influence?	○	
22. Do you receive new ideas with skepticism rather than enthusiasm?		○
23. Do you deal with mistakes and frustrations as learning experiences?	○	

continues

Exhibit 15–7 Continued

		Yes	No
24.	Do you avoid experimenting with new ways for fear of repercussions?		○
25.	Do you feel impatient when problems persist over time, and do you want to proceed to solve them?	○	
26.	Do you spend a lot of time "trying" and much less time "finishing"?		○
27.	Do you track results in order to hold yourself accountable?	○	
28.	Do you overanalyze a problem instead of moving to solve it?		○
29.	Do you take the initiative to seek out other managers to help you with problems or projects?	○	
30.	Would other managers characterize you as territorial?		○
31.	Do you confront other managers whose actions or inactions have a negative impact on the effectiveness of your department?	○	
32.	Are you cynical about solving problems that cut across departmental lines?		○
33.	Do you ask your staff for suggestions and ideas about how the organization, not just your department, can be strengthened?	○	
34.	Do you withhold information from employees about the organization's status for fear it will dampen their morale?		○
35.	Do you accept responsibility for helping staff understand difficult administrative decisions so that they retain their faith in the organization's leadership?	○	
36.	Do you find yourself resenting decisions that affect your staff negatively even if you know these decisions are wise for the organization?		○
37.	Do you help others see the good in the organization?	○	
38.	Are you more likely to complain than to take action to make things better?		○
39.	Are you a positive force for change in your organization?	○	
40.	Do you see change and experimentation as threats more than as adventures?		○

Scoring: Count the number of checks inside the circles. The highest possible score is 40. The higher your score, the more you already *think* in ways that support the customer-driven process of continuous quality improvement.

Source: Adapted from *Health Care Managers in Transition: Shifting Roles and Changing Organizations,* by W. Leebov and G. Scott, with permission of Jossey-Bass, Inc., © 1990.

easy way for managers to identify their areas of strengths and potential development to become true leaders in the quality movement.

Make It Work

Recognizing and appreciating quality service is one thing. "Delivering it consistently is something else" (Krupp and Klein 1991, 9). It is ultimately a manager's responsibility to ensure that the quality program is more than just a slogan. Managers must follow up to ensure that strategies are implemented and that staff accountability is ensured. If the strategy involves a systems change it will be more cumbersome to implement. Yet staff might actually be more comfortable addressing a complex system issue because they do not view a system as personally threatening. The reverse may be true when the change involves addressing the performance-related issues of an individual within the system. The manager must act as a coach, not a critic, when working with a struggling staff member to improve performance.

The manager ensures that quality-improvement activities are done. In quality initiatives, as in any other activity, this means getting them done through people, not controlling every step of the process at the top. Continuous quality improvement is truly an ensemble performance, not a one-person show. The ultimate reward for a clinical manager's efforts is not being named the year's "best director." Rather, the manager's goal is for the cast to achieve "best picture" recognition.

Empower Employees

The term *empowerment* tends to be overused and the concept underpracticed. In some companies, "empowering" people really means "dumping" on them. "Empowerment and delegation are not the same thing" (Kennedy et al. 1992, 93). If a staff is empowered, it has the freedom and ability to fix problems as they occur. The meaning of empowerment also tends to be confused with the concept of employee equality. Employees at every level of an organization should be empowered. However, the quality actions taken by front-line workers may be very different from those at a manager's level. One of the primary responsibilities of a clinical manager is to ensure that employees are given the independence and resources they need to take care of the special situations they encounter in their day-to-day jobs.

Numerous benefits accrue when employees join the trek on a quality journey. This means using a team approach and including employees as part of quality teams. Some advantages of the team approach are the following (Moving managers 1992):

- Employees begin to think for themselves. Because they are closest to the problem, employees often know what should be done to correct a problem. They are, in fact, happy to make changes to improve quality when management empowers them to make their own decisions.
- The group works to its capacity. Quality problems are often complex and cross-functional. By including many individuals on quality teams, the group benefits from the insights and experiences of others. The group expands and becomes a team—more than the sum of its parts.

- Teamwork boosts productivity. Empowered employees respect management and give more energy and enthusiasm to their work. Managers who share power with their employees show that employees are valued; they reap the benefits of having employees involved in and excited about their work.
- Managers don't always have the solution to a problem. By using the team approach, the manager can depend on employees to think through alternative solutions and decide on the best mode of action.
- The team approach frees time for high-priority tasks. Empowered employees make many of their own decisions; the manager need not be involved in every transaction. Thus managers are free to focus on important functions such as new quality initiatives, cost-reduction strategies, or cost-benefit studies.
- Quality is a long-term program; teams give a better return on investment than time and effort spent on one employee alone. When employees are involved in the quality process, they make a personal commitment to it. Although team projects are time-consuming and sometimes cumbersome, employee commitment pays off in the long run.

It takes skill and practice for a manager to get everyone working as a team and to use personal skills and expertise to solve problems. Employee training is needed both to develop new skills and to help employees who are used to working independently function as part of a dynamic team. Employee training topics might well include such topics as interpersonal skills, giving and using feedback, and how to influence others. Managers may need to hone their own skills for coaching and group facilitation. Also, giving quality top priority in a department means that the clinical nutrition manager will need to delegate some tasks to others in order to free time for the quality journey.

Meet Accreditation Standards

The Joint Commission distinctly addresses the quality-related duties of department directors in Volume 1 of the *1993 Accreditation Manual for Hospitals* (Joint Commission 1992). In addition to providing effective leadership, the department manager is responsible for

- the integration of the department into the primary functions of the organization
- the coordination and integration of interdepartmental and intradepartmental services
- the development and implementation of policies and procedures that guide and support the provision of these services
- recommending a sufficient number of qualified and competent persons to provide care and service
- determination of the qualifications and competence of department personnel
- the continuous assessment and improvement of the quality of care and services provided
- the maintenance of quality control programs
- the orientation and continuing education of all persons in the department
- recommendations for space and other resources needed by the department

Terms that incorporate the word *quality* are specifically stated in three of the standards. But each of the manager's other responsibilities are quality related as well. Providing leadership and resources, determining manpower requirements, and ensuring the competence and development of staff are all essential to quality improvement. Incorporating quality management responsibilities directly into managerial functions will help integrate quality-improvement efforts into day-to-day operations. This is much more desirable than having quality-improvement duties exist as an entity, distinct and separate from other management functions.

Educate and Train

Staff education and training are two factors most critical to the success of a quality-improvement program. Let your staff know what is expected of them. We do this for virtually every aspect of the job from hand washing to calculating parenteral solutions. Yet it is quite uncommon for a department to provide an in-service on its quality program (Gardner 1990). Managers should commit at least as much time to disseminating the information gathered in quality-monitoring studies as is spent gathering data.

Departments have used some informal yet effective methods to keep quality issues in the forefront of daily practice. For example, staff members can rotate the responsibility to research and select a quality-related article. This article can then be circulated among staff or presented at a journal club.

Sharing the status of ongoing quality activities and results of completed projects is one of the easiest ways to advance the staff's knowledge of the quality-improvement program. When a dietitian or dietetic technician is participating as a team member, have him or her present a brief progress report at staff meetings. Encourage the staff member to explain the tools the team is constructing or analyzing. This demonstration will increase the familiarity of each staff member with the quality-improvement process.

The department should also encourage continuing education and participation in quality monitoring and evaluation at the professional level. Often the continuing education programs that dietitians elect to attend tend to concentrate around those that relate to advancing clinical knowledge and skills. By extending departmental support to attend quality-related programs, the department is investing in education that could potentially benefit the department across the board. It is important that this support be demonstrated in a substantive manner. If adequate funding is available, travel and registration expenses can be reimbursed. Other methods of financial support, such as providing paid or compensatory time off to attend seminars, are usually at the manager's discretion. This support is usually much easier to justify if there is direct potential for the department to benefit from the information provided.

Build a quality-improvement module into the plan for professional staff development. This can be accomplished by having established criteria for determining which continuing education programs being offered best fit with the department's strategic plan, goals, and quality objectives. Planning ahead will increase the chances of being able to provide financial support for the staff member selected to attend designated programs. Have the attendee provide department peers with a summary of the program's key points. This summary should specifically address how the information gained could be applied at the department level. Finally, follow up on the suggestions. Incorporate those that are appropriate into the quality-improvement process. If this final step is achieved, the department's resource investment of dollars and employee hours will be more than justified.

ROLE OF THE DIETITIAN AS TRAINER

The trainer is the acting coach. The coach needs to assist the players in performing so convincingly that everyone who observes them believes their actions to be genuine. And with adequate training, that is in fact what occurs: the players are no longer rehearsing a role, they are performing a service.

Some nutrition services departments are fortunate enough to have their own trainer on staff. Some share a trainer position with another department. Others coordinate training programs with the hospital's educational services department. Yet others have all dietitians act in the training mode and rotate responsibility according to the calendar, or they assign training responsibilities according to the individual's skill and areas of expertise.

Dress rehearsal comes before opening night. Providing training upfront is synchronous with the quality-improvement axiom "do it right the first time." In-depth orientation for new employees and provision of ongoing training and self-development programs for long-term employees are ways to guarantee the delivery of quality care to patients and clients (Harris 1989). Just as employees are involved in determining key aspects of care, their input should be solicited to determine their own training needs. A trainer can lead an employee group in a brainstorming session to come up with a list of ideas for potential training programs. For example, dietetic technicians may identify a need for more information on the diet instructions they provide. The trainer could assemble a list of these instructions and develop a knowledge survey (Exhibit 15–8) to be completed by the dietetic technicians. The results of this survey will assist in prioritizing training needs. A self-assessment component can be built into the survey, making it a valuable planning tool for individual professional staff development.

A trainer can assume a variety of roles in the quality-improvement process. Because trainers have educational expertise and may not have day-to-day operational responsibilities, they are often the persons best suited to performing the role of facilitator. Other training responsibilities may include coordinating the departmental quality committee or its peer review component. The trainer is often a key person in the "do" phase of the FOCUS-PDCA cycle (see Chapter 13). This is especially true if the plan of action relates to a knowledge issue or requires an operations change. The trainer can coordinate or conduct the in-services required or arrange the retraining sessions to accommodate the revised employee job duties.

According to Katz and Green (1992), the individual who serves as trainer and coordinator of the departmental quality committee must possess 12 critical competencies:

1. proficiency in the scientific approach
2. currency in accreditation requirements and processes
3. computer literacy
4. critical thinking and problem-solving abilities
5. strong negotiation skills
6. sales and marketing skills
7. assertiveness
8. ability to organize information, people, and resources
9. ability to work collaboratively with and through others
10. strong writing skills
11. strong presentation skills
12. team-building skills

DIETITIANS AND DIETETIC TECHNICIANS AS STAR PERFORMERS

Interviews with registered dietitians and dietetic technicians demonstrate that these professionals are truly committed to the quality process. They have many suggestions on how to improve quality in their areas of practice. Which quality tasks are performed by which group of employees depends on how the roles and responsibilities between dietitians and dietetic technicians are divided. Roles vary from place to place and are a function of an institution's type, size, and other characteristics. The following ideas for dietitians and dietetic technicians will benefit patients, employees, and other customers, as well as promote quality practice in the nutrition profession.

- *Support excellence in practice with sensitivity in interactions.* Patients tend to correlate technical skills with in-

Exhibit 15–8 Knowledge Survey Form, Nutrition and Dietetics: Diet Instruction Knowledge Survey

Skill Levels:
1 = In-depth knowledge
2 = Above average knowledge
3 = Meets job knowledge criteria
4 = Training level
5 = No knowledge of task
6 = Desires to learn task
7 = Should learn task

Expectations and Requirements for Dietetic Technicians

Knowledge Level: 1 2 3 4 5 6

General Information — **Knows How To:**
1. Select & prepare the written materials in advance.
2. Introduce her/himself and explain the general purpose of the diet to the patient.

Low Simple Sugar — **Knows How To Choose:**
1. Foods to avoid (including "traps" such as fruit, yogurt, fructose, alcohol, etc.).
2. Serving sizes according to ADA guidelines.
3. Alternatives to sugar.

4 gram Sodium — **Knows How To Explain:**
1. Label reading to determine sodium content.
2. Alternatives to sodium and can provide examples.
3. Foods to avoid (including "traps" such as Oriental foods and sauces).
4. OTC medications which may contain sodium.

Blenderized Diet for Wired-Jaw Surgery — **Knows How To:**
1. Judge appropriate consistency of food and how to liquify foods.
2. Find choices to meet calorie and protein needs including a variety of foods and sources for additional calories.
3. Add seasoning for flavor.
4. Accommodate/adjust for egg/milk allergies.
5. Explain how items are strained from foods such as seeds and fibers.

Source: Adapted from Harris, R.D., Returning to the Basics, *Hospital Topics*, Vol. 67, p. 22, Heldref Publications, 1989, and from Standards of Practice, Department of Nutrition and Dietetics, Ohio State University Medical Center, Columbus, Ohio.

terpersonal skills (Press et al. 1992). The courtesy and friendliness a dietitian conveys during a counseling session may positively influence a patient's judgment of clinical skills.

- *Maintain visibility in rounds,* in the medical records, during nonroutine working hours, and especially with patients. Results of one survey showed that although 82 percent of consumers thought dietitians were helpful and informative, only slightly over 50 percent felt that they were visible to patients during their hospital stays (DeLuco and Cremer 1990). Clearly this is an area where dietitians have the opportunity to improve patient perceptions.
- *Double-check diet orders,* especially those that pose problems in meeting patients' food choices.
- *Take the extra step* to incorporate a patient's special request or make the necessary diet adjustment so that the very first meal received is satisfactory.
- *Make meal rounds.* This was suggested for both dietitians and dietetic technicians, not necessarily on a daily basis, but at least occasionally. Making meal rounds for patients who are eating is just as important as making medical rounds for patients receiving parenteral and enteral nutrition. This sets an excellent example for other staff. It also provides the dietitian the opportunity to observe firsthand how patients are served. Often the dieti-

tian is called in to trouble-shoot when a problem arises and will be better prepared to intervene if she or he is familiar with meal service procedures.
- Enable the dietitian to *accept verbal orders* for nutrition-related services. This will help overcome the obstacles that an inappropriate diet order poses for quality nutrition care.
- *Identify your department by name* each time you introduce yourself or answer a phone.
- *Turn problems,* dilemmas, and topics of professional debate or disagreement *into quality-improvement opportunities* or research proposals.
- *Ask patients what their goals are* for today's counseling session. Be willing to forgo the ones you have set to help them achieve theirs.
- *Follow up, follow up, follow up!* Check to see that your nutrition recommendations are implemented. A dietitian's obligation to the patient only begins when he or she records the nutrition care plan in the medical record. His or her true responsibility is to see that it is achieved.
- *Promote the role of the dietitian and dietetic technician as the nutrition experts* by educating fellow employees on nutrition topics. If feasible, discount nutrition services to fellow staff members or provide complimentary wellness programs. Publish nutrition-related public service announcements in the employee newsletter or utilize other in-house communications to share nutrition information.
- *Support the department's quality-improvement efforts.*

Excellence in clinical practice can be achieved only with full dietitian support. One of the hallmarks of traditional quality assurance programs was the peer review process. Clinical managers recognized how awkward staff felt with the process. When a dietitian performed the review function she or he was concerned about encountering a problem when evaluating a colleague. The dietitian was then faced with the dilemma of either potentially alienating the peer or compromising personal standards of professionalism. Dietitians whose performances were being reviewed expressed concern that their professional judgment was being questioned or that their confidentiality could be violated if a performance-related issue surfaced. Some institutions now prefer to describe the peer review process by another name, such as medical record review or rotating review, to de-emphasize the employee-to-employee performance evaluation.

Another manager described how, when it was audit time, the staff would attempt to justify why the peer review process should not be conducted now. Typical comments were as follows:

> There were too many people on vacation last month. (Translation: We're concerned the results might not look good.)

> There are too many people on vacation this month. (Translation: We don't have enough time to complete the audit because we're covering extra patient care units.)

The transition from traditional quality assurance to continuous quality improvement requires a change in focus from individual performance to systems improvement. If dietitians recognize this shift and act accordingly it should remove some of the discomfort associated with the peer review aspect of the program.

How Can One Person Improve Quality?

Every member of the department at every level is key to the quality-improvement process. Each employee can make use of the following tips (Leebov 1991) to leave a unique mark on the quality quest.

- *Do it right the first time.* Correcting errors frustrates the customer, costs money, and wastes time. Every error or job left half-done is a quality problem. Managers do not desire to look over their employees' shoulders constantly, and dietitians don't welcome this scrutiny. By taking ownership and responsibility for maintaining the quality of their own work, dietitians promote autonomy along with efficiency.
- *Confront poor quality when you see it.* Don't walk past a problem, error, or example of shoddy service without comment or action. Offer your help. Instead of letting poor quality happen, step in and do what you can to make things right. When you notice a problem but say or do nothing, you're part of the problem. When you speak up or act, you're part of the solution.
- *Pursue quality excellence in yourself.* Stretch, don't settle. When you stop actively improving yourself and your work, quality doesn't stay the same, it slips. Some people do just enough to get by on the job, but that "getting by" attitude threatens quality improvement. Adopt a "good is never enough" attitude. In words and actions, reach for increasingly higher standards for yourself and those you influence. Expand your skills so that you can be more effective in making quality happen. Stretch your own capabilities and contributions by choosing continuing education opportunities to maximize quality potential.
- *Seize opportunities to get involved.* You will have increasing opportunities to become involved in making quality improvements. You can make a difference, but only if you assert yourself, make suggestions, and join quality teams. You do have something valuable to contribute. Draw on your firsthand experience with your clients and patients. People who seize opportunities to become involved find it stimulating and revitalizing. It's a

chance to break the monotony in the job routine, get to know other people, see and be part of the "bigger picture," and make your mark in your organization.

- *Look for solutions.* Negative activities such as blaming and griping have proved to be among the most debilitating obstacles to positive change. They drain energy from the forces working for change. From each person's vantage point, there are problems and frustrations that seem to go on day after day. It's easy to become cynical or skeptical about the possibilities for change. Perhaps your hesitation is based on valid observations of past failures. Even so, try to suspend your disbelief and give change a chance. It is better to offer your support so that you can use your potential influence positively. Some veteran employees who have "seen it all" find it refreshing to pair up occasionally with a new employee. In exchange for mentoring the new hire, the experienced staff member may benefit from working with a person who doesn't yet know what "can't be done."

- *Appreciate quality when you see it.* Don't take quality for granted. Don't say something only when things go wrong. Appreciation focuses attention on what's happening right. It takes energy to keep quality high. Without reinforcement, people have trouble maintaining energy for quality work. This should not be limited to the supervisor-employee relationship. Rather, compliment your colleagues, your boss, and other team members. In traditional quality assurance peer review programs, corrective actions were taken only when someone's performance resulted in deviations from standards. The continuous quality-improvement philosophy allows peer recognition when things are done right. It provides the opportunity for an individual to become a positive force who reinforces quality in others.

A simple way to promote collegial appreciation is a "quality alert" program. Note cards or fliers with the quality mission statement or slogan are convenient to each work area. When an employee notices a fellow worker performing a quality service, he or she can jot down what has been seen and send it to the colleague. Encourage all employees to use the cards. Customarily, supervisors take the responsibility to recognize employees for a job well done. Quality alerts can be used to accomplish this, but they also make it convenient for employees to recognize each other and their managers. This type of program reinforces the fact that quality improvement is everyone's responsibility.

- *Become a quality advocate.* "Dare to go public with your commitment to quality" (Leebov 1991, 36). Change, even change for the better, can be tiring. It truly takes commitment. Be willing to express your commitment and say "I'm glad this is happening; I believe this will work!"

Quality management places the challenges and rewards of assessment and improvement squarely in the domain of each organization, and hence its people. The fact that quality improvement fills the organization's need to meet external accreditation requirements is incidental to the central effort (American Hospital Association 1991). Managers, trainers, staff dietitians, and other staff members who use the strategies cited in this chapter will personally step forward and become a reflection of quality and a full-fledged contributor in the organization. Not only will patients, clients, and organization benefit, but individuals will no doubt be energized by both personal involvement and increased value to the organization.

REFERENCES

American Hospital Association and National Society for Patient Representation and Consumer Affairs. 1991. *Defining, measuring and meeting customer requirements: The meaning of quality in a CQI environment.* Teleconference. Chicago.

Betts, P.J., and N. Baum. 1992. Borrowing the Disney magic. *Healthcare Forum Journal* 35, no. 1:61–63.

Bevsek, S.A., and J.H. Walters. 1990. Motivating and sustaining commitment to quality assurance. *Journal of Nursing Quality Assurance* 4, no. 2:28–36.

DeLuco, D., and M. Cremer. 1990. Consumers' perceptions of hospital food and dietary services. *Journal of the American Dietetic Association* 90:1711–1715.

Flower, J. 1990. Managing quality. *Healthcare Forum Journal* 33, no. 5: 64–68.

Frankmann, C. 1990. Agenda for change: JCAHO's focus on quality. *Clinical Nutrition Management* 9, no. 4:1–4.

Gardner, S. 1990. Quality assurance: Keys to success in implementing your plan. *Clinical Nutrition Management Newsletter* 9, no. 4:4–5.

Gordon, W.I. 1985. Gaining employee commitment to quality. *Supervisory Management* 30, no. 11: 31–33.

Harris, R.D. 1989. Returning to the basics: The bottom line to quality improvement. *Hospital Topics* 67, no. 2:19–23.

Hopkins, J.L. 1990. Evaluating the QM program. *QRC Advisor* 7, no. 1:1–9.

Joint Commission on Accreditation of Healthcare Organizations. 1992. *1993 Joint Commission Accreditation Manual for Hospitals.* Oakbrook Terrace, Ill.

Jurdi-Haldeman, D. 1992. *The innovator.* Metrohealth Medical Center, Ohio: National Society for Healthcare Foodservice Management.

Katz, J., and E. Green. 1992. *Managing quality: A guide to monitoring and evaluating nursing services.* St. Louis, Mo: Mosby–Year Book, Inc.

Kennedy, M., et al. 1992. A roundtable discussion: Hospital leaders discuss QI implementation issues. *Quality Review Bulletin* 18, no. 3:78–96.

Krupp, S., and M. Klein. 1991. Enthusiastic leadership: Building a service culture. *Common Goals*, no. 9. Warrendale, Pa: Hospital Shared Services.

Leebov, W. 1991. *The quality quest: A briefing for health care professionals.* Chicago: American Hospital Publishing, Inc.

Leebov, W., and C.J. Ersoz. 1991. *The health care manager's guide to continuous quality improvements.* Chicago: American Hospital Publishing, Inc.

McCabe, W.J. 1992. Total quality management in a hospital. *Quality Review Bulletin* 18, no. 4:134–140.

McLaughlin, C.P., and A.D. Kaluzny. 1990. Total quality management in health: Making it work. *Healthcare Management Review* 15:7–14.

Moving managers and employees toward the team approach. 1992. *Briefings on CQI* 2, no. 4:1,8.

Press, I., et al. 1992. Patient satisfaction: Where does it fit in the quality picture? *Trustee* 45:8–10, 21.

Scholtes, P.R. 1988. *The team handbook.* Madison, Wis: Joiner Associates, Inc.

Staff. 1991. *The total quality management story.* Roundtable discussion with John J. Hudiburg and a select group of business leaders. Chicago: Harris Trust and Savings Bank.

Chapter 16

Managing Change To Achieve Quality Improvement

Quality improvement calls implicitly for changes in both individual behavior and organizational functioning. Every aspect of health care brings together three interacting components: practitioner, organization, and patient (Jessee 1982). These three components are the basis for quality indicators in the realms of process, structure, and outcomes. Health care comprises the dynamic interactions among the process of care (preventive, curative, or palliative) provided by the practitioner; the patient's own attitudes, behaviors, and health status; and such organizational variables as equipment, personnel, policies, procedures, and standards (Jessee 1982). When monitoring reveals a variance between standards and performance, change may be required in any one or all three components of health care.

Burke and co-workers (1991) measured managers' knowledge of how to bring about organizational change. Analysis of results revealed "an alarming lack of knowledge of these issues." For example, only 61 percent of managers understood the complexity of individual response to change, 68 percent knew general information about the nature of change, and 69 percent knew how to manage personnel through the change process. The highest score received (79 percent) related to managing the organization side of change, perhaps the least-threatening aspect of engineering change and innovation.

This chapter is designed to help dietetic professionals better understand the change process, enhancing their ability to plan and implement successfully the changes needed for quality improvement. Topics include the framework for change, the general nature of change, individual response to change, types of problems encountered in quality management and how to address them, and using practice-based research as a beginning point for changing dietetic practice.

FRAMEWORK FOR CHANGE

In his book *Teaching the Elephant to Dance: Empowering Change in Your Organization,* Belasco (1990) asserted that an organization can "learn to break its chain to the past" but that many small steps, subsumed into four major categories, must be taken to create change.

1. Create a vision. This implies translating quality into specific actions, as noted in the important aspects of care and quality indicators. In dietetics, new visions might encompass the redesign of care delivery systems to be compatible with patient-focused care strategies or the use of clinical pathways for cross-functional diagnoses.

2. Establish the future. Strategies must be determined and resources made available to reach the desired goals. For example, from the start promote dietitians as integral members of cross-functional teams and involve dietitians in planning key processes for care delivery.

3. Prepare for change. Such things as attitudes, policies, procedures, standards, job descriptions, and work groups must be altered in anticipation of new approaches to the work of an organization. Also, barriers to change must be envisioned and minimized to reduce resistance among employees.

4. Empower change. Work with both individuals and the organization to both motivate and empower personnel at all levels to help bring to reality a shared vision of the future.

Dietitians, in carrying out their responsibility for high-quality nutrition services, need to create a positive environment and develop a personal quality vision. Then dietetic staff members can be empowered to make changes necessary to strive for quality improvement.

GENERAL NATURE OF CHANGE

Quality management in health care involves three processes: (1) monitoring, (2) assessment, and (3) improvement or change. Of these three, the least well understood and most poorly implemented is the change process (Kaye 1983). For change to occur, individuals must give up current behaviors, try new ways of doing things, and continue to perform their duties in the newly adopted manner. These three elements of change, first described by Kurt Lewin (1951), are commonly referred to as "unfreezing," "changing," and "refreezing."

Unfreezing

Unfreezing usually requires an attitude change. To accomplish this, personnel engage in a sometimes lengthy process in which they move from total disregard for innovation to acceptance of it. This happens through a series of steps, recounted by Schiller et al. (1991), in which the individual

- seeks information about the innovation
- examines the personal dimensions of the change
- notes that the change cannot be avoided
- recognizes how the change affects daily activities
- collaborates with others in similar circumstances
- refocuses attention positively toward implementing the change, and even improving on it.

Any change in behavior or attitudes tends to be resisted because embracing change requires a person to acknowledge that previous performance or attitudes were wrong or inadequate, a conclusion repugnant to most individuals, especially those working in the health care field. If change is to occur, then, it must be preceded by an awareness that the current system that supports present behavior and attitudes may be in some way deficient. For example, much of dietetic practice is based on custom or trial and error. In some cases, however, research utilization can bring a higher degree of excellence to patient care.

Unfreezing can be accomplished by making announcements or holding meetings to discuss with staff current research findings and how practice might be changed and, as a result, quality of care improved (Goode and Bulechek 1992).

It is also important to meet with staff to solicit their ideas on how the use of research or other innovations could improve their work. Launching a campaign in a departmental newsletter or on a staff bulletin board can also help to disseminate knowledge about research utilization and unfreeze the situation.

Quality care may also be improved by reduction in cycle time. For example, can patients choose today's selective menu today? Can screening be completed at the time of patient admission, rather than 24 or 48 hours later? Can nutrition care documentation be integrated with progress notes? Thawing of these and similar knotty issues can be initiated by posing "what if" questions to sensitize staff members to potential valuable changes in the status quo.

Change

Change, the dynamic transition between unfreezing and refreezing, requires seeking out, processing, and using information for the purpose of integrating within oneself new values, perceptions, attributes, attitudes, and behaviors. In quality management, the change process requires taking initiative to improve or update a satisfactory situation to remain out front in quality service. It may also mean pinpointing root causes of problem areas to be changed, cooperation of key staff in defining a desirable new course of action, and, finally, acceptance by those involved in improvement activities. Kaye (1983) summarized five modes of intervention for effecting change.

1. Try intervention into the physical structures upon which behavior depends, such as clinical departments, the dietetic office, or organization of the unit. For example, if better use of a dietitian's time is desired, it may be necessary to transfer the dietitian to a different unit, change to a consultation-based service, and/or reassign some tasks to a dietetic technician.

2. Try intervention into the ongoing function of the individual. For example, if a dietitian consistently makes mistakes in calculating nutrient needs or glosses over nutritional problems, additional training may be needed before performance will improve. Also, a journal club may help to alert dietitians to innovative ideas and new opportunities for achieving quality care.

3. Use verbal interventions. For conscientious personnel, all that may be required is pointing out to them that their performance or productivity is below standard. This in itself may lead to the desired change. Others may need further discipline, such as counseling, verbal warning, written warning, or more serious intervention.

4. Use tactics designed to change overt patterns of behavior. It is sometimes desirable to upgrade employee friendliness or sensitivity toward patient expectations in an effort to achieve desired levels of patient satisfac-

tion. In this case, public relations training may be needed to uproot sullen or hostile demeanor and replace them with greater openness, affability, and cooperation.

5. Modify or control the situational circumstances that surround the person and that, in turn, govern the occurrence of his or her behavior. Sometimes, productivity levels are not met because dietitians are slow in doing calculations or because students demand an inordinate amount of their time. The purchase of hand-held computers will expedite calculations; changes in productivity standards may be needed when staff members regularly work with students or new house staff.

The challenge to dietitians, quality assurance coordinators, and supervisors is to identify creatively which strategies will work best to bring about change in any particular situation. Qualities, characteristics, and motivations of the work force, as well as organizational attributes, will determine the best intervention method to use in a given circumstance.

Refreezing

Attention needs to be exerted to sustain change once it has been implemented. This period of "refreezing" is fostered by sound management, close supervision, regular monitoring to encourage continued adherence to new behaviors and procedures, and follow-up to be sure the desired change is sustained over time. Since managers generally have more difficulty dealing with managing the "people side of change" than with the "organization side of change" (Burke et al. 1991), clinical managers are encouraged to give priority to personnel matters in attempting to improve quality performance.

Group motivation is necessary to sustain quality improvements (Bevsek and Walters 1990). As described by these authors, three strategies can be used to help create an environment conducive to motivation and change.

1. Provide relevant and meaningful information to all employees. Such information should include trends and innovations, the basics of quality care, how quality is integral to each individual's job, and the relationship of quality practice to job performance.
2. Provide support during all phases of planning and carrying out quality-related activities. This may involve helping to select quality problems of keen interest to staff members, guiding the collection and organization of data, using techniques to include appropriate staff members in the process, focusing on quality improvement as a source of job satisfaction, offering feedback as a way to monitor and guide the quality process, setting priorities to allow time for conducting quality-management activities, giving rewards and affirmation for quality improvements, and reducing environmental barriers that impede quality services.
3. Provide sufficient resources to achieve quality goals. Resources may include adequate clerical staff, computer support, educational materials and tools, support for attendance at pertinent conferences and seminars, and outside consultation when necessary.

THE HEALTH CARE ENVIRONMENT

It is helpful to view health care as a conceptual model as a way of viewing the interconnectedness of various facets subject to change. Such an approach invites consideration of the whole, not just one of its parts.

Such a model, as described by Batalden (1991) and shown in Figure 16–1, incorporates the practitioner, the locus of care and services, the social context of health care, and the individual patient or client. Professionals—with their personal attributes, competence, technical skills, motivation, knowledge, and interpersonal relationships—are at the heart of the health care process. These professionals provide services within a particular institution or agency that provides equipment and materials, support personnel, and those policies and procedures needed to standardize the operation. The social context creates the milieu for service delivery, including social policy, accreditation and other regulations, fiscal and economic policy, and professional licensing and development opportunities. The patient enters this maze of interacting forces, receives appropriate care and services, and returns to society with his or her needs being met.

Change in any one aspect of this model is likely to affect other components of the system. It is also apparent that change in the social context or locus of care delivery can have a major impact on patient care even without modifications in the professional's behavior, knowledge, or competence. Systematic application of change within this model offers a significant agenda for change as organizational leaders seek to increase the value and quality of services provided within the context of total health. While the social context contributes important elements that affect care, the players closest to quality-improvement programs are the patient, the professional or other support personnel, and the hospital or organization systems through which services are delivered.

Health care reform and the move toward managed care will have a major impact on dietetic practice. New fiscal policies may modify reimbursement practices for nutrition services. With greater emphasis placed on disease prevention, dietitians are challenged to develop innovative counseling and education programs for children as well as adults. Professional roles may expand to include fitness and smoking cessation (Finn 1993). Enhanced leadership and creative thinking skills may facilitate integration of the challenges presented by health care reform.

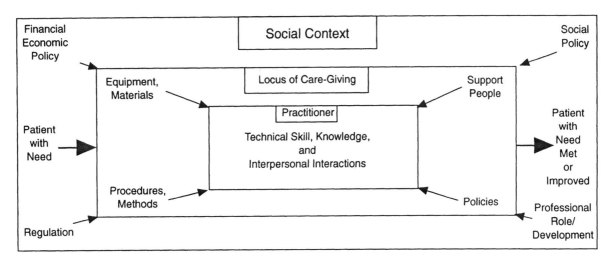

Figure 16–1 Health Care Delivery Model: A Context for Change. *Source:* Reprinted from Batalden, P.B., Organizationwide Quality Improvement in Health Care, *Topics in Health Record Management*, Vol. II, No. 3, pp. 1–12, Aspen Publishers, Inc., © 1991.

BARRIERS TO CHANGE IN HEALTH CARE INSTITUTIONS

As indicated earlier, change in health care settings requires an adjustment in at least one area of the health care triad: the health care practitioner, the patient, or the organizational structure. Luke and Boss (1983) offered ten barriers to change as an integral component of quality improvement.

1. Health care practitioners, by the very nature of their professionalism, expect autonomy and functional independence. Thus their utmost cooperation is required before any substantial change will occur, particularly with regard to personal behaviors and attitudes.
2. Health care institutions structure behavior through adherence to standards of care, standards of practice, procedural manuals, accreditation mandates, and the like. A collective stability becomes entrenched as everyone within the institution subscribes to the expected behaviors. This collective force creates an enormous barrier; unfreezing entrenched behaviors is no small feat.
3. Individuals resist change for personal reasons. Among these might be perceived loss of power, influence, or status; self-appointed responsibility to "protect" quality when proposed changes are viewed as a "threat" to quality; unwillingness to put forth the psychic energy needed to engage in changed behavior; and inability to see the big picture and to think creatively about how to engage oneself in improving it.
4. Hospitals foster desired and even "programmed" behaviors through such things as divisions of labor, recruitment and selection practices, reward structures, training programs, and activities to build cohesiveness among staff members. It is difficult for personnel, particularly professional staff members, to set aside these symbolic expressions of organizational preference. Development of patient-focused care systems demands tearing down departmental walls and working without boundaries for the patient's well-being (Brider 1992).
5. Health professionals tend to focus attention on the tasks that they perform, sometimes concentrating so intently that all else seems unimportant in comparison. Thus any attempt to improve overall quality performance in a unit or within the institution may appear as an intrusion. Even attempts to enhance the work environment, such as installing computers for clinical use, are seen as disruptive. Such individuals are well-intentioned; they simply have tunnel vision that causes them to lose sight of the bigger picture.
6. Scarce resources limit the options available to enhance quality care. In some instances, health care professionals would gladly take on new modes of behavior, but decreased resources and increased regulation deter real change.
7. Previous investments in certain high-cost systems, equipment, or training programs limit the hospital's ability to abandon major assets in favor of other alternatives, even if the new systems are more conducive to quality care. For example, assume that patient satisfaction scores for dietary services are low and that the hospital has a cook-chill patient food system. The plan to improve low scores is not likely to include replacement of the system. Capital costs are too high, even if such a change might improve satisfaction scores.
8. Over time the accumulation of statutes, rules, regulations, policies, criteria, specifications, professional

skills, and the like serve as behavioral constraints. Willingness to adhere to institutional norms can, in the long run, be a barrier to change. Herein lies a primary deterrent to cross-functional team care or "departments without walls." Health care professionals are strongly socialized to departmental cultures and hospital mores; old traditions are not easily replaced.

9. Influential health care professionals (e.g., physicians) often determine, albeit informally, which changes will be successfully implemented. Through support or opposition to proposed changes, physicians or other informal leaders can wield their power over quality-improvement initiatives. For example, clinical dietitians in some hospitals document in progress notes, increasing the likelihood that physicians will notice and use nutrition information. If one or more influential physicians oppose the adoption of this practice when proposed at other hospitals, its use is doomed to failure.

10. Interinstitutional arrangements and contracts may deter change. For example, if a hospital uses shared purchasing or if a single menu is used throughout a hospital system, food specifications or menus for one institution cannot be modified. Arrangements that lower costs and streamline operations can, on the other hand, inhibit alternatives for quality improvement.

RESISTANCE TO CHANGE

Resistance to change is often considered a major deterrent. Oromaner (1985) enumerated several factors that can motivate lack of cooperation on the part of an individual:

- fear of losing control over one's position or activities
- threatened changes in informal social relationships
- impending loss of status or esteem
- inconvenience required by imposed changes
- personal perception that the decision is "wrong" and that there would be a better way to approach the problem.

Bushy (1991) noted that "resistance to change increases in direct response to pressure to change" but that resistance can be decreased when change is "reinforced by trusted others, such as high-prestige figures, those whose judgment is respected, and those of like mind." Although employees may have a tendency not to cooperate when they are pressured to improve performance, often they can be influenced to come around when key leaders in the department are both interested in the innovation and committed to making it happen. Thus nutrition care managers need to reinforce continually their vision of quality care; support the efforts of supervisors, individual employees, and the quality committee; and participate in activities designed to improve or reward quality performance.

Other general strategies can be used within the department to reduce resistance to change, such as the following:

- Involve those affected in planning the change.
- Keep everyone informed, not only about what will happen but why.
- Conduct a trial run of new procedures to work out the bugs before wholesale implementation of an innovation.
- Protect social relationships and status as much as possible.
- Allow personnel to vent their opinions, even negative ones, as they work through the change-acceptance process.
- Remain open to new ideas and be ready to modify planned change if the alternative fails to meet expectations.

Jessee (1982) offered seven specific rules for promoting change needed to improve health care performance and outcomes.

1. Be sure of the facts. Know why a problem exists and deal with the underlying problem before expecting personnel to demonstrate improvements.
2. Take appropriate action. Avoid actions that may produce hostility or defensiveness, such as writing a letter to point out deficiencies and mandate that performance comply with established standards. Rather, use a personal approach to maintain human dignity, using data to point out deficiencies, and together seek ways to address the problem.
3. State the problem specifically. Be sure those involved have a clear understanding of the opportunity or problem, its magnitude, how it affects quality, and the consequences of failing to address it.
4. State the desired solution. Goals, stated in operational, behavioral, or quantitative terms, need to be established so that employees know what is expected of them. If resources are unavailable, or circumstances are such that a desired solution is impossible, it should not be pursued until its accomplishment is feasible.
5. Be prepared to deal with resistance. Not everyone will be immediately willing to change. What are the consequences of continuing present behavior or level of service? If there is not a clearly negative impact on patient care, risk, or resource utilization, it is usually not worth the effort, in both emotional and human terms, to insist on change.
6. Do not force the issue. Exerting one's authority or personal power only increases resistance to change. It is better to use a logical approach by pointing out facts and consequences of failure to meet established norms

and standards. Most employees and health care professionals are eager to perform well; they will go along with programs designed to upgrade both individual and group performance when they see the advantages of changing their behaviors.

7. Use concurrent monitoring to ensure that new behaviors are maintained. Continuous quality improvement strategies incorporate this concept; they are based on monitoring present activity rather than retrospective analysis. However, when changes are made, adequate time should be given to integrate the change into performance before a subsequent audit of the indicator is conducted.

IMPROVING PERFORMANCE AND OUTCOMES

Quality improvement may be the result of improved performance on the part of health professionals or other hospital employees, better patient compliance, or institution of organizational changes that facilitate delivery of improved patient care.

Improving Performance of Health Care Practitioners

Once resistance is overcome, several methods can be used to achieve behavioral change among health care personnel. Donabedian (1991) offered six options.

1. Direct intervention: Use administrative authority to halt or redirect a practice. For example, a hospital management team may introduce a new policy requiring clinical dietitians to bring in sufficient revenue to support their salaries. Such a policy would create changes in what activities would be performed and how dietitians would spend their time. The policy may also lead to improved quality care by giving priority to those patients who would benefit most from nutritional services.

2. Reminders: Find effective but simple ways to reduce inattentiveness created by stress or complexity of tasks performed. Standard forms for nutritional assessment, planning specialized nutrition support, a pocket-size manual of standards, and the like reduce the likelihood that important points will be omitted. Such written documents serve as a reminder for expected levels of performance.

3. Feedback: Give information on monitoring and survey results. It is assumed that when objective comparisons are made between personal performance and group norms, health professionals will be motivated to bring their performance in line with the rest of the group. Accordingly, when the work of one clinical dietitian causes departmental performance to vary significantly from thresholds, one way of handling the problem is to share with the offending dietitian raw data showing how individual noncompliance affects overall audit results. Such feedback will be most effective if it is provided face to face, by a respected colleague, and based on comparison with peers.

4. Education and training: Offer new knowledge and opportunity to develop skills when ignorance and ineptitude lead to deficient care. About 15 percent of all quality deficiencies are attributable to a lack of education or knowledge (Patterson 1989). When a knowledge deficit is identified, general techniques for improving practice, such as those suggested by Elrod (1991) and outlined in Exhibit 16–1, may be used to correct the problem. These general methods may be insufficient to ensure that a particular individual has learned those things necessary to correct performance inadequacies. Eisenberg (cited in Donabedian 1991) noted that, at least in the case of physicians, educational programs will be more effective if

- the practitioner accepts personal inadequacies and the need to learn new techniques
- the educational content is specific and geared directly toward the identified deficiency
- the training is conducted face to face
- one-on-one training opportunities are provided

Exhibit 16–1 Possible Actions To Solve or Improve Knowledge Deficiencies

- Attend workshop/seminar
- Revise orientation information
- Provide mentors/facilitators
- Provide an in-service
- Attend a course or class
- Provide continuing education
- Establish mutually set goals
- Develop policy, procedure, standards, or protocols
- Confer with staff to determine the best method to develop knowledge/skill
- Offer 1:1 training/education
- Arrange for "expert" guest speaker
- Provide information from books or articles
- Stress expectations or consequences
- Provide opportunities to practice deficient skills
- Develop self-learning packets
- Use audiovisual material
- Utilize skill inventories

Source: Reprinted from Elrod, M.E.B., Quality Assurance: Challenges and Dilemmas in Acute Care Medical-Surgical Environments, in *Monitoring and Evaluation in Nursing*, by P. Schroeder, ed., pp. 27–56, Aspen Publishers, Inc., © 1991.

- the educational program is conducted by someone the practitioner trusts and respects.

5. Incentives and disincentives: Use rewards or punishments to entice the desired behaviors. To be effective, this technique requires that the incentives and disincentives be directly linked with desired changes in behavior. In theory this may sound good, but in practice it is difficult to achieve and the practice may backfire, creating even greater problems.

6. Using consumers to influence practitioners: Let patients play a role in quality improvement by using their expectations to help shape standards. Patient food preferences, attitudes toward nutrition counseling, and desired nutrition services can be used to pressure changes in menus, food availability, counseling schedules, and the like.

McConnell (1993) cautions that corrective discipline is an inappropriate strategy for dealing with poor performance. Discipline is rightly used only for problems of conduct. Those who exhibit substandard work have not "broken the rules." They have, however, failed to meet expected levels of performance.

Employees who deliver poor-quality work need to learn how to perform properly. As described by McConnell (1993), the supervisor should do the following:

- Review job standards with the employee. Be sure the task is clearly understood. Coach the employee to make a commitment to meet expected standards.
- With the employee, develop an action plan for meeting quality standards. Include a timetable. Document such conferences, including desired outcomes. Give the employee a copy of the completed plan.
- Conscientiously monitor employee performance. Be sure the employee makes regular progress as planned. Give additional guidance and assistance as needed.
- Remove any apparent deterrents to employee success when possible. Help the employee to gain confidence; give affirmation and help build self-esteem.

Exhibit 16–2 illustrates some general techniques for improving performance of both individuals and groups (Elrod 1991). Often a change in the work environment will enhance the outcomes of individual efforts toward work improvement.

Improving Patient Compliance

Quality indicators, particularly outcome measures, can be enhanced when patients respond positively to therapeutic or preventive regimens. Patient compliance can be improved by using techniques such as the following (Jessee 1982):

- Alter patients' perceptions and beliefs. Patients have come to recognize the value of low-fat/low-cholesterol diets in reducing risk of heart disease. Also, diabetic patients who recognize the integral links among diet, exercise, and insulin have better control over their diabetes. Those convinced of the need to exercise and make sustained lifestyle changes are more successful in maintaining weight control than those who yo-yo diet without altering other aspects of behavior. Often patients' compliance is directly related to their health beliefs and how they respond to both internal and external cues to action. When practitioners understand the factors that enhance patient compliance, such as motivation, health attitudes, access to care, economic status, previous treatment, and social pressures, they can develop individualized treatment plans with which the patient is more likely to comply.

- Use behavior modification techniques. However, it is incumbent on professionals to work closely with patients when designing goals, appropriate behaviors, contracts, and the like to be sure these are consistent with the patient's values and interests.

- Use principles of adult learning and effective counseling to facilitate the change process. Time spent in nutrition counseling will meet standards of care only if desired patient outcomes are achieved: weight loss, reduced blood lipids, controlled blood glucose, management of renal disease, and so on. Use of certain techniques is known to increase patient compliance, such as establishing a behavioral plan, identifying appropriate behavioral strategies, and making the patient accountable for follow-up and compliance (Roach et al. 1992).

Exhibit 16–2 Possible Actions To Solve or Improve Performance Deficiencies

- Encourage people to develop potential
- Communicate expectations and goals for unit and individuals
- Coach or counsel those who do not meet expectations
- Observe performance personally
- Provide prompt positive/negative feedback
- Find ways (rewards) to strengthen desired behavior
- Communicate priorities
- Remove rewards for poor performance
- Use negative consequences for nonperformance
- Remove any obstacles causing nonperformance
- Arrange some time off for personal problems, if indicated
- Establish employee recognition programs
- Role model expected performance
- Maintain open and honest communication with staff at all times
- Post positive letters from patients on bulletin boards
- Transfer or terminate
- Develop detailed job descriptions
- Suggest counseling by a professional for personal problems

Source: Reprinted from Elrod, M.E.B., Quality Assurance: Challenges and Dilemmas in Acute Care Medical-Surgical Environments, in *Monitoring and Evaluation in Nursing*, by P. Schroeder, ed., pp. 27–56, Aspen Publishers, Inc., © 1991.

Improving Organizational Functioning

Some of the factors associated with organizational functioning are individual diversity, group dynamics, the interdependency of human and technological factors, and the levels of authority that constitute any given organization. As noted by Jessee (1982), hospitals are particularly resistant to organizational change because within the health care setting goals are diverse, authority is diffuse, professionals have independent roles, and there are few performance measures. Despite these factors, several techniques, such as the ones shown in Exhibit 16–3, can be used to foster improved performance through changing systems (Elrod 1991).

Another way to bring about change to achieve quality improvement is through the use of conflict resolution (Pastor 1992). One of the greatest sources of conflict in quality improvement initiatives is differing opinions and expectations on how quality can and should be addressed within different professional groups. Such conflicts are most likely when cross-functional quality teams are involved. When potential conflicts are recognized and resolved in early stages of working together, full-blown antagonisms can be prevented. Dealing with conflict requires use of communication, listening, and negotiating skills. Following are the steps of dealing with conflict as outlined by Pastor (1992).

1. Prepare. Take time quietly to identify your own thoughts, feelings, attitudes, biases, and feelings regarding the situation.
2. Determine the problem. In dialogue, open the lines of communication to explore how others perceive the situation. Be sincere, open, and frank. Paraphrase and offer feedback in honest conversation. Try to identify the true problem. Since this phase of conflict resolution is often difficult, many hospitals engage outside facilitators to help professionals clarify, share, and agree on problems standing in the way of improved patient care.
3. Seek common agreement. Focus on areas of agreement while delineating only the major areas of disagreement. Since most health professionals are truly interested in high-quality patient services, progress can be made by emphasizing areas of common agreement. Sources of disagreement are usually minor; these can be openly addressed and resolved once the numerous areas of agreement are recognized.
4. Take responsibility. At this point, each party needs to acknowledge how she or he has contributed to the problem. Consider, for example, a cross-functional team on a surgical unit in which one indicator of care is the patient's ability to describe planned home care, including diet restrictions. Dietitians may routinely neglect to document evidence of patient competence, and nurses may be lax about communicating discharge plans. Members of each group need to assume responsibility for how they contributed to the problem and to express, without being personal, how the conduct of others impedes quality care.
5. Resolve the conflict. Having re-established openness and trust, each party can make suggestions as to how the conflict can be resolved. Brainstorming is an effective technique at this point because it involves everyone, allows piggybacking of ideas, offers numerous alternatives, and diffuses tension by focusing on new issues and constructive ideas. The discussion should end with selection of a solution and agreement on specific actions to rectify the problem.

Donabedian (1991) suggested two other techniques for change within organizational systems: fostering change in individual performance and making it easier to comply with institutional standards and expectations. Professionals and others are more likely to work toward improved quality care if organizations do the following:

- Facilitate change by releasing time, providing equipment available, or supporting new behaviors. For example, nutrition care managers can reduce the resistance of dietitians to conduct nutritional assessments by employing dietetic technicians to assume many routine tasks ordinarily performed by dietitians. Resistance can also be reduced by the purchase of hand-held programmed calculators or personal computers to facilitate nutritional assessments and calorie counts.
- Routinize and guide required activities. For example, both nutrition screening and assessments can be facilitated by the use of checklists and printed forms, including labels for use in medical record charting (Fatzinger et al. 1992). The use of established protocols and guidelines

Exhibit 16–3 Possible Actions To Solve or Improve System Deficiencies

- Redesign delivery systems
- Create/revise forms
- Purchase/lease equipment
- Adjust inventories/supplies
- Formulate/revise policy or procedure
- Adjust staffing levels or ratios
- Seek services of a consultant/expert
- Revise job descriptions
- Redesign organizational structure
- Alter communication systems
- Eliminate unnecessary policies or other hindrances
- Computerize systems

Source: Reprinted from Elrod, M.E.B., Quality Assurance: Challenges and Dilemmas in Acute Care Medical-Surgical Environments, in *Monitoring and Evaluation in Nursing*, by P. Schroeder, ed., pp. 27–56, Aspen Publishers, Inc., © 1991.

also aids the completion of many patient care procedures. For example, development of a formulary will make it easier for dietitians to make recommendations for appropriate enteral products and supplements to be used in specific circumstances by reducing the number of options while still meeting the needs of patients (Schiller et al. 1991).

RESEARCH UTILIZATION AS A BASIS FOR CHANGE

In their efforts to improve quality, professionals may ask certain questions (Batalden 1991).

- Who are the customers of what I do? Who is the immediate recipient of what I produce? Who ultimately benefits from my services?
- What are the major processes that I work in to produce given outcomes for my customers? Are the outcomes those expected?
- What are the steps involved in those processes? How do these processes vary from one customer or one professional to another?
- How might I improve those processes to reduce waste and needless complexity? What could be done to improve results or outcomes?
- Who else does things similar to what I do? Do they do it better? What methods do they utilize? Are they different from mine? Do they have different outcomes? Do they attract different customers? Why?

Practice-based research evolves when dietitians consider questions such as these. A cursory review of professional journals reflects a variety of studies whose results could be used to change dietetic practice for the better. To continually prepare themselves for change, every dietitian should regularly read professional journals (Vickery and Cotugna 1992). Many departments conduct nutrition rounds, hold journal clubs, and circulate readings to increase dietitians' awareness of current research and its potential applications.

Not every dietitian will engage in research, but every dietitian should participate in research utilization. In addition, the clinical nutrition manager must help the department create a climate for basing practice on scientific knowledge. Research utilization requires individual involvement and both depart-

---

Exhibit 16-4 Comparison of Individual and Departmental Approaches to Research Utilization

Clinical Problem: Threat of losing clinical dietitian positions because of budget shortfalls; dietitians spend most of their time engaged in activities that could be performed less expensively by dietetic technicians or other support personnel. A dietitian reads a research study that shows how one department increased revenue when registered dietitians charged patients for legitimate nutritional services.

Individual Process	**Organizational Process**
Reads research article.	Reads research article.
Discusses article with peers in the department.	Discusses article with clinical nutrition manager or quality improvement coordinator.
Approaches clinical nutrition manager about initiating changes in the department to modify dietitians' job descriptions and establishing a reimbursement system for dietetic services.	Committee proceeds with research utilization and asks staff dietitian to participate.
Believes job descriptions of dietitians should be changed. Unsure about how to develop reimbursement system.	Research utilization activities involve (1) gathering studies related to roles of dietitians and reimbursement practices; (2) critiquing articles; (3) establishing a research base; (4) translating findings into a new research-based procedure for role delineation and reimbursement procedures; (5) providing in-services for dietitians, medical record administrators, physicians, accountants, nurses, and administrators to explain new procedure and reason for practice change; (6) implementing new procedures; and (7) evaluating the outcome.
Assumes responsibility and accountability for implementation and for evaluating outcomes.	

Source: Adapted from Goode, C. and Bulechek, G.M., Research Utilization: An Organizational Process that Enhances Quality of Care, *Journal of Nursing Care Quality*, Special Report, pp. 27–35, Aspen Publishers, Inc., © 1992.

mental commitment and organizational change. The interest and commitment of individual dietitians alone are not sufficient. The change process requires the structuring of mechanisms to implement research findings and to provide resources necessary to maintain these mechanisms over time (Crane and Horsley 1983). Specifically, mechanisms and resources, as outlined by Crane and Horsley (1983), must be provided to ensure the following:

- There is a systematic way to identify patient care problems that need new or more effective solutions.
- Relevant scientific knowledge is identified, assessed, and selected for use in solving patient care problems.
- Scientifically sound practice innovations are adapted from the research base and designed to meet the particular requirements of the hospital or service agency.
- The practice innovation is initially introduced on a trial basis, which includes an evaluation of its effectiveness in solving patient care problems.
- A reasoned decision is made to adopt, modify, or reject the innovation based in part on data generated during the trial period.
- Adopted innovations are diffused to other units as appropriate.
- The quality of the practice innovation is maintained over time.

Those desiring to use research utilization as a basis for organizational change may refer to Crane and Horsley (1983) for specific suggestions on how to integrate these mechanisms with a quality-improvement program.

Research utilization is an organizational responsibility and is best accomplished when there is organizational commitment to scientifically based practice (Goode and Bulechek 1992). Each dietitian has an important role to play, but research utilization is too overwhelming for one individual to carry the burden alone. The clinical dietetic manager, quality-improvement coordinator, or other individual should be designated to provide departmental leadership for the process. Individual dietitians and dietetic technicians can then contribute their time and expertise in an integrated fashion. A comparison of individual and departmental roles is shown in Exhibit 16–4.

Many but not all quality improvements will evolve from monitoring and evaluating care. Only those elements monitored at a particular time will provide an assessment of performance and opportunities to compare performance with established standards. Constant vigilance and openness to new research and innovations in dietetic practice, as well as commitment to research utilization, are essential if the department is to remain current and, in fact, offer services of the highest quality.

REFERENCES

Batalden, P.B. 1991. Organizationwide quality improvement in health care. *Topics in Health Record Management* 11, no. 3:1–12.

Belasco, J.A. 1990. *Teaching the elephant to dance: Empowering change in your organization.* New York: Crown Publishers, Inc.

Bevsek, S.A., and J.A. Walters. 1990. Motivating and sustaining commitment to quality assurance. *Journal of Nursing Quality Assurance* 4, no. 2:28–36.

Brider, P. 1992. The move to patient focused care. *American Journal of Nursing* 92, no. 9:26–33.

Burke, W.W., et al. 1991. Managers get a "C" in managing change. *Training and Development* 45, no. 5:87–90.

Bushy, A. 1991. Changing practice: An overview. In *Monitoring and evaluation in nursing,* ed. P. Schroeder, 253–270. Gaithersburg, Md.: Aspen Publishers, Inc.

Crane, J., and J. Horsley. 1983. Research innovations in nursing: Implications for quality assurance programs. In *Organization and change in health care quality assurance,* eds. R.D. Luke, et al., 171–182. Gaithersburg, Md: Aspen Publishers, Inc.

Donabedian, A. 1991. Reflections on the effectiveness of quality assurance. In *Striving for quality in health care: An inquiry into policy and practice,* eds. R.H. Palmer, et al., 59–128. Ann Arbor, Mich: Health Administration Press.

Elrod, M.E.B. 1991. Quality assurance: Challenges and dilemmas in acute care medical-surgical environments. In *Monitoring and evaluation in nursing,* ed. P. Schroeder, 27–56. Gaithersburg, Md: Aspen Publishers, Inc.

Fatzinger, P., et al. 1992. Development and use of preprinted forms and adhesive labels in medical record charting. *Journal of the American Dietetic Association* 92: 982–985.

Finn, S.C. 1993. Nutrition services give managed care organizations an edge in managed competition. *Journal of the American Dietetic Association* 93:533–534.

Goode, C., and G.M. Bulechek. 1992. Research utilization: An organizational process that enhances quality of care. *Journal of Nursing Care Quality* Special Report:27–35.

Jessee, W.F. 1982. Achieving improved health care performance and outcomes. In *Teaching quality assurance and cost containment in health care,* ed. J.W. Williamson and Associates, 165–181. San Francisco: Jossey-Bass Inc., Publishers.

Kaye, R. 1983. Quality assurance—A strategy for planned change. In *Organization and change in health care quality assurance,* eds. R.D. Luke, et al., 157–169. Gaithersburg, Md: Aspen Publishers, Inc.

Lewin, K. 1951. *Field theory in social science.* New York: Harper & Row.

Luke, R.D., and R.W. Boss. 1981. Barriers limiting implementation of quality assurance programs. In *Organization and change in health care quality assurance,* eds. R.D. Luke, et al., 147–155, Gaithersburg, Md: Aspen Publishers, Inc.

McConnell, C.R. 1993. Behavior improvement: A two-track program for the correction of employee problems. *Health Care Supervisor* 11, no. 3:70–80.

Oromaner, D.S. 1985. Winning employee cooperation for change. *Supervisory Management* 30, no. 12:18–23.

Pastor, J. 1992. Managing change and resolving conflict for efficient implementation of TQM. *QRC Advisor* 8, no. 6:1–4.

Patterson, C. 1989. Standards of patient care: The Joint Commission focus on nursing quality assurance. *Nursing Clinics of North America* 23, no. 3:625–638.

Roach, R.R., et al. 1992. Improving dietitians' teaching skills. *Journal of the American Dietetic Association* 92, no. 12:1466–1473.

Schiller, M.R., et al. 1991. *Handbook for clinical nutrition services management*. Gaithersburg, Md: Aspen Publishers, Inc.

Vickery, C.E., and N. Cotugna. 1992. Journal reading habits of dietitians. *Journal of the American Dietetic Association* 92, no. 12:1510–1512.

Chapter 17

Patient-Focused and Cross-Functional Teams

Traditionally, hospitals have been organized around functional areas or disciplines such as nursing, pharmacy, dietetics, and laboratory, and they have employed a hierarchical decision-making model in which issues are dealt with by channeling them up through carefully structured lines of authority. At the end of the line, decisions are made by those at higher levels of management. Ultimately this approach leads to poor quality care because managers may not be in touch with root problems, they are under time constraints and pressures, and they tend to focus on individual departments rather than collaboration and cooperation (Fargason and Haddock 1992).

Some startling statistics surfaced as analysts took a closer look at departmentalized hospitals (Borzo 1992).

- Only 16 percent of hospital services were spent on delivering patient care at Bishop Clarkson Memorial Hospital, Omaha, Nebraska. A whopping 29 percent was spent on documenting work, 18 percent on waiting or downtime, and 14 percent on scheduling and coordinating care.
- A stroke patient travels an average of 47 miles from service to service, department to department, during a seven-day stay at Mercy Hospital in San Diego, California.
- In a typical hospital, clerical employees outnumber patients. Most large hospitals have upwards of 300 different job classifications. Most job titles have two to four incumbents, but for over half of these only one person holds the title.
- Most patients receive services from 40 to 50 different employees from nearly 100 different departments during a three-day hospitalization.

These factoids highlight the astonishing fragmentation caused by compartmentalization and departmentalization. Many hospitals are taking a closer look at how they deliver care. Some institutions are completely overhauling care by implementing patient-focused systems. Others are taking a less dramatic approach such as the use of cross-functional teams.

The Joint Commission's *1994 Accreditation Manual for Hospitals* (1993) gives a sharper focus to its Agenda for Change by redirecting quality assessment toward "processes" as well as departments. This new focus of quality care targets patients and what happens to them. Major sections in the new manual are (A) patient care: patient rights, admission, patient evaluation, nutritional care, treatments and procedures, patient and family education, and coordination of care; (B) organizational functions: leadership; human resources, information, and environmental management; and quality assessment and improvement; (C) structural components: governing body, management and administration, medical staff, and nursing; and (D) department-specific requirements.

This shift in content connotes that quality is not the sole responsibility of any one department or person. Quality is essentially customer-driven, and in health care it is the patient and family, rather than professionals and their various services, whose treatment and welfare are the targets of quality care improvement.

The new focus is on a multidisciplinary and key process approach to care. New accreditation manuals de-emphasize departmental activities and advocate a restructuring of services around self-contained units staffed by multidisciplinary caregivers who can provide much of what patients need in their rooms (Borzo 1992). Dietitians must ensure that nutrition services remain an integral component of patient-focused care.

Patients and their families are not the only customers in health care, however. Other customers may include the facility staff, physicians, discharge planners, referral agencies, federal and state surveyors, volunteers, and the community (Tishman 1992). Thus most hospitals will continue to have departmental quality committees to assess and continually improve non-patient-related outcomes, processes, and structures.

The focus of this chapter is interdisciplinary cross-functional patient care teams. Content includes a summary of approaches to delivering and evaluating patient-centered care; definition, roles, function, and suggestions for establishing cross-functional teams; case management; several examples of cross-functional teams; factors contributing to the success of teams; pitfalls to be avoided; and some particular concerns of dietitians.

EVALUATING PATIENT-CENTERED CARE

During the mid-1980s, various authors recognized that patient care evaluation could be improved through a patient-centered approach to quality management. At the time, three different types of quality-assurance audits were commonly used, and in some cases these models are still followed (Krause et al. 1984).

1. Departmental audits are audits in which the entire quality assessment and improvement process is conducted independently by each department. Krause et al. (1984) assert that this type of quality program is "a waste of paper, lacks completeness, intimates a basic distrust of the fellow professions, and is time consuming on the part of the Governing Body, the Executive Committee and the Hospital Administrator."
2. Parallel audits are monodisciplinary studies performed simultaneously on the same topic by two or more disciplines. The same patient population may be used, but the criteria are developed and applied independently. When these audits are conducted by medical and nursing personnel, as they usually are, allied health professions are often lost in the middle or—worse yet—omitted entirely from consideration.
3. Multidisciplinary audits include at least one representative from pertinent disciplines to evaluate the care of a specific group of patients. This streamlined approach to the evaluation of patient care provides a composite set of criteria. The multidisciplinary team works together to write criteria, make judgments, and recommend actions (Mosleth 1984). Each discipline, however, reviews only its own discipline, makes its own judgments, and recommends its own actions, except in areas of shared responsibility.

Patient-centered evaluations go one step further. All involved share information regarding the planning and implementation of patient care. Each discipline defines its activities, delineates its responsibilities, and states its expectations. The end result is mutual problem solving. Not only does the patient benefit through improved care, but patient-centered audits limit duplication of effort and apply the team approach to delivery of health care (Krause et al. 1984).

Each of these "audits of care" approaches, although a step in the right direction, left something to be desired. The audits of care failed to "identify the basic organization wide problems that interfered with the provision of high quality care" (Carroll 1992, iii). Addressing this deficiency, true interdisciplinary team delivery and evaluation of quality patient care is the ultimate objective of cross-functional teams. Although nutrition is not addressed as an issue, Laughon et al. (1993) described the transition from unit-based quality assurance to multidisciplinary quality improvement in a coronary care unit.

While some hospitals are implementing cross-functional teams, a few hospitals are experimenting with patient-focused teams and case management approaches to care. These new initiatives suggest a dynamic future for health care delivery. They should also alert clinical nutrition managers to the need for involvement in such new initiatives to ensure that high-quality nutritional care is incorporated into new models of patient care.

PATIENT-FOCUSED CARE

When a hospital converts completely to patient-focused care, the delivery system is radically transformed. Key components of this system (Borzo 1992) are as follows:

- cross-training of staff and creation of multidisciplinary care teams
- decentralization of services such as nursing, pharmacy, X-ray, and housekeeping, and movement of them to patient floors whenever feasible
- use of critical pathways and computerized exception-reporting medical records.

At Mercy Hospital in San Diego, California, a 523-bed, full-service, not-for-profit hospital, the decision was made to convert the entire facility to patient-focused care (Brider 1992). In this new arrangement only four health care worker roles are recognized.

- clinical partner—a registered nurse, pharmacist, medical technologist, dietitian, respiratory therapist, or other licensed professional; responsible for assessing and planning patient care within each discipline

- technical partner—a licensed vocational nurse, certified nursing assistant, dietetic technician, pharmacy technician, or physical therapy technician; works under the supervision of a clinical partner
- service partner—a unit inventory specialist who orders supplies, stocks and cleans patient rooms, and provides patient transportation and other services
- administrative partner—a job combining the functions of unit receptionist, secretary, admitting clerk, medical record coder, and financial analyst.

All of these positions report to a single unit director rather than to their previous department heads.

Patient-focused caregivers provide the full range of services, including nursing care, routine ancillary procedures, patient support, and patient-related clerical and administrative services.

Hospitals that switched to patient-focused care report some dramatic results (Brider 1992). For example, there are fewer incident reports. Average length of stay is down. Patients receive care from fewer individuals; instead of 50 caregivers, the patient now deals with only 13. The average turnaround time for test results has dropped from 157 to 48 minutes.

St. Vincent Hospitals and Health Services, Indianapolis, Indiana, reported other outcomes from a trial patient-focused care unit (Farris 1993). Patient satisfaction increased. Physicians thought nursing care was much better on the trial unit. Staff members were pleased with their increased responsibility, increased efficiency, and greater involvement in problem solving. Quality care indicators showed improved patient outcomes. Costs were down 10 percent to 18 percent. A care team spent 60.2 percent of its time on direct patient care, as opposed to the 46 percent nurses provided in the previous system of care. Also, each patient received an average of 4.7 hours of direct care per day, whereas the traditional unit provided only 2.8 hours of direct care per day. It is safe to assume that, with these positive results, other hospitals will move quickly to implement similar systems.

The Joint Commission is watching patient-focused care systems to ensure that quality of care is maintained (Brider 1992). The goals of patient-focused care cannot target only efficiency. Quality care must also be maintained; measures within structures, outcomes, and processes need to show that quality standards are upheld. For example, accreditation standards require that professional services be provided by licensed individuals as prescribed by state laws. Nursing diagnoses, nursing care prescriptions, and nursing interventions are restricted to registered nurses. In states where dietitians are licensed, nutrition assessments and counseling may be carried out only by licensed dietitians. Thus, as cross-training of professionals occurs, the scope of practice for licensed individuals must be protected.

CASE MANAGEMENT

Another interdisciplinary approach to quality improvement is case management. Case management may be a process for monitoring services or for limiting their volume. It can also denote an entire continuum of services, including interdisciplinary planning and delivery of care in terms of defined outcomes. Strictly speaking, "case management is a model of patient care delivery which restructures and streamlines the clinical production process so that it is outcome based" (Giuliano and Poirier 1991). Managed care is also "a method of organizing care delivery that emphasizes communication and coordination of care among health care team members" (Mosher et al. 1992).

St. Peter's Medical Center, New Brunswick, New Jersey, developed a case management approach to patient care. Goals of this innovative approach, as delineated by Manco et al. (1992), were to

- improve patient perceptions of quality
- improve selected quality indicators including patient education and discharge planning
- reduce lengths of stay
- assure appropriate use of resources
- reduce costs of care
- increase patient satisfaction
- promote provider satisfaction with case management
- introduce a high-touch remedy into a high-tech environment.

In this system, care managers oversee patient care and monitor resource usage (Manco et al. 1992). The case managers act as facilitators, maintaining a multidisciplinary focus. They concentrate on process analysis and systems improvements. They continually challenge all care providers to "examine and reexamine quality and resource use issues."

Critical pathways are established to ensure that care processes are standardized and result in a defined outcome. A *critical pathway* is a map used to track a patient's hospitalization. Critical pathway maps are developed for high-volume diagnosis-related groups (DRGs). They describe which interventions should take place on any given day of hospitalization to achieve the standard outcomes within the given DRG length of stay. Cross-functional teams are convened to develop the pathways. Usually the pathway is set up so that it can be integrated with an institutional database for tracking and monitoring and for quality management.

Appendix A–8 shows an example of a critical pathway developed for DRG 106, coronary artery bypass graft. The critical pathway is accompanied by a quality-monitoring work sheet delineating key monitors and expected outcomes. The work sheet allows tracking of variations due to delivery systems, patients, or practitioners. Types of variances can be tracked. The clinical path implementation and reporting flowchart, also shown in Appendix A–8, provides for problem identification, analysis, and action planning. When variances occur, the quality team analyzes the need for improvements in the path, the system, or caregiver performance.

Patients play a key role in progressing through a critical pathway. Guidelines may be provided for patients so that they know what to expect and how to participate in their own care during each day of hospitalization (Mosher et al. 1992). For example, patient daily guidelines may describe dietary preparation for surgery, diet for day of surgery, and progression to regular or modified diet.

DEFINITION AND ROLES OF CROSS-FUNCTIONAL TEAMS

Cross-functional teams are interdepartmental quality work groups. They are an integral part of a total quality-management system in hospitals. They can accompany any system of care: patient-centered, case management, or the conventional departmental approach. Such teams address problems, evaluate progress, and work toward meeting customer expectations. As noted by Tishman (1992), cross-functional teams are "empowered to identify and prevent problems, determine new procedures and processes, evaluate their progress, and report to a quality council." Typically, such teams include representatives of disciplines involved in the care of a designated group of patients. Such teams are not "limited to supervisors representing various functions or to subordinates who report to a common supervisor; rather they cross traditional hierarchical lines of authority" (Fargason and Haddock 1992).

To illustrate, the cross-functional team at Memorial Health Alliance, a long-term care facility, consists of the assistant administrator, director of nurses, assistant director of nurses, medical director, a physician, a social worker, and staff nurse (Tishman 1992). Given the complex nature of hospitals, each will have several cross-functional teams, one for each major patient population, such as cardiology, surgery, oncology, and so on (McEachern et al. 1992). Regardless of the nature and number of teams, they are developed and implemented in steps similar to those shown in Figure 17–1. This team approach is not as radical as the patient-focused teams described earlier. However, the teams are truly interdisciplinary, and each team member works with the group to plan, assess, deliver, evaluate, and improve care processes.

The typical nutrition support team consisting of one or more physicians, dietitians, nurses, and pharmacists would be considered a cross-functional team. It is not, however, a patient-focused team, since the work of a nutrition support team is limited to nutrition. As dietitians encounter innovation and restructuring in health care organizations, they will need to adapt roles, procedures, and processes to remain vital partners in health services delivery.

STARTING A CROSS-FUNCTIONAL PROBLEM-SOLVING TEAM

The Joint Commission mandate to initiate cross-functional teams will facilitate their establishment. Ordinarily, it is one individual who provides the impetus for starting a team. McEachern et al. (1992) describe three different approaches used at one hospital to develop direct patient care teams.

1. Early on, before interdisciplinary teams were common practice, any individual with a keen personal interest in cross-functional collaboration would approach the quality-improvement consultant for assistance in setting up a team. Often the initiator was a team leader who wanted to learn more about work processes in order to make them function better.

2. As the idea of cross-functional teams caught on, hospital administration encouraged caregivers to organize teams around body systems or major functional processes. A Pareto chart (see Chapter 8) was prepared showing the hospital's top 25 DRGs by volume. These were analyzed and grouped where possible. Twelve cross-functional teams were developed, each responsible for a high-volume group of patients: obstetrics, neonate, psychosis, rehabilitation, diabetes, back pain, gastrointestinal, pulmonary, cardiovascular, gynecological, connective tissue disease, and central nervous system. These teams represented 91 percent of the hospital's top 25 DRGs.

3. Direct patient care teams were also developed by clinicians themselves to help them organize their daily work activities. The hospital found this to be the most effective method of establishing ongoing, productive, patient care teams. Two teams developed by using this approach were the human immunodeficiency virus (HIV) team and the chest pain team.

Who should be included on the team? Fargason and Haddock (1992) suggest that teams be composed of those individuals who can have greatest impact on solving the problem at hand. Rather than calling on those who have the greatest prestige or hierarchical influence, effective teams often include nonprofessional staff such as housekeepers, students, house staff, dietary aides, and others. Both external and internal customers of a process may also be invited to help solve problems of particular interest to them. Dietitians can influence the makeup of teams dealing with problems related to such things as tray delivery, nutritional screening, infection control, nourishments, pantry supplies, and selective menus by encouraging cross-functional teams to include nonprofes-

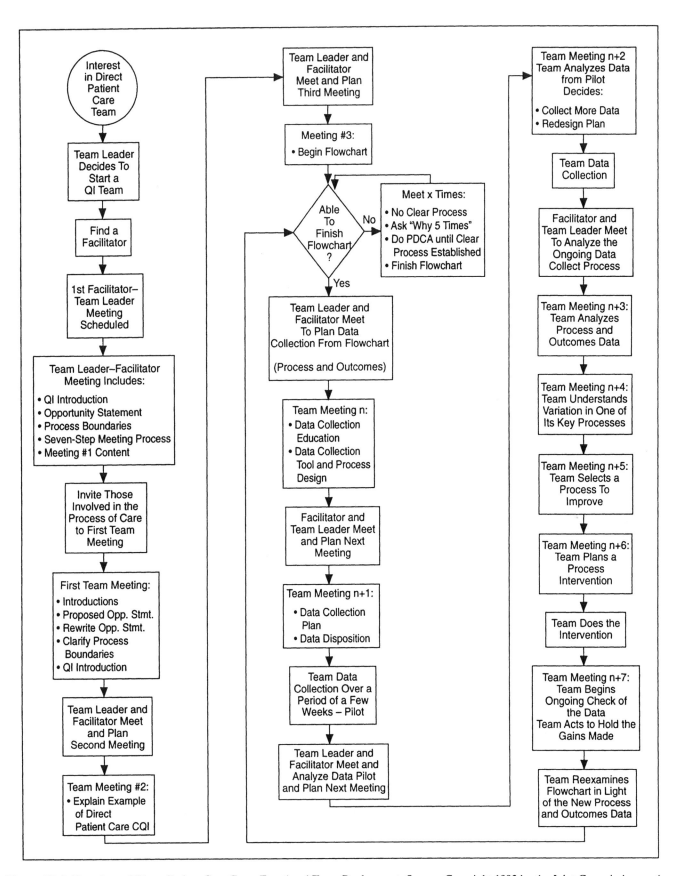

Figure 17–1 Flowchart of Direct Patient Care Cross-Functional Team Deployment. *Source:* Copyright 1992 by the Joint Commission on Accreditation of Healthcare Organizations, Oakbrook Terrace, Illinois. Reprinted from the June 1992 *Quality Review Bulletin* with permission.

sional staff, interns and residents, former patients, and others who might not otherwise be included.

Heterogenous problem-solving groups should not be formed simply by adding a few members to existing committees (Fargason and Haddock 1992). When groups exceed 12 members they often experience poor communication, inadequate coordination, and inability to make decisions or take action. Turnover and absenteeism tend to increase as the group becomes larger. Job satisfaction increases up to a team size of about five members and decreases as members are added.

The size of a group depends on its purpose. Large, even-numbered groups (12 or more members) are best if the purpose is fact-finding and deliberation (Fargason and Haddock 1992). On the other hand, odd-numbered groups of five to seven members are best for decision making, problem solving, and taking action. Before selecting members, it is helpful to develop a flowchart of the process to be addressed and form the quality-improvement group from among those identified in the process. Such an approach helps define the process under consideration and helps the group remain focused on the problem at hand.

USE OF A FACILITATOR

Both Mosleth (1984) and McEachern et al. (1992) recommend the use of a facilitator to assist in the development and deployment of multidisciplinary and cross-functional teams. Sometimes the facilitator is the institution's quality improvement coordinator. The role of the facilitator is to help move "the ubiquitous tension present in health care today toward peaceful, productive, and creative resolution" (McEachern et al. 1992). A facilitator would be a valuable addition to any cross-functional team, but in the experience of many dietitians, nutrition support teams were initiated without the benefit of such an individual. Looking back, a nutrition support team facilitator could have helped prevent many problems, such as turf protection, poor communication, egomania, infighting, and publicly criticizing one or another individual for weaknesses in the patient care process.

EMPOWERMENT THROUGH COLLABORATION

Oie and Recker (1992) described their experience in developing a team quality assurance model. As in many instances, the need for a multidisciplinary patient care team arose when critical care nurses were frustrated by (1) the complex and highly variable processes needed to resolve quality issues and (2) the inability of health care professionals to reach consensus over methods to best address multidisciplinary problems.

In the situation described by Oie and Recker (1992), one of the first barriers encountered when switching from departmental and unit-based quality committees to a multidisciplinary approach was the realization that the term *multidisciplinary* held different meanings. Physicians viewed the term as meaning various clinical services within medicine; among nurses the term meant the various professionals caring for the patient. One-on-one conversations among staff nurses, quality committee members, and physicians uncovered additional resistance to developing a multidisciplinary approach to quality care, particularly a feeling among physicians that they lacked sufficient knowledge about quality management and that they were being criticized for personal shortcomings. A proposal was developed and refined to give a format that was concise, specific, and logical but with examples to clarify points.

A two-part solution was offered:

1. A team approach would be used to evaluate patient care provided by all caregivers.
2. The focus would be on process instead of on outliers, noncompliant persons, punitive measures, and the like.

A model, shown in Figure 17–2, was used to depict the numerous services involved in direct patient care. It was agreed that some of the team members shown should be permanent committee members while others would be invited on an ad hoc basis. A flowchart (Figure 17–3) was developed to illustrate the team approach to quality improvement. The ten-step monitoring and evaluation process was used; because of physician unfamiliarity with this process, each step needed to be described in detail and supported with examples.

The multidisciplinary team selected the aspects of care that would be monitored and evaluated. The first topics chosen for study were ones that

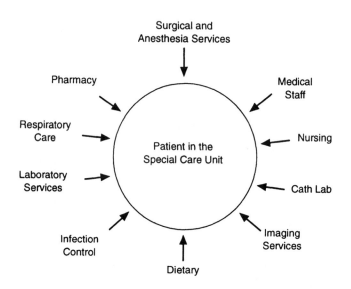

Figure 17–2 Team Members Providing Care for Critically Ill Patients. *Source:* Reprinted from Oie, M. and Recker, D., Empowerment through Collaboration: Implementing a Team Quality Assurance Model, *Journal of Nursing Care Quality*, Vol. 6, No. 2, pp. 32–40, Aspen Publishers, Inc., © 1992.

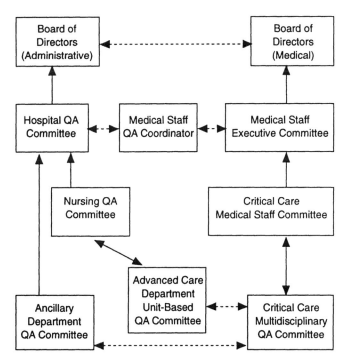

Figure 17–3 Flow of Information Using Team Approach to Quality Management. *Source:* Reprinted from Oie, M. and Recker, D., Empowerment through Collaboration: Implementing a Team Quality Assurance Model, *Journal of Nursing Care Quality*, Vol. 6, No. 2, pp. 32–40, Aspen Publishers, Inc., © 1992.

- various team members had already identified as important aspects of care
- lent themselves to multidisciplinary participation
- included ideas and suggestions for indicators provided by the literature
- physicians were interested in

Nurses assumed leadership for development of the process. Standards were developed with input from physicians, respiratory therapists, dietitians, and infection control personnel. Evaluation of the process was positive, primarily because communication was improved, differences were accepted, terms were clarified, and mutual understanding was enhanced.

This scenario offers several lessons for dietitians:

- Multidisciplinary or cross-functional patient care teams will frequently be spearheaded and led by nurses, although other caregivers may certainly provide leadership as well.
- Physicians are an important part of quality monitoring and evaluation; close collaboration will foster better working relationships.
- Dietitians may not be included in all cross-functional teams. However, nutrition screening, assessment, and intervention must be provided and integrated with patient-focused and cross-functional groups when appropriate.

EXAMPLES OF CROSS-FUNCTIONAL TEAMS AND INDICATORS

Most hospitals are relatively new at the business of interdisciplinary patient care teams, particularly teams that include nutritional care personnel. Following are a few examples of multidisciplinary or cross-functional teams described in the literature.

A Study of Height and Weight Measurements

At Rochester Methodist Hospital, an 800-bed teaching hospital in Rochester, Minnesota, nurses and dietitians came together to conduct a joint study of completion and accuracy of height and weight measurements (McMillin and Jasmund 1985). Although each department had its own quality-assurance representative who was responsible for quality improvement, the departments desired to integrate their quality-assurance activities whenever possible.

Dietitians and nurses jointly developed a height and weight study consisting of

- documentation of admission heights and weights
- documentation of routine weights
- routine weight needs assessment
- assessment of significant weight changes
- perceptions regarding height and weight measurements

The study, as described in detail by McMillin and Jasmund (1985), includes study findings, the actual questionnaire used to determine perceptions of nurses and dietitians, quality goals, corrective actions to help achieve goals, and study results. Because heights and weights are an important component of nutritional status assessment and monitoring, other groups may wish to replicate this study. To facilitate such a study, the questionnaires used to survey attitude towards height and weight measurements are reprinted in Exhibits 17–1 and 17–2.

Nutrition Support Team

The organization and function of nutrition support teams are familiar to many dietitians. However, many teams neither perceive themselves as cross-functional teams nor have they delineated a quality-improvement plan for the team. If this is the case at your institution, the nutrition support dietitian might take the initiative to direct the group toward a cross-functional orientation. Srp et al. (1991) illustrated how the Joint Commission ten-step process can be applied to a nutrition support team engaged in either hospital or home care. Powers et al. (1991) delineated important aspects of care, clinical indicators, and criteria that may be adapted for any nutrition support team.

Exhibit 17–1 Questionnaire Used To Determine Perceptions Regarding Height and Weight Measurement: Dietetics Department

---

Perceptions Regarding Height and Weight Measurement: Dietetics Department

1. Physicians respond to your requests for daily weight orders:
   _____ all of the time
   _____ usually
   _____ occasionally
   _____ rarely

2. How are height/weight data communicated to you? (Check all that apply.)
   _____ verbally
   _____ nurse
   _____ physician
   _____ physicians' progress notes
   _____ nurses' notes
   _____ nursing patient care planning card (cardex)
   _____ reviewing flow sheet
   _____ patient
   _____ other: please list
   _____

3. Indicate your experience with the following problems:
   1 = frequently   2 = occasionally   3 = rarely/never
   _____ weights not recorded as per order
   _____ no notification received of significant weight changes
   _____ unclear weight record
   _____ questionable weight data
   _____ metric versus English units used
   _____ discrepancies in patient care planning card (nursing cardex) versus medical record

4. What do you see as major concerns regarding the collection, communication, or use of weight measurements. Be descriptive.

5. How frequently do you review daily weights for patients with a compromised nutritional status?
   _____ daily
   _____ every other day
   _____ every 3–5 days
   _____ other: please list _____

6. How do you monitor weight from one day to another?

7. Rank the following indicators in order of importance to you in screening patients:
   1 = very important   3 = important in certain situations
   2 = moderately important   4 = rarely or never used
   _____ diagnosis   _____ age   _____ diet
   _____ height   _____ medications   _____ nurses' notes
   _____ weight   _____ past medical history _____ progress notes
   _____ intake/output

8. Rank the following indicators in order of importance to you in monitoring patients' nutritional status:
   1 = very important   3 = important in certain situations
   2 = moderately important   4 = rarely or never used
   _____ diagnosis   _____ age   _____ diet
   _____ height   _____ medications   _____ nurses' notes
   _____ weight   _____ past medical history _____ progress notes
   _____ intake/output

9. Do you have any suggestions for improving the line of communication between the dietetics and nursing departments regarding significant weight changes?

10. In your opinion, are nurses generally aware of your need for height/weight data?
    _____ no   _____ yes
    Comments:

Source: Copyright 1985 by the Joint Commission on Accreditation of Healthcare Organizations, Oakbrook Terrace, Illinois. Reprinted from the February 1985 *Quality Review Bulletin* with permission.

Exhibit 17–2 Questionnaire Used To Determine Perceptions Regarding Height and Weight Measurement: Nursing Department

---

Perceptions Regarding Height and Weight Measurement: Nursing Department

1. When weight measurements are ordered for a patient, how often do you believe that this is an appropriate order?
   ____ all of the time   ____ most of the time   ____ some of the time   ____ none of the time
   Please give some examples of inappropriate weight orders that you have encountered:

2. What are the weight measurements for?

3. Who reviews the recorded weights?

4. What is the reason for admission height/weight measurements?

5. Which of these people use admission height/weight measurements in their professional assessment of patients? (Check all that apply.)
   ____ pharmacists   ____ physical therapists
   ____ dietitians   ____ nurses
   ____ physicians

6. Which of these people use (in their day-to-day assessments of patients) the recorded weights ordered throughout a patient's hospital stay? (Check all that apply.)
   ____ pharmacists   ____ physical therapists
   ____ dietitians   ____ nurses
   ____ physicians

7. If you weigh a patient and there is a wide fluctuation of weight from one day to the next, what would your nursing assessment and action be?

8. Which of these people have you called, left a note for, or talked to directly regarding a significant weight change?
   1 = most often   2 = sometimes   3 = seldom   4 = never
   ____ pharmacists
   ____ dietitians
   ____ physicians
   ____ physical therapists
   ____ nurses

9. What other methods of communicating significant weight changes have you used? (Please explain.)

10. How is a significant weight change communicated to other professionals?
    1 = most often   2 = sometimes   3 = seldom   4 = never
    ____ verbally
    ____ shift report
    ____ nursing notes
    ____ patient care plan card
    ____ flow sheet
    ____ other

Comments:

Source: Copyright 1985 by the Joint Commission on Accreditation of Healthcare Organizations, Oakbrook Terrace, Illinois. Reprinted from the February 1985 *Quality Review Bulletin* with permission.

Medical-Surgical Environments

Important aspects of care delineated by many nursing units will include a reference to nutrition assessment, intervention, or patient counseling. For example, "Management of the nutritional needs of the diabetic patient" is included as a high-volume, problem-prone item in Important Aspects of Care for a Diabetic Unit (Elrod 1991). It is appropriate for a dietitian to be an integral member of a cross-functional team responsible for diabetic patients. The same may be true of teams associated with the care of patients in cardiovascular, gastrointestinal, pediatric, surgical, renal transplant and dialysis, neurological/neurosurgical, and oncology units.

In *Monitoring with Indicators: Evaluating the Quality of Patient Care,* Carroll (1992) provides several indicators for both disciplinary and cross-functional groups. Two indicators specifically for dietitians are modified diets and uncontrolled diabetes.

1. number of patients discharged on modified diets without evidence in the medical record of dietitian's interview, counseling, and dietary education/number of patients discharged on modified diets during the relevant period
2. number of diabetic patients discharged after a stay of more than two days without evidence in the medical record of dietitian's interview, counseling, and dietary education/number of patients discharged with diagnosis of diabetes mellitus during the relevant period

In *Monitoring with Indicators* (Carroll 1992), nutrition services are addressed in cross-functional indicators related to outpatient diabetes care, chronic hemodialysis, inpatient diabetes mellitus, intravenous nutrition support, and anorexia nervosa. The lack of attention to nutrition in other important conditions included in this book, such as burns, surgery, gastric ulcers, and iron-deficiency anemia, suggest that dietitians need to take the initiative to establish strong communication with personnel on units where nutritional care is of vital importance. These and other personal observations verify that dietitians must exert themselves to ensure that nutritional care is included in cross-functional quality monitoring and evaluation processes.

Rehabilitation

Summers et al. (1988) described an interdisciplinary effort that satisfied the requirements of program evaluation, business planning, and quality assurance within the context of quality management. The purpose of this effort was to develop a rehabilitation service with defined outcomes. Although the program described did not specifically mention nutritional care, Huyck and Rowe (1990, 123) delineated standards of nutrition care for a rehabilitation unit. These standards could aptly be integrated with a cross-functional rehabilitation team providing care for stroke, general rehabilitation, brain injury and spinal cord injury patients.

Pressure Ulcers

A multidisciplinary team was formed at Meadowlands Hospital Medical Center, Secaucus, New Jersey, to improve the treatment of nosocomial pressure ulcers (Hegelein 1992). The team consisted of representatives from administration, medical staff, clinical nutrition, nursing, and infection control. The goals of the team were to update the hospital's protocol for decubitus ulcers, reflect the importance of nutritional care in the prevention and treatment of pressure ulcers, and to decrease the formation and progression in acuity of pressure sores. Data analysis for 1991 showed an increased awareness of the importance of nutritional care in the treatment and prevention of decubitus ulcers, a decrease in ulcer progression, and a significant cost savings to the facility.

Neonatal Intensive Care

Ying et al. (1992) reported the procedure used to develop a nutrition-related outcome indicator for use by a multidisciplinary care team in a neonatal intensive care unit. The dietitian's responsibilities and important aspects of care were reviewed. The nutritional goal for low-birth-weight infants is to provide adequate calories for weight gain. Therefore, weight gain in conformance with growth standards set for premature infants was selected as an indicator. Daily weights were plotted for low-birth-weight infants. An 85 percent conformance with the growth curve was achieved during the last three quarters of 1991. Since these results were congruent with other published data, a threshold of 85 percent was set for the indicator. The proposed indicator and threshold were submitted to the hospital's quality-assurance committee and the perinatal committee for approval and implementation.

FACTORS CONTRIBUTING TO TEAM SUCCESS

Cross-functional teams are relatively new and outside the experience of many dietetic professionals. However, a few hospitals have successfully initiated such teams. McEachern et al. (1992) cited several factors, shown in Exhibit 17–3, that accelerated the development process.

Puta (1991) noted several key elements of collaborative interdisciplinary teams, including

- mutual trust, required for honest, open discussions and evaluation of findings from monitoring activities
- open communication, including willingness to disclose feelings, share interpretations of study findings and

Exhibit 17-3 Accelerators to Success with Cross-Functional, Direct Patient Care Teams

---
- Understanding that patient care involves not only decisions about care, but also the actions and communications of every individual involved in the delivery of care
- Having teams examine the complete process of care as it actually happens
- Listening to patients and other customers—a tremendous opportunity to learn about where improvements can be made
- Strong, directed facilitation as needed to keep the team on track
- Teaching team members team tools they need to use at the meetings
- Having one physician on the direct patient care team who has spent time working on other non-patient care teams
- Diligent reading and facilitated learning about continuous quality improvement (CQI) by interested parties
- Convenient meeting times (with food provided)
- Helping team leaders with team leadership skills between meetings
- Providing administrative help with data analysis

Source: Copyright 1992 by the Joint Commission on Accreditation of Healthcare Organizations, Oakbrook Terrace, Illinois. Reprinted from the June 1992 *Quality Review Bulletin* with permission.

---

analysis, ask questions, identify areas needing improvement, and suggest action strategies for change

- effective communication skills, especially use of constructive feedback and listening skills
- clinical competence, a key component to mutual trust and respect, the basis for meaningful insights and judgments into the causes of quality-related problems, and appropriate actions to resolve quality issues
- responsibility and accountability among both professionals and nonprofessionals, because change will happen only if and when individuals take responsibility for their decisions and actions
- assertiveness to ensure that all participating disciplines adequately contribute input to the selection of important aspects of care to be monitored, the review and evaluation of findings, and identification of areas for improvement and planning of corrective actions.

Dietitians can foster expression of these traits within the department to ensure that they possess the attributes necessary to be effective participants of cross-functional teams.

Establishing cross-functional teams is useful only if the institution makes a long-term commitment to the quality-improvement process. The inclusion of dietitians, dietetic technicians, and food service workers on multidisciplinary teams is useless if decisions continue to be made by physicians, nurses, and key administrators. Raising the expectations of allied health personnel and then ignoring their contributions to group decision making only leads to "disappointment, bitterness, and deterioration of staff morale" (Fargason and Haddock 1992).

PITFALLS AND PROBLEMS

McEachern et al. (1992) delineated several pitfalls inhibiting the success of interdisciplinary patient care teams, shown in Exhibit 17-4. These and other common problems can be avoided (Fargason and Haddock 1992) by

- ensuring senior management's overriding commitment to delivery of high-quality health care
- providing formal recognition of teams, team members, and results of cross-functional team efforts
- setting aside time during regular work hours for team meetings
- focusing on clearly defined goals (Goals are different for each cross-functional team and flow from the primary objective for forming the team.)
- strategically including those most involved in the process under consideration, even if they appear disinterested in quality improvement
- using group management skills (brainstorming and nominal group processes) to enhance creativity and arrive at integrative solutions
- employing principles of group dynamics to handle emotional issues, hidden agendas, disagreements, and individual behaviors that may disrupt group cohesion and team decision making
- training team members in the techniques of group processes, quality improvement, data collection and analysis, reporting, and new methods of work performance
- developing team leaders to facilitate effectively cross-functional teams and group decision making.

CONCLUSION

Both interdisciplinary patient-focused care teams and cross-functional quality teams depend on the involvement of

---

Exhibit 17-4 Examples of Pitfalls Associated with Cross-Functional Teams

- Starting teams without significant organizational readiness
- Placing only doctors and nurses on teams without involving all people who work in the process of care
- Starting a team without significant cross-functional team experience
- Not keeping a storyboard of progresses made
- Not sharing lessons learned
- Too much help with the team's clerical work—the team must own the data
- More than eight people on a team
- Involvement by many different people with competing interests

Source: Copyright 1992 by the Joint Commission on Accreditation of Healthcare Organizations, Oakbrook Terrace, Illinois. Reprinted from the June 1992 *Quality Review Bulletin* with permission.

committed individuals who possess the motivation and skills needed to participate fully in the process. No amount of managerial advocacy, conscientious effort in appointing committees, or making sure that teams have the right composition can substitute for investment in the *people* who make quality happen at every step of the process. Individuals must begin to improve their own ability to manage themselves and understand how their behaviors and attitudes affect others negatively or positively (Strasser 1992). Managerial innovations are facilitative, but the key to quality care and improvement lies with individual health care professionals. High-quality people are the heart and soul of high-quality health care services.

REFERENCES

Borzo, G. 1992. Patient-focused hospitals begin reporting good results. *Health Care Strategic Management* 10, no. 8:16–22.

Brider, P. 1992. The move to patient-focused care. *American Journal of Nursing* 92, no. 9:26–33.

Carroll, J.G. 1992. *Monitoring with indicators: Evaluating the quality of patient care.* Gaithersburg, Md: Aspen Publishers, Inc.

Elrod, M.E.B. 1991. Quality assurance: Challenges and dilemmas in acute care medical-surgical environments. In *Monitoring and evaluation in nursing,* ed. P. Schroeder, 27–56. Gaithersburg, Md: Aspen Publishers, Inc.

Fargason, C.A., and C.C. Haddock. 1992. Cross-functional, integrative team decision making: Essential for effective QI in health care. *Quality Review Bulletin* 18, no. 5:157–163.

Farris, B.J. 1993. Converting a unit to patient-focused care. *Health Progress* 74, no. 3:22–25.

Giuliano, K.K., and C.E. Poirier. 1991. Nursing case management: Critical pathways to desirable outcomes. *Nursing Management* 22, no. 3:52–55.

Hegelein, S. 1992. Utilizing the CQI process to improve the treatment of nosocomial decubitus ulcers. *Journal of the American Dietetic Association* 92 (suppl), no. 9:A-65. Abstract.

Huyck, N.I., and M.M. Rowe. 1990. *Managing clinical nutrition services.* Gaithersburg, Md: Aspen Publishers, Inc.

Joint Commission on Accreditation of Healthcare Organizations. 1993. *1994 Accreditation manual for hospitals.* Oakbrook Terrace, Ill.

Krause, C.G., et al. 1984. *Patient-centered audit.* St. Louis, Mo: Warren H. Green, Inc.

Laughon, D., et al. 1993. From unit-based quality assurance to multidisciplinary continuous quality improvement in the coronary care unit. *Journal of Nursing Care Quality* 7, no. 3:19–27.

Manco, B., et al. 1992. "Business managers" of mini product lines characterize case management at St. Peter's Medical Center, New Brunswick, New Jersey. *Strategies for Healthcare Excellence* 5, no. 9:1–12.

McEachern, J.E., et al. 1992. How to start a direct patient care team. *Quality Review Bulletin* 18, no. 6:191–200.

McMillin, B.A., and J.M. Jasmund. 1985. A quality assurance study of height and weight measurements. *Quality Review Bulletin* 11, no. 2:53–57.

Mosher, C., et al. 1992. Upgrading practice with critical pathways. *American Journal of Nursing* 92, no.1:41–44.

Mosleth, R.R. 1984. A practical guide to multidisciplinary auditing. In *Nursing quality assurance: A unit based approach,* ed. P.S. Schroeder and R.M. Mailbusch, 191–203. Gaithersburg, Md: Aspen Publishers, Inc.

Oie, M., and D. Recker. 1992. Empowerment through collaboration: Implementing a team quality assurance model. *Journal of Nursing Care Quality* 6, no. 2:32–40.

Powers, T., et al. 1991. A nutrition support team quality assurance plan. *Nutrition in Clinical Practice* 6:151–155.

Puta, D.F. 1991. Interdisciplinary quality assurance: Issues in collaboration. In *Issues and strategies for nursing care quality,* ed. P. Schroeder, 141–162. Gaithersburg, Md: Aspen Publishers, Inc.

Srp, F., et al. 1991. Quality of care concepts and nutrition support. *Nutrition in Clinical Practice* 6:131–141.

Strasser, S. 1992. The Achilles' heel of quality management: The human quotient. *Quality Review Bulletin* 19, no. 5:156.

Summers, P.M., et al. 1988. Quality management: Program design: An interdisciplinary approach. *Nursing Clinics of North America* 23, no. 3:665–670.

Tishman, E. 1992. Total quality management: The bridge to customer satisfaction. *Provider* 18, no. 10:30–42.

Ying, M., et al. 1992. The development of an outcome oriented nutrition indicator in the neonatal intensive care unit. *Journal of the American Dietetic Association* 92 (suppl), no. 9:A-70. Abstract.

Chapter 18

Evaluating Quality Management Programs

Is this program worth it? If not, what can we do better? This philosophy provides a basis for evaluating quality management programs (Hopkins 1990). As for many other aspects of clinical management and dietetic practice, valuable insights can be gained when managers and practitioners stand back and look at both the process and results of their quality-related activities. This chapter reviews the purpose and essential components of an evaluation system designed to improve quality improvement processes and to meet external accreditation requirements.

PURPOSE OF EVALUATION

There are four important reasons to evaluate the quality management program.

1. Evaluation is required for compliance with hospital accreditation standards (Joint Commission 1992).
2. Evaluation facilitates the program's growth, integrity, and effectiveness (Sawyer-Richards 1991, 181). The ultimate goal of continuous quality improvement is to effectively change the practice of clinical nutrition to improve patient care. Evaluation is the control mechanism that determines whether this goal has been met.
3. Evaluation assists with the reduction of potential problems in risk management.
4. Evaluation directly supports the maintenance of important professional values such as credibility, autonomy, and accountability (Sawyer-Richards 1991). By providing a framework for regulating nutrition practice, evaluation enables clinical nutrition departments to meet obligations to the hospitals' governing bodies and to the public.

Most important, when evaluation of a quality management program demonstrates effectiveness, it instills greater confidence and satisfaction among nutrition professionals (Sawyer-Richards 1991). Success brings further success, making it easier to motivate individuals to maintain high standards, implement needed changes, work collaboratively with others to identify and solve complex problems, and put forth the efforts needed to meet customer needs and expectations continually.

PLAN EARLY TO EVALUATE LATER

Even the most progressive hospital nutrition departments are probably just getting up to speed with the Joint Commission Agenda for Change. Many are still in the transition phase, planning and establishing continuous quality improvement programs. Evaluation of the system may seem a long way off, but deciding how to measure the success of a quality management program in advance can provide direction for the implementation process and prevent problems down the road. Therefore, it is not too soon to draft program evaluation procedures while still in the program development stage.

Identifying quality program goals in advance will make it much easier to determine whether expectations have been met. Simply beginning the quality monitoring process, then waiting a year to see what, if any, results are achieved, may have a devastating effect. Desired results will be evident only if procedures are set in place and steps are taken to evaluate achievement of desired outcomes.

A quality program should take into account customers' needs, present and future (Lehr and Strosberg 1991). Quality goals, like the mission statements of which they are often a part, tend to be relatively stable from year to year. However, the particular objectives selected to meet the goals may change annually to keep abreast of changes in customer mix, needs, and services. New objectives are set as former ones are achieved (Meisenheimer 1985).

Goals or targets, as they are also known, are different from thresholds. A threshold often defines the lower limit of what is acceptable practice. A goal, however, may be set above the threshold. Goals will always be above the current level of practice if performance lags behind stated thresholds. For example, the threshold for patient satisfaction rating with clinical nutrition services may be set at 90 percent. The department achieved an average rating of 91 percent in this area last year. One of the department's goals for the coming year is to achieve a 93 percent rating. Employees developed and implemented a variety of quality-related objectives and programs designed to reach this goal.

Exhibit 18–1 provides a format for reviewing progress made on current quality objectives and planning over a two-year period. The form capsulizes the quality efforts of the Ohio State University Medical Center's department of nutrition and dietetics.

Goals in three areas of emphasis for quality initiative remain constant: food, service, and staff development. But note how the proposed objectives or strategies for each area build on what has been accomplished in the past year. If objectives are written in specific and measurable terms at the onset, they will streamline the evaluation process afterward (Meisenheimer 1985).

The work involved in conducting the evaluation review is streamlined if the forms used throughout the monitoring process are designed to facilitate the evaluation process. One helpful format for this purpose is the "picture book" style shown in Figure 18–1. The picture book illustrates the major components of the project on a single page. Information for the picture book is compiled and recorded by the departmental quality coordinating team or a task force. Several different forms may be required when tracking milestones for large, complex problems. Data from these forms are used by the team to complete the summary form or picture book.

The picture book becomes a valuable tool for the annual evaluation even if a quality task force disbands. The form identifies the rationale for project selection and summarizes the data collection and analysis process. Actual charts, graphs, and other tools may be attached. The actions taken and goals are stated. The last box in the picture book, the "evaluation" box, may be the most helpful when it is time to evaluate the program as a whole. In this box, the team notes which of its actions or decisions were the most or least effective. It lists any obstacles encountered in the FOCUS-PDCA cycle (see Chapter 13) and suggests areas for improvement.

Sawyer-Richards (1991) described six criteria that can be used to evaluate the effectiveness of quality programs:

1. Changes and improvements in the department and in patient care occur at the behavioral and organizational level.
2. Problems are followed to resolution.
3. There is demonstrated evidence of improved patient care services.
4. Desired outcomes are achieved and maintained.
5. There are comprehensive monitoring and evaluation activities for the entire scope of services offered.
6. Conclusions and recommendations are communicated within the department and reported to the organizationwide quality management program as specified in the overall quality plan.

The Annual Appraisal form found in Appendix D–7 includes many of the evaluation criteria cited above. If the hospital does not have a standardized process for departmental quality evaluating programs, such a form can be completed quickly, yet provides an opportunity to assess the relevance and effectiveness of the plan currently in place.

EVALUATE PURPOSE, OBJECTIVES, AND OUTCOMES

The purpose of any quality management program is to improve both the quality of services and customer satisfaction with the products and services provided by the food and nutrition department. Annual quality objectives flow from specific goals and actions designed to improve both quality and customer satisfaction. The first step in an annual evaluation of the quality management program is to review stated purposes, goals, and objectives to be sure that they reflect high standards and movement toward improved outcomes.

A quality management program is effective when tangible, working results are observed. The quickest and most obvious method of evaluation is simply to compile a list of the problems resolved and improvements made during the past year (see Appendix A–5 and A–6). What kinds of results can you expect? Quality management can promote improvement in areas where an increase is desirable, such as in patient satisfaction, productivity, and market share (Dougherty 1992). Several such advances have been achieved by dietitians as they implement quality improvement programs. Following are

Exhibit 18-1 Nutrition and Dietetics Quality Improvement

AREA OF EMPHASIS	OBJECTIVES 1991–1992 (Previous Year)	MET	NOT MET	PROPOSED OBJECTIVES (Next Year 1992–93)
Food	Recruit Chef/Production Manager	X		Complete Phase I of Menu Revision Develop Menu Philosophy, Cycles, Items, Recipes.
Service	Achieve 93% Patient Satisfaction Rating	X		
	Increase patient contact by initiating Patient Visitation Program by:	3rd Qtr		
	Patient Services Staff	X		
	Management Staff		X	Management Staff will participate in Patient Visitation Program.
	Promote Future Orientation. Develop RFP & Award Bid for Consultant Services for development of a Master Plan for Nutrition Services	X		Propose Top Priorities as identified in Master Plan to Administration
	Streamline Patient Menu system. Complete Phase I of Computerized Diet Office/Menu Program		X	Complete Phase I of Computerized Diet Office/Menu Program
Staff	Enhance expertise of Frontline Managers: Begin inservice program for Operations (front-line managers) Staff	X		Maintain opportunities for Skill Enhancement. Continue schedule of inservices.
	Convert Food Service position to training position. Recruit department trainer.	X X		Ease staff into jobs, reduce turnover—implement department orientation program.
Staff	Focus on Peer Review and Attendance components	X		Reorganize recognition programs along new department lines.
	Initiate Department Employee of the Quarter and Excellent Attendance Awards	X		

Source: Adapted from Meisenheimer, C.G., Program Evaluation and Nursing Quality Assurance, in *Issues and Strategies in Nursing Care Quality*, P. Schroeder, ed., p. 190, Aspen Publishers, Inc., © 1991 and from the Department of Nutrition and Dietetics, Ohio State University Medical Center, Columbus, Ohio.

some examples of advancements reported in the literature (Gardner 1990):

- increased accuracy of diet orders, menu editing, and nutrient intake analyses
- improved timeliness of documentation of nutritional risk of patients who receive nothing orally (NPO) and have no other form of nutrition support for more than three days
- improved communication between dietitians and medical and nonmedical staff
- increased consistency of nutritional assessments of patients initiated on enteral feedings, including accurate assessment of hydration requirements and tolerance to tube feedings
- improved documentation of patient care activities
- closer monitoring of nutrition intake in the geriatric population
- increased appropriateness in use of total parenteral nutrition (TPN)
- increased appropriateness in the use of cholesterol-lowering medications
- earlier initiation of nutrition supplements and/or tube feedings in radiation oncology outpatients
- increased efforts by dietitians to be the recognized primary source for answers to nutrition questions
- enhanced patient transfer and discharge planning (Laramee 1992)
- increased implementation of dietitians' recommendations with the use of preprinted order labels (Hester and Ford 1992)

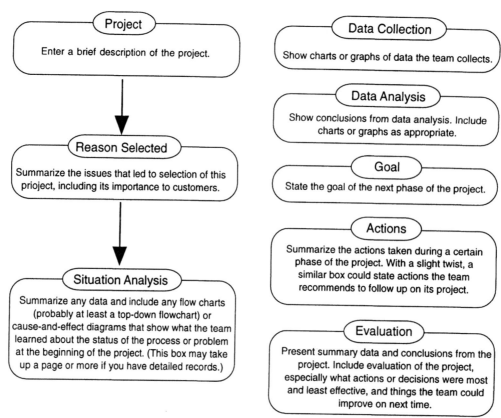

Figure 18-1 The Picture Book Format. *Source:* Reprinted from *The Team Handbook,* by P.R. Scholtes, et al., pp. 4-14, with permission of Joiner Associates, Inc., © 1988.

A quality management program can also result in decreases in areas where reductions are preferable, such as in the number of errors, amount of wasted product, or rate of system failures (Dougherty 1992). Examples of such successes include

- decrease in the severity and overall incidence of nosocomial pressure ulcers (Hegelein 1992)
- decreased total tray assembly error rate (Gardner 1990)
- decrease in the number of negative comments on the patient satisfaction questionnaire
- reduction of the need for crisis intervention, incidents of "fire fighting," and reactive style of management (King 1992)

The list of annual accomplishments is the major factor demonstrating program effectiveness. Reviewing the list is both satisfying and motivating, and it can be the most enjoyable part of the process. But the evaluation process does not end here. If evaluation is limited to an analysis of outcomes and neglects taking a look at the quality monitoring process itself, the full value of a quality management program is never realized (Hopkins 1990).

EVALUATE EFFICIENCY

In addition to evaluating the objectives and results of quality efforts, evaluation should also examine whether the quality management program is efficient. One of the best models available for this purpose was proposed by Hopkins (1990), who outlined the steps in evaluating a hospitalwide quality program, then focused in on departmental and certification issues that are especially relevant to clinical nutrition. A synopsis of Hopkins's (1990) review process with application to clinical nutrition services follows.

Managerial Support

Managerial support for a quality management program is critical to its success. Genuine support is more than just verbal; it must be concrete and demonstrable. Assessing commitment at this level takes into consideration the following questions. Do department managers

- address quality in the mission statement?
- actively participate in quality-related activities?
- ensure that all areas participate in the process?
- consider quality when discontinuing, adding, or modifying services, or when making financial and program decisions?
- challenge recommendations when insufficient information exists?

- follow up to ensure resolution of identified problems?
- critically review, discuss, and sign off on reports?
- orient new staff members to the quality-monitoring and evaluation process?
- increase their knowledge about continuous quality improvement or other approaches to quality management?
- incorporate quality indicators and risk reduction into performance and self-evaluations?
- hold their supervisors accountable?
- define responsibilities for quality management in job descriptions?

There are several ways these questions can be answered. A form can be developed by the quality committee and administered to managers to ascertain managerial support. Managers may also use the questions for self-evaluation by answering yes or no to each query. Each yes answer should be justified and verified. For example, if a manager answered yes to the question related to increasing personal knowledge of the quality evaluation process, a list of relevant programs attended or articles read should accompany the response. Managerial commitment can also be evaluated by support staff. The department quality committee, supervisory staff, or other staff members can be surveyed as to their assessment of how managers are supporting the quality improvement program. If a particular question is consistently answered no, that aspect of the program should be analyzed, and strategies should be developed to address the issue in the coming year.

Managers are responsible for providing the resources needed to make the program work. Examine the resources provided for staff support. If a valid process has been used to select quality improvement projects, the availability of resources will have been determined prior to the analysis and action phases. Check to see whether the inability to implement any quality recommendations was a result of the failure to provide, or lack of, adequate resources. Departments still can achieve significant quality improvements if they realistically assess the resources at their disposal and charter projects accordingly.

Quality Management Structures

It is useful to check the effectiveness of the departmental quality committee. Has the committee been productive? Who serves on it? Is it time for a change? Rotating membership on a staggered schedule will help infuse fresh ideas into an experienced group with an established framework. Has the group met its deadlines? If not, determine whether the schedule, the membership, the leader, or some other factor was unrealistic or dysfunctional. Adjust the committee structure accordingly when planning for future program success.

Many of these same questions can be asked about departmental quality circles, teams, or task forces. What issues were addressed by these groups? Were meetings held? Was time used effectively? Can these groups demonstrate measurable outcomes? Are any new task forces needed? Should any quality teams be discontinued, having outlived their usefulness? Should additional or new staff members be appointed to ongoing quality teams? A formal annual evaluation time offers the unique opportunity to address these and similar questions. When changes result from regular program evaluations, individual employees or groups are less likely to feel put down or threatened if their quality team roles change. They view such changes as a part of normal operations.

Quality Functions

The annual quality report might not reflect some components of quality management that can benefit from periodic, if not annual, review. Such components include staff qualifications, monitoring issues, data processing, staff involvement, strategic problems and opportunities, meetings, communication flow, and reports. Evaluation of these factors is described briefly here.

Professional Qualifications

Conventional quality assurance focused on individual performance as the root cause of poor quality care or services. The transition away from quality assurance to quality improvement has shifted attention away from this good-apple/bad-apple syndrome. Attention is now directed almost totally on systems; problems are more likely traced to system failures than to poor individual performance. Thus it is imperative to have systems in place to promote quality practice among individuals.

Practicing clinical nutrition is both a responsibility and a privilege. A review of hiring policies helps to ensure that this privilege is extended to reputable practitioners who are ready to accept their responsibilities and who have demonstrated commitment to quality nutrition. The review process should demonstrate that each new or returning practitioner has maintained certification. Any gaps in training or experience should be justified and rectified as soon as possible. If exceptions are made to standard appointment guidelines, the number of exceptions and rationale should be recorded. Regular review of all dietetic staff will help managers recognize overall deficits and facilitate delineation of specific qualities or skills needed when new or vacant positions are filled.

Monitoring Issues

If the ten-step process or a similar model was used in the development and implementation phase of a quality monitoring program, there should be an affirmative answer to each of the following key monitoring questions:

- Was a departmental quality plan written?
- Has the department defined its scope of responsibility,

important aspects of care, quality indicators, and thresholds?

- Do the indicators evaluate all critical aspects of care?
- Are the indicators appropriate?
- Are pertinent findings acted upon?
- Have monitoring results been communicated internally and to other affected departments, services, and disciplines?
- Have relevant findings been incorporated into system changes to improve individual practices?

These questions can be answered easily by members of the department quality committee. Another, more valuable, approach would be to ask these same questions of staff member(s) who have been on the periphery of the process. If non–committee members have a working knowledge of the process and are able to answer the questions accordingly, this is a good indication that the reach of the quality program has indeed extended past those who have been "assigned" to monitor quality.

Data Processing

A voluminous amount of information can be accumulated over the course of a year. Gathering and processing data require time. Manpower is a major resource needed for any quality program. When evaluating this program aspect, consider the ease with which information is obtained and the usefulness of the information, as well as where it comes from and where it is going.

Evaluate whether time is used efficiently by examining how data are gathered and stored. If a manual system is used, the annual evaluation time is a good opportunity to re-examine whether resources are available now that were not in place a year ago to computerize the process. When the following questions were posed to a group of dietitians and managers, they were able to suggest a number of improvements that either streamlined the collection process or increased the validity of the data.

- Are data collected by the individual(s) with easiest access to the information? *Suggestion*: The admitting department sent a patient census report to the nutrition department's purchasing group to compute month-end financials. This information was also needed by the nutrition department's assistant director in another departmental area for productivity calculations. Every month the assistant director made several calls to a friend in the admitting department to obtain a verbal report of the information she needed. Neither the nutrition department nor the admitting department was aware that the same information was needed by nutrition services in two different areas for different purposes. Once the duplication in communication flow and the overlap in report calculations were identified, the nutrition department centralized all month-end reports within a single area.
- Is data collection timely enough to be useful? *Suggestion*: The morning admission census was used in determining staff assignments for the entire day. Often this information was irrelevant by the time the evening meal was served. This resulted in inefficient staffing patterns, last-minute readjustments, and late dinner service. The problem was corrected by having a supervisor get updated census data from the admitting department prior to the beginning of the second shift.
- Are records centralized and conveniently accessed? *Suggestion*: A multidisciplinary team was studying the enteral feeding infusion process. The group needed information related to patient tolerance of the feeding, medications mixed with formula, and accurate intake and output calculations. A major obstacle encountered was the inordinate amount of time required to gather data to analyze the system. Each piece of information was kept in a different part of the medical record. Furthermore, some information was kept at the nursing station and some at the patient's bedside. The team designed a form that consolidated all needed information on a single sheet. With the assistance of the nurse and medical record team members, the group was able to conduct a trial of the condensed format in the intensive care unit that had the highest incidence of enteral feedings. The form not only provided the team with the information they needed to study the process accurately, it also became a permanent part of their solution to improve the process as a whole.
- Are all data essential? *Suggestion*: Program evaluation should address critically the essentiality of each piece of data. Data should be available for each targeted major aspect of care. When quality programs are initially set up, it is often difficult to part with some perennial favorite indicators that are holdovers from traditional quality assurance programs. If there is conflict initially among the group about whether to keep or eliminate a particular indicator, it may be best to leave it in. The group can return to the issue during the evaluation process, and if in fact the information has improved the process, it can be retained. If not, it can be deleted at that time and the group should be more comfortable with the decision. Any extraneous information that does not contribute to the quality management process should be eliminated at this point so that time spent on quality improvement can be allocated more effectively in the future. For example, during a mock review a group of dietitians debated deleting two questionable pieces of data collected when monitoring consultations:

1. Medical record documentation includes time in appropriate military format.

2. Patient charge information is entered into the computer within 24 hours of completing the consultation.

The manager thought the first monitor was important because it helped determine whether required time frames were met (to the hour) in responding to consultations. A dietitian thought the second monitor should be deleted because it had nothing to do with the quality of the nutrition recommendations made. Via mutual agreement, the group opted to cease gathering information on military time but continue monitoring billing punctuality. The rationale was that the first piece of information was not deemed essential, as it had little or no impact on patient outcome or satisfaction. However, information on patients' bills was very important to those on the receiving end, whether it be the patient or a third-party payer. Delayed charges often did not make it on the primary bill. A separate, follow-up bill had to be sent that required duplication of work on the part of the fiscal department, confusion on the part of the patient, and increased chance for denial on the part of third-party payers. The need to "do it right the first time" justified maintaining this monitor.

Feedback from Staff

Dietitians, technicians, and other clinical staff members who have participated in quality monitoring and evaluation can provide valuable insight regarding their experiences with the process. Staff can rate the effectiveness of the indicators they have monitored for identifying potential areas for improvement. Since it is these employees who conduct peer reviews and who may have rotating monitoring responsibilities, their input can be especially relevant in evaluating the effectiveness and efficiency of the process.

Input from staff dietitians can be obtained anecdotally or by survey. Formats similar to those in Exhibits 18–2 and 18–3

Exhibit 18–2 Quality Program Assessment

This assessment has been designed for you to evaluate our Program. As you know, Quality Improvement is only as effective as we make it. Please place an " X " in the "yes"/"no" columns with explanations, suggestions of who, what, where, and when in the "Comment" section.

A. Does the Quality Improvement Plan:

	Yes	No	Comments
1. Refer to the organization's philosophy?			
2. Refer to the organization's goals/objectives?			
3. Describe the purpose of the Quality Improvement Program?			
4. Clearly define its goals/objectives?			
5. Define the program's scope including integration with hospital and medical quality improvement programs?			
6. Address authority?			
7. Identify person(s) responsible for coordinating the program?			
8. Describe the Quality Improvement Committee: a. purpose/functions? b. membership? c. chairperson?			
9. Identify committees interfacing with Quality Improvement Committee?			
10. Describe a monitoring approach (i.e., problem-focused)?			
11. Delineate lines of communication?			
12. Address confidentiality?			
13. Include a plan for periodic evaluation?			

B. As a result of Quality Improvement efforts, do you think:

	Yes	No	Comments
1. You appropriately utilize standard(s) established by: a. the Department? b. Accreditation requirements? c. Health Department requirements?			
2. Your practice has improved?			
3. Your documentation has improved?			
4. Patient care has become more cost effective?			
5. Patient satisfaction has increased?			
6. Job satisfaction has increased?			
7. Staff performance appraisals are more accurate?			
8. Your time and involvement is worthwhile?			
9. Continuing education programs are more relevant?			

C. Additional Comments:

Source: Adapted from *Quality Assurance: A Complete Guide to Effective Programs,* by C.G. Meisenheimer, pp. 76–77, Aspen Publishers, Inc., © 1985.

Exhibit 18-3 Annual Assessment of Monitoring and Evaluation Activities

Name of Organizational Unit:

Assessment Year:

Date of Assessment:

Individual Performing Assessment:

Please review your monitoring and evaluation (M&E) activities for the past year and assess the effectiveness of the M&E process within your organizational unit.

A. Indicators

List all indicators used in monitoring and evaluating your service during the past year.
1.
2.
3.
4.
5.
6.
7.
8.
9.
10.

B. Assessment of Effectiveness*   Indicator 1 2 3 4 5 6 7 8 9 10

1. Implementation Status
   a. Not implemented
   b. Data collected only
   c. Data collected, analyzed, and compared to thresholds
   d. Cases evaluated and findings reported
   e. Reports communicated
   f. List number of months implemented.

2. Methodology
   a. Were any problems encountered with the M&E methodology? (Y=Yes; N=No) If YES, indicate
   a. Implementation was not practical/problematic.
   b. Data collection process was not possible/problematic.
   c. Data collected did not adequately measure indicator, changes needed.
   d. Criteria (if used) not reliable/objective.
   e. Indicator did not address quality or appropriateness (eg, measured quality control).
   f. Threshold too high or low.

3. M&E Outcome
   a. Did the indicator identify problems and/or opportunities to improve? (Y or N)

   Indicator 1 2 3 4 5 6 7 8 9 10

   b. Will this indicator be continued next year? (Y or N) If YES, will there be changes made to the
      1) data collection process?
      2) criteria?
      3) threshold for evaluation?
   c. Were changes made in selecting important aspects of care for monitoring because they were no longer high risk, high volume, or problem prone? (Y or N)

4. Overall Assessment
   a. Identifying problems and/or opportunities to improve care, this indicator was
      1) most effective
      2) somewhat effective
      3) not effective
   b. In measuring quality and appropriateness of care/service, this indicator was
      1) most effective
      2) somewhat effective
      3) not effective

5. Documentation of M&E Effectiveness

Please list any opportunities to improve care or services and describe any problems identified and resolved or addressed through the M&E process for the past year.

* For each indicator, assess its effectiveness.

Source: Reprinted from Hopkins, J.L., Evaluating the QM Program, *QRC Advisor*, Vol. 7, No. 1, p. 5, Aspen Publishers, Inc., © 1990.

may be useful for gathering this information. They provide feedback on the methodology of data collection, such as its availability and validity. Evaluation should be geared to elicit whether staff favor continuing, altering, or deleting particular indicators during the next cycle.

An example of how the nutrition care service at the University of Utah Hospital and Health Sciences Center utilized an evaluation form is shown in Exhibit 18–4. Three staff members completed the assessment for eight indicators monitored over the past year. Determining the efficacy of indicator num-

Exhibit 18–4 Annual Assessment of Monitoring and Evaluation Activities

Name of Organizational Unit: Nutrition Care Service
Assessment Year:
Date of Assessment:
Individual Performing Assessment:

INDICATORS

LIST ALL INDICATORS USED IN MONITORING AND EVALUATING YOUR SERVICE DURING THE PAST YEAR

1. Identification of patients requiring nutritional intervention by questionnaire and/or screen form according to Nutritionist
2. Accuracy of calorie count implementation
3. Timeliness of calorie count implementation
4. Accuracy of implementation of diet order
5. Proper performance of nutritional assessment for consultation/care plan
6. Accuracy of administration of enteral and TPN (i.e., tube feedings and TPN)
7. Patient surveys which indicate a need for improvement in service provided
8. Identification of patients on unsupplemented NPO or clear liquid diets to assure appropriate nutritional intervention

IMPLEMENTATION STATUS	1	2	3	4	5	6	7	8
Not implemented								
Data collected only								
Data collected, analyzed and compared to thresholds	X	X	X	X	X	X	X	X
Cases evaluated and findings reported	X	X	X	X	X	X	X	X
Reports communicated	X	X	X	X	X	X	X	X
List number of months implemented	6	12	12	12	12	12	10	6
METHODOLOGY								
Were any problems encountered with the M&E methodology? (Y = YES, N = NO) If Y indicate:	N	Y	N	Y	N	Y	Y	N
Implementation was not practical/problematic		X		X		X	X	
Data collection process was not possible/problematic				X		X	X	
Data collected did not adequately measure indicator, changes needed							X	
Criteria (if used) not reliable/objective							X	
Indicator did not address quality or appropriateness (e.g., measured quality control)						X		
Threshold too high or low						X	X	
M&E OUTCOME								
Did the indicator identify problems and/or opportunities to improve? (Y or N)	N	Y	N	Y	Y	Y	Y	N
Will this indicator be continued next year? (Y or N)	Y	Y	N	Y	Y	Y	Y	N
If Y, will there be changes made to the:								
data collection process	Y	Y		Y	Y	Y	Y	
criteria	N	N		N	N	N	N	
threshold for evaluation	N	N		N	N	Y	Y	
Were changes made in selecting important aspects of care for monitoring because they were no longer high risk, high volume or problem prone? (Y or N)	Y	N	Y	N	N	Y	N	Y
OVERALL ASSESSMENT								
Identifying problems and/or opportunities to improve care, this indicator was:								
most effective		X		X	X		X	
somewhat effective								
not effective	X		X			X		X
In measuring quality and appropriateness of care/service, this indicator was:								
most effective	X	X	X	X	X			X
somewhat effective							X	
not effective						X		

Courtesy of Nutrition Care Services, University of Utah Health Science Center, Salt Lake City, Utah.

ber six, related to the accuracy of the administration of total parenteral nutrition (TPN) solutions, was rated especially problematic. Based on this information, the department elected to evaluate further why it was difficult to assess this indicator. As part of their investigation they developed a cause-and-effect diagram detailing the causes of inaccurate TPN administration. As a result of their efforts the departmental quality committee took four actions.

1. Initiated formation of an interdisciplinary task force for the units with highest TPN utilization, to track areas of variance. Task force members included a dietitian, a head nurse, a pharmacist, and a physician.
2. Developed and piloted guidelines for properly filling out TPN order forms.
3. Participated in the revision of intake and output (I & O) sheets.
4. Educated nursing staff in high TPN usage areas regarding importance of nutrition as a vital medical therapy.

Note how the action plan incorporated actions, such as the revision of the I & O sheets, which will facilitate the monitoring and evaluation process in the future.

Strategic Problems and Opportunities

If a group has not been successful in meeting its goals or has spent an inordinate amount of time on tangential issues, it may have been following the wrong course from day one. Perhaps it was attempting to resolve a massive issue rather than a defined project. A team can spend an inordinate amount of time studying a sideline issue that is not part of the problem. The annual review provides the opportunity for the department to assess how well it is doing on selecting opportunities that it has the time, resources, and wherewithal to improve effectively.

Meetings

The effectiveness of team meetings is critical to the success of quality management and should be evaluated as part of the annual review. One type of meeting evaluation format is found in Exhibit 18–5. For this assessment to be meaningful, the information must be gathered while the teams are still in session. The team should critique itself and should select the evaluation criteria based on what the members feel is important. "The most convenient setting for evaluation is at the end of meetings, but this is often difficult because people are often tired and don't feel like challenging themselves" (Scholtes 1988, 4-33). An alternate recommendation is for the team to rate itself at the midpoint of the meeting rather than at the end. That way, if a change of direction is needed, it can be incorporated during the remaining time.

Exhibit 18–5 One Type of Meeting Evaluation Format

To use this format, the team leader or other facilitator asks team members to rate the meeting on criteria important to the team. Often team members work individually first, then share and discuss answers with their teammates. Other evaluation formats are less structured.

Our meeting today was:
Wonderful	1 2 3 4 5 6					Lousy
Very Focused	1 2 3 4 5 6					Rambling
Energetic	1 2 3 4 5 6					Lethargic

•
•

I would characterize our methods as:
The Scientific Approach	1 2 3 4 5 6					Shooting From The Hip
Cooperative	1 2 3 4 5 6					Divisive

•
•

Source: Reprinted from *The Team Handbook,* by P.R. Scholtes, et al., pp. 4–34, with permission of Joiner Associates, Inc., © 1988.

Individual staff members may spend large blocks of time in quality-related meetings. This represents an enormous investment on the part of the department. A review of the meeting evaluation forms provides the opportunity for the department to assess how well meeting time is being utilized and judge the value of the return on its investment. Meeting evaluations also provide valuable feedback on the skills and effectiveness of team leaders, facilitators, and members. This information can be used to select topics for future continuing education programs and to facilitate a change in group dynamics, when appropriate.

Communication Flow

Is information supplied to only one group or shared with others? The tendency to conduct quality assurance activities as covert operations is self-defeating in the quality management framework. Program review should include a component that evaluates whether information has been sent as outlined in the information flowchart. Inquire whether those on the receiving end of the information flow have in fact received the data and whether they were received in a timely manner.

Interruption of the information flow can thwart the effectiveness of the entire quality management process. One dietitian described how a standardized form was developed for ordering enteral feedings. The form was designed to improve the care of patients by initiating early dietitic intervention and to promote consistency in monitoring and follow-up activities. It was developed through the joint efforts of the appropriate multidisciplinary staff members. It received the stamp of approval from medical records and was then printed and distributed to nursing units.

Although the form had worked well during a trial period, after it was formally adopted the expected improvements were not realized. An informal poll revealed the vast majority of floor staff was not aware of the form's existence. The major obstacle to success in this instance was not in the problem-solving phase but in the information flow. There had been inadequate communication regarding the availability and purpose of the new order sheet. The staff needed to emphasize communication strategies to achieve their goals fully.

Reports

Reports should be user friendly. If someone unfamiliar with a particular indicator can review a report and make an appropriate judgment as to what the results show, it probably meets this requirement. Results should include an analytical component, such as a graph of the relevant information, rather than just contain a compilation of raw data. Include a standard of reference so that outcomes can be compared with thresholds and prior performance. Reports should be edited to exclude nonessential information. Charts or visuals, if used appropriately, may condense or clarify long lists of numbers. Make sure that reports include clear-cut recommendations for required actions. Then check to make sure the reports go directly to those who need to act.

Reports and meeting minutes should be written correctly the first time, using a format that has been predetermined to be acceptable. Often facilities use the same format throughout the house for consistency. The Joint Commission prohibits the falsification of information by adding, deleting, or fabricating quality-related documents. This includes rewriting or reformatting previously prepared quality assessment records. It is important that there is not even the slightest appearance that data have been adjusted for any reason. The Joint Commission policy was enacted to address complaints accusing facilities of altering records to satisfy accreditation requirements. Penalties for falsification can be as severe as the loss of accreditation for the institution (Staff, *Trustee* 1992).

Overall Quality Management Evaluation

Finally, analyze the program from an all-encompassing standpoint. Does the process articulate a complete cycle that begins with problem identification and ends with resolution and sustained improvement (Meisenheimer 1985)? Randomly select one of the department's quality monitoring projects. Examine the steps that were taken to achieve the desired outcome. Analyze the components that worked well and those that did not. Were there trouble spots? What part of the process was most frustrating for participants? Where were the most roadblocks encountered? In the examples provided throughout this book, many actual and potential barriers in developing and implementing the quality management process in clinical nutrition have been cited. Some groups have difficulty scheduling meetings at times when all members can attend. Others may find that, despite their best efforts, traditional hierarchies tend to dominate group dynamics. Group members may become disenchanted with the quality management process if they feel that the same team members are always delegated the same tedious chores (Scholtes 1988). Others are frustrated because their recommendations are never fully implemented because of inadequate resources.

Each department will inevitably discover a set of challenges unique to its own organization. The focus of quality evaluation is to provide a mechanism to identify these issues and develop strategies and alternative approaches that teams can pilot over the next year. Then, at the end of the next cycle, the changes can be re-evaluated to determine whether the process is running more smoothly.

Expect obstacles, but do not make them the focus of the evaluation. To maintain momentum early on, it is important that the evaluation highlight the program's strengths as well as its weaknesses. Include both points in the report summary. Concluding with a list containing only "what went wrong" can sabotage next year's efforts. Identify those procedures that do work well, then apply them across the board. In effect, apply the continuous quality-improvement process to the program evaluation process. Appendix A-7 contains a completed quality program evaluation report including many of the concepts discussed throughout this chapter.

EVALUATION AND ACCREDITATION SURVEYS

The two primary considerations with respect to quality care are the management of an organization and the care that is directly delivered to the patient. Historically, internal and external evaluation of these two areas has been studiously avoided over the years (O'Leary 1991). The Joint Commission shift in focus from quality assurance to continuous quality improvement has redirected attention to management and care issues. Meeting accreditation standards should not be the sole determining factor in pursuing quality improvement. In reality, the need to comply with external review requirements has driven the quality improvement process in many hospitals. An evaluation program should be designed both to achieve internal goals and objectives and to meet externally imposed criteria. Cornett (1992) lists the most frequent problems identified by accreditation surveyors. These are delineated in Exhibit 18-6. Two of these problems relate to the evaluation process, particularly that program evaluation has not been assessed or is incomplete.

The final standard in the chapter on quality assessment and improvement in Volume 1 of the *1993 Accreditation Manual*

Exhibit 18–6 Problems Frequently Identified by Joint Commission Site Surveyors

> Both the JCAHO and CARF compile a list of the most frequent problems identified by surveyors. These problems include:
>
> - Lack of documentation of patient care services, management activities, and processes.
> - Lack of evidence of involvement of the patient and family in goal-setting, treatment, and discharge planning.
> - Goals of rehabilitation are not measurable, or described in functional/behavioral terms.
> - QI plans and monitoring processes are incomplete; there is little evidence of follow-up of problems.
> - Effectiveness of QI monitoring and evaluation program has not been assessed.
> - The program evaluation system is incomplete/inadequate.
> - Safety monitoring and inspections are inadequate.
> - Monitoring and review of case records are inadequate.
> - Employee performance appraisals lack performance expectations and evaluations do not relate to job descriptions.
>
> *Source:* Reprinted from Cornett, B.S., Quality and Accreditation Standards: JCAHO and CARF Surveys, *Quality Improvement Digest*, Spring 1992, pp. 1–5, with permission of the American Speech-Language-Hearing Association, © 1992.

for Hospitals (Joint Commission 1992) defines and addresses requirements for evaluation: "The objectives, scope, organization and effectiveness of the activities to assess and improve quality are evaluated at least annually and revised as necessary" (QA 4.4).

To satisfy this requirement, the scoring guidelines (Joint Commission 1992, Vol 2, 20) recommend that the organization

- provide evidence in the way of reports detailing quality improvement activities
- evaluate the program, at least annually
- be able to demonstrate that changes were made if indicated by the review

All quality-related documentation should be accessible for the surveyor's perusal, including the most recent evaluation report. The director or manager meeting with the surveyor should be able to explain comfortably what the department has accomplished as a result of quality monitoring and evaluation activities (Krasker et al. 1992). The report should clearly identify and support this discussion. Information should include specifics on what has been done over the most recent 12 months. It may be helpful to select several activities and assemble the complete cycle of information related to studying the process. Organize the documents in a fashion that will illustrate the progress made from the problem-identification stage to its resolution. Be able to provide evidence that shows that gains in improvement have been maintained.

EVALUATING BENEFITS OF CONTINUOUS QUALITY IMPROVEMENT

Evaluating and improving the quality management process in the field of clinical dietetics may result in more than what was originally expected when the program was first set in motion. There are potential spinoff benefits. It is great if goals are met. It is even better if some have been achieved that the department did not consciously set out to attain. Consequences related to improved staff morale and job satisfaction are possible as a result of effective quality monitoring and evaluation. Workers may feel less stressed if they have fewer patient complaints to deal with. This may reduce staff turnover as well as recruiting and training costs (Zucchelli and Zucchelli-Sorenson 1991).

Participating in quality monitoring and evaluation provides increased opportunities for professional staff to meet their needs for responsibility, achievement, and recognition. The pleasure of being recognized as a person whose input has made a difference provides a powerful incentive to support the program (Bevsek and Walters 1990). Distributing a staff questionnaire, similar to the one in Exhibit 18–7, can help assess the program's fringe benefits as perceived by employees.

A major advantage of quality management is that it provides a platform in which everyone is communicating on the same level (Robert et al. 1990). For small working groups, it provides a straightforward method of communication that is based on facts. This helps allay the emotional side of the problem-solving process. For large, complex organizations, it creates a framework for multidepartmental interaction. It provides the opportunity, probably for the first time, for both individuals and groups to seek input from others and ask how they can be of help. In industries in which quality management has been established for longer periods of time, managers report that rather than being task-oriented, people "are thinking more about the meaning of their jobs. And, that creates improvements all the way down the line" (Robert et al. 1990, 61).

PLANNING FOR THE FUTURE

The focus of quality management is improvement, and improvement takes time. Therefore, quality programs should be aimed at customers' needs, present and future (Lehr and Strosberg 1991). For this to happen, quality initiative must be incorporated into the department's strategic plan. An example of how this has been accomplished by the nutrition services department of the Cleveland Clinic Foundation is provided in Exhibit 18–8. Based on a five-year vision statement, milestones are projected for the department's financial, marketing, logistics, operations, employee development, and education endeavors for intervening years. Quality improvement components are built in throughout the strategic milestone projections. Note, for example, how the elements pertaining to

Exhibit 18-7 Implementing Quality Service Initiative: A Self-Assessment Tool

---

Use this assessment tool to help identify areas of strengths and opportunities. You and/or your Quality Service Initiative team can diagnose the service culture in your area and devise ways to strengthen it.

	Yes	No
1. We seek feedback on how other departments (our internal customers) perceive our service.		
2. We regularly seek feedback from our key external customer groups to understand their service expectations.		
3. Employees regularly give their input on service problems and service improvements.		
4. We monitor our service quality over time.		
5. We cross department lines to solve service problems.		
6. Customer feedback is regularly shared with staff.		
7. Employees value our department reward and recognition efforts.		
8. Employees participate in planning for changes that will affect them.		
9. Accomplishments and improvements that contribute to customer satisfaction are shared with staff.		
10. I actively advance our priority on QSI by allocating time for implementing service improvements.		
11. Employees know the service behaviors expected of them in their jobs.		
12. The steps to build cooperative team relationships with physicians and/or other depts. are actively taken.		

	Yes	No
13. Special events are conducted to energize staff with respect to service.		
14. Staff receive frequent pats on the back for positive service performance.		
15. We seek reactions from customers, including employees, to improvements we plan as a result of their issues or suggestions.		
16. Staff are actively encouraged to take advantage of service training opportunities.		
17. Management staff solicit, listen and respond to staff concerns.		
18. Time and energy is devoted to continuous improvement.		
19. Employees are regularly informed about the financial situation and what is being done to succeed.		
20. Managers in our area take the time to nurture and support staff.		
21. Management in our areas consider employees as customers whose satisfaction is key to the success of RMH.		
22. Employees work as a team to provide exceptional service.		

Involve staff to celebrate the "yes" answers. Discuss the Quality Service Initiative accomplishments and reinforce employee commitment to service. Include the "base hits" as well as the "home runs." If you answered "No" to any item, brainstorm ways you and your staff can make improvements. It's one way to initiate new energy and direction for the Quality Service Initiative teams.

Source: Reprinted with permission of Riverside Methodist Hospitals, Columbus, OH. All rights reserved.

---

benchmarking, building liaisons, and improving the repertoire of clinical products and services are addressed in light of how the future will affect department operations.

Any quality management system is geared to produce results over the long haul. Many of the Japanese companies that adopted the quality management principles process in the 1950s had to wait until the 1970s to reap the benefits (Mathews and Katel 1992). To ensure continuity and relevance, the department's quality program should mesh with both the hospital's and the department's long-term vision and strategies. A few strategies, described below, have been suggested for the hospital of the future (Staff, *KSA Perspective* 1991). If clinical nutrition professionals desire to have a part in that future, they should plan to adapt and integrate these proposals into their own strategic plans.

One strategy is to develop integrated services to provide patients with the right treatment at the right time and place, and at their convenience. The successful nutrition services department must diversify to offer a full range of services, along a continuum from primary to acute care. For most organizations this is business as usual already. What may change in the future, however, is the location where these services are provided. Provision of services must be specialized to customer needs, distributed throughout the geographic area served, and designed for easy customer access.

A second strategy is increased nutrition services in non-hospital settings. Traditionally, proportionately more dietitians are employed by hospitals than by other institutions. In most hospitals, the majority of services are aimed at inpatient care. "While inpatient volume will continue to grow because of aging and population growth, the most significant growth in health care volume will be in ambulatory care" (Staff, *KSA Perspective* 1991, 4). More patients will be treated without ever entering an inpatient facility. To remain competitive, dietitians must continue to improve nutrition services in the acute care setting while further developing nutrition care delivery systems for the chronically ill patient in the home or ambulatory centers such as health clinics, neighborhood service centers, senior citizen housing projects, and congregate feeding or recreation centers. To do this successfully and efficiently, nutrition professionals will need to form alliances with other health care providers and become part of mobile delivery care teams.

Enhanced value will result from integrating nutrition services and shifting delivery sites to the patient's locale. Constantly revising quality-improvement objectives based on customer needs will assist dietitians in moving forward to accommodate this change.

Exhibit 18–8 Five-Year Vision Statement with Three-Year and Two-Year Milestones Showing Incorporation of Quality Management into a Strategic Plan

Nutrition Services—1993 Strategic Plan

THREE-YEAR STRATEGIC MILESTONES (1995)

- **Financial:** Profit margin 5–10%; increases in expenses have paralleled increased revenues; billable services include screens for nutritional risk.

- **Marketing:** Nutrition services are included as an important part of managed care partnerships that the Foundation has forged with major self-insured corporations and other third-party payors.

- **Logistics:** Clinical nutrition network includes all practitioners, including off-site satellite offices in the Grogan and West Side Dialysis buildings; order entry and all internal patient files are maintained on the network.

- **Operations:** Volunteer program is fully implemented; staff is participating in cross-functional clinical teams; benchmarking of major products and services is in place.

- **Employee Development:** All performance evaluations include a peer and/or subordinate feedback component; use of quality management tools and methods are routinely used to improve clinical products and medical services.

- **Education:** The dietetic internship is modified to coincide with the supervised practice recommendations by ADA's Critical Issues Task Force, possibly necessitating more liaisons with didactic programs in dietetics.

TWO-YEAR STRATEGIC MILESTONES (1994)

- **Financial:** Profit margin 5–10%; increases in expenses have paralleled increased revenues; financial support for clinical nutrition services is included in national healthcare reform legislation.

- **Marketing:** Nutrition services are offered in a "packaged" format as an incidental aspect of physician care to offset RBRVS cutbacks and enhance managed care initiatives within the Foundation.

- **Logistics:** Centralized operations in the hospital have moved from M-17 into M-18; the clinical nutrition network includes all main campus practitioners, including offices in the Crile, Rehab and Cancer Center buildings.

- **Operations:** Pilot program using volunteers is launched; intradepartmental communications occur using electronic mail through the clinical nutrition network; the interface between A/D/T and Diet Office is implemented.

- **Employee Development:** Managerial and dietitian performance evaluations include a peer and/or subordinate feedback component; quality management tools and techniques are used to improve support services and some clinical products.

- **Education:** The mid-accreditation Program Evaluation Document is submitted to maintain approved internship standing with the ADA; specialized nutrition practitioners can competently handle general nutrition inquiries from internal and external customers; graduate students from OSU are assisting with clinical research analysis and publications.

Courtesy of The Cleveland Clinic Foundation, Cleveland, Ohio.

REFERENCES

Bevsek, S.A., and J.A. Walters. 1990. Motivating and sustaining commitment to quality assurance. *Journal of Nursing Quality Assurance* 4, no. 2:28–36.

Cornett, B.S. 1992. Quality and accreditation standards: JCAHO and CARF surveys. *Quality Improvement Digest* Spring:1–5.

Dougherty, D.A. 1992. JCAHO/agenda for change one year later. *Clinical Nutrition Management* 10, no. 3:9–10.

Gardner, S. 1990. Q and A on QA. *Clinical Nutrition Management Newsletter* 9, no. 5:6–8.

Hegelein, S. 1992. Utilizing the CQI process to improve the treatment of nosocomial decubitus ulcers. *Journal of the American Dietetic Association* 92 (suppl), no. 9:A–65.

Hester, D.D., and D.B. Ford. 1992. The use of preprinted order labels. *Journal of the American Dietetic Association* 92 (suppl), no. 9:A–66.

Hopkins, J.L. 1990. Evaluating the QM program. *QRC Advisor* 7, no. 1:1–9.

Joint Commission on Accreditation of Healthcare Organizations. 1992. *1993 Joint Commission Accreditation Manual for Hospitals*. Oakbrook Terrace, Ill.

King, P. 1992. A total quality make-over. *Food Management* April:97–107.

Krasker, G.D., et al., presenters. 1992. *Executive Briefing at the Ohio State University Hospitals: 1993 AMH Standards and Survey Process*. In-service, October. Oakbrook Terrace, Ill: Joint Commission on Accreditation of Healthcare Organizations.

Laramee, S.H. 1992. Evaluation of nutrition referral form. *Journal of the American Dietetic Association* 92 (suppl), no. 9:A–85.

Lehr, H., and M. Strosberg. 1991. Quality improvement in health care: Is the patient still left out? *Quality Review Bulletin* 17:326–329.

Mathews, J., and P. Katel. 1992. The cost of quality. *Newsweek*, September 7, 48–50.

Meisenheimer, C.G. 1985. Designing a quality assurance program. In *Quality assurance: A complete guide to effective programs*, ed. C.G. Meisenheimer, 73–89. Gaithersburg, Md: Aspen Publishers, Inc.

O'Leary, D.S. 1991. Accreditation in the quality improvement mold—A vision for tomorrow. *Quality Review Bulletin* 17, no. 3:72–77.

Robert, M., et al. 1990. A worker's mind is a terrible thing to waste. *Quality Progress* 23:59–61.

Sawyer-Richards, M. 1991. Program evaluation and nursing quality assurance. In *Issues and strategies in nursing care quality*, ed. P. Schroeder, 179–198. Gaithersburg, Md: Aspen Publishers, Inc.

Scholtes, P.R., and other contributors. 1988. *The team handbook.* Madison, Wis: Joiner Associates, Inc.

Staff. 1991. The hospital of the future. Atlanta: Division of Kurt Salmon Associates. *Hamilton/KSA Perspective* July:1–6.

Staff. 1992. JCAHO prohibits rewriting records. *Trustee* 45, no.7:22.

Zucchelli, L., and S. Zucchelli-Sorenson. 1991. Put the patient first, or else your hospital may not last. *Healthweek,* June 21.

Chapter 19

Quality Management in Hospital-Affiliated Services

Much of the activity related to quality monitoring and evaluation is focused on acute care hospitals. However, as hospitals expand their scopes of service to meet the health care needs of targeted population groups, hospital-based dietitians must extend quality nutrition services to diverse settings. Concern for quality is paramount in these health and nutritional care delivery settings for several reasons.

- Consumers are demanding more value for their health care dollar, in whatever setting services are delivered.
- The public, disenchanted with high costs and proliferation of unnecessary procedures, is beginning to hold health care professionals accountable for the outcomes of comprehensive health care.
- The trend toward both early hospital discharge and the expansion of outpatient services, home health care, and long-term care facilities have contributed to the movement for extension of quality assessment and improvement beyond the hospital.
- Managed care programs provide a mechanism to track patient care services and outcomes within a network of delivery systems.

Hospitals, cognizant of the de-emphasis on inpatient services, look to alternative strategies for maintaining viability and growth. Many hospitals have converted inpatient space to other health care uses. Other hospitals have developed off-campus health care options under the jurisdiction of hospital-based personnel. Quality and efficiency are concerns here just as they are in the acute care setting.

This chapter addresses quality programs in hospital-affiliated programs and settings, including guidelines and standards, description of various settings where nutritional services may be provided, sample quality indicators, organizational structures for quality management, and application of the monitoring and evaluation process.

GUIDELINES AND STANDARDS

Standards of quality for non–acute care nutritional services are provided by the Joint Commission, Health Care Financing Administration, governmental agencies, and professional organizations. When these services are administered by the hospital, quality is monitored as part of the overall quality management program.

Joint Commission Standards

The Joint Commission has had a long history of involvement in evaluating quality in long-term care. The *Long Term Care Standards Manual* (Joint Commission 1988a) is used to evaluate quality care in skilled nursing facilities that undergo voluntary accreditation. This manual parallels the one for hospitals and is organized in similar fashion, except that the standards apply specifically to long-term care settings. One section of the manual is devoted to standards for dietetic service. Other sections of the manual also relate to dietetic services, such as "Organizational Structure and Management," that addresses policies and procedures and quality assurance requirements; and "Patient/Resident Care Management," which includes standards and guidelines for patient assessment, care planning, nutrition intervention, and evaluation of effectiveness of care.

Joint Commission standards for home care (1988b) and ambulatory care (1987) do not include specific sections for nutrition services. However, these manuals include standards for patient/client care and quality assurance activities.

Omnibus Budget Reconciliation Act of 1987 (OBRA)

The Health Care Financing Administration (HCFA) published regulations that cover requirements for long-term care facilities that provide services funded through Medicare and Medicaid (OBRA '87 update 1991). These regulations focus on such things as resident rights, quality of life, resident assessment, dietary services, administration, and quality of care. Dietitians are held accountable for high-quality nutrition care and the elimination of malnutrition. Quality-of-care provisions of particular concern to dietitians are care of residents

- with pressure sores to promote healing, prevent infection, and prevent development of new sores
- receiving enteral nutrition support to ensure compliance with nutritional needs; prevention of diarrhea, vomiting, dehydration, and metabolic abnormalities; and, if possible, to resume normal feeding
- who require a therapeutic diet to ensure that optimal nutritional status is maintained unless the resident's clinical condition demonstrates that this is not possible
- to ensure that proper hydration is maintained
- receiving parenteral fluids or nutrition support to ensure proper treatment in line with assessed needs

These regulations also give residents the right to choose where, when, and what they eat. Residents also have the right to refuse a therapeutic diet. Nutrition assessments must be conducted upon admission and periodically thereafter.

State and Local Agencies

Quality standards for food services often come under the purview of local health departments. Examples of such institutions and programs sometimes managed by hospitals are daycare facilities, school lunch programs, psychiatric facilities, and home-delivered meal programs.

When institutions and agencies are not covered by a system of voluntary accreditation, appropriate standards of quality for nutrition and other health services may be established by state licensing agencies or health departments. Health maintenance organizations; home care agencies; hospice programs; and women, infants, and children (WIC) programs fall in this category.

Professional Associations

Generic standards of quality were developed by the American Dietetic Association (1986). These six standards must be adapted in specific terms for each institution or agency. For example, these standards were used to define 27 outcome-oriented guidelines for quality management in long-term care facilities (Gilmore et al. 1993). The standards can be aptly applied in other situations as well.

The American Dietetic Association regularly publishes position papers on topics related to quality nutritional care and services. Contents of these papers can be used as the basis for establishing practice guidelines and quality indicators in specific settings. Pertinent position papers are cited in the appropriate sections of this chapter.

Through collaborative efforts of the American Academy of Family Physicians, the American Dietetic Association, and the National Council on the Aging, Inc., guidelines for nutrition screening were developed and published (Nutrition Screening Initiative 1991). Although the guidelines were established primarily for older Americans, they can be adapted for use in any health care setting.

Through another collaborative effort, a consensus panel of dietitians, physicians, nurses, and others developed practice guidelines for the nutritional care of outpatients with type II diabetes mellitus (Franz 1992). These guidelines, including expected outcomes related to nutrition counseling and therapy, can be incorporated into quality management systems in a variety of outpatient settings.

The American Society for Parenteral and Enteral Nutrition publishes guidelines as they relate to specialized nutritional support. For example, this professional group published standards for home nutrition support (1992), standards for nutrition support for residents of long-term care facilities (1989), and standards of practice for nutrition support dietitians (1990). These documents are particularly useful as a basis for developing a quality management program in the non–acute care setting.

Under the sponsorship of the National League for Nursing, Mitchell and Storfjell (1989) developed *Standards of Excellence for Home Care Organizations*. Only the hospice section of this manual deals with nutritional care directly. Provisions are described later in this chapter.

QUALITY MANAGEMENT IN LONG-TERM CARE

Many hospitals or health care systems have developed extended care units or acquired nursing homes as part of a strategic plan. Usually hospital-based dietitians assume responsibility for nutritional care services in these settings.

Nutritional well-being of the elderly is an integral component of the health, independence, and quality of life (American Dietetic Association 1987b). The aging process is often accompanied by chronic diseases and other disabilities, making it difficult for the older person to remain fully independent. One study showed that 39 percent of patients admitted to a skilled nursing facility were malnourished (Nelson et al. 1993). Thus eldercare provided in a step-down hospital unit or skilled nursing facilities is an important phase of health care and represents a unique challenge for dietitians (Posner et al. 1987).

Nursing homes differ from hospitals in several important ways (Kane 1981):

- Quality programs in long-term care (LTC) facilities must monitor quality of life as well as quality of care.
- LTC facilities are smaller and have lower staff-patient ratios and higher personnel turnover. Thus LTC facilities generally lack the technology and research base necessary for an elaborate quality assessment and monitoring program.
- The quality of record systems in LTC facilities varies, and even good record systems rarely contain quality-of-life information, such as psychosocial well-being.
- Physicians do not generally take as active a role in LTC as in hospitals, tipping the delicate balance of power and quality control in favor of administrators.
- Residents in LTC facilities usually have multiple chronic conditions and complex psychosocial needs that affect their functional status, making it impractical to use diagnostic classification systems as the basis for developing quality indicators. Diagnoses alone only partially suggest the kind of care required by various members of medical, nursing, and allied professionals.
- Because LTC lengths of stay are longer than those in hospitals, outcomes of care are more easily monitored.
- Unlike hospitals, where physician performance figures predominately in quality evaluations, quality in LTC facilities depends heavily on non–physician personnel, whose efforts can go a long way toward improving both quality of care and quality of life. By the same token, these same staff members can affect the patient or resident negatively through inadequate or incompetent care.

Because of these differences between hospitals and LTC facilities, quality-improvement programs in these types of institutions are different. In the past, quality assurance efforts within long-term care facilities have shown only limited success in improving quality (Dimant 1991). Thus major changes in quality assessment and monitoring are afoot in LTC because of OBRA guidelines, new Joint Commission standards, and the application of continuous quality improvement (CQI) principles. Many long-term care administrators have become familiar with key CQI principles and are beginning to implement them. Principles of particular importance used by administrators and summarized by Gustafson (1992) are the need to do the following:

- Train both managers and employees, including emphasis on commitment to the *mission* of the facility.
- Identify one's *customers* and their unique needs, and design and provide services that address those needs.
- Recognize that any process can be improved, and then commit oneself to *continuously* working to improve the process.
- *Empower employees* by giving them the authority, training, and resources necessary to deliver quality care and services.
- Believe that most problems are the fault of systems and processes, not the staff members, and direct most of the quality-improvement effort toward *identification and modification of inadequate processes,* instead of finding and punishing poor work performance.
- Commit oneself and the institution to the use of *data* to manage and improve processes.
- Develop *long-term partnerships* with collaborators who are likewise committed to the delivery of high-quality goods and services.
- Demonstrate top administrators' *commitment* to continuous quality improvement by creation of an environment where quality care can flourish.
- Learn from others and use *benchmarks* of the best as the basis for establishing quality processes and thresholds.

These principles and a process for quality monitoring and evaluation can aptly be applied to dietetic services. Because the perceived roles of full-time dietitians, part-time dietitians, or consultants may differ (Gilbride and Simko 1986; Finn et al. 1991), these principles will need to be adapted to suit specific facilities and situations.

Exhibit 19–1 gives examples of quality indicators for use in long-term care settings. As noted, these indicators address a range of services, including direct patient nutrition assessment and intervention, nutrition education, food services, customer satisfaction, and employee safety. One monitor that could be added to these indicators is healing of pressure sores, especially among residents receiving enteral formulas (Jackobs 1992). Both sentinel and rate-based indicators are included. Also, process changes can be used to improve outcome indicators. For example, Case-McAleer and Hopkins (1992) showed that a liberalization of the therapeutic diet in long-term care resulted in decreased weight loss, increased compliance with diet restrictions, and increased patient satisfaction.

Resident satisfaction is an important element in quality assessment and is one of the factors mandated by OBRA. McChesney (1991) delineated elements of quality usually evaluated by residents, including the following:

- Reliability—The facility should dependably and accurately deliver what it promises.

Exhibit 19-1 Quality Indicators for Long-Term Care Facilities

Clinical Indicators for Long-Term Care Facilities

Function	Outcome Indicators	Process Indicators
A. Nutrition Intervention	1. Nutritionally compromised residents with ineffective treatment (Sentinel Event) 2. Residents involved and satisfied with nutrition intervention	3. Residents with nutrition assessments within 7 days of admission or change in medical condition 4. Residents with appropriate nutrition intervention 5. Nutritionally compromised residents with follow-up on a monthly basis 6. Residents with appropriate diet prescriptions
B. Diet Therapy	1. Residents with adequate nutrient intake 2. Residents satisfied with diet therapy 3. Residents with foodborne illness (Sentinel Event)	4. Diets conforming to the RDA's and diet prescriptions 5. Residents with accurately implemented diet orders 6. Rate of compliance with meal service protocols 7. Rate of compliance with standards for nutrient retention 8. Rate of compliance with food safety and infection control standards
C. Nutrition Education	1. Residents having knowledge and compliance with their therapeutic diets	2. Residents receiving education within 7 days of the determined need or the physicians' orders
D. Cafeteria Services	1. Customer satisfaction with cafeteria service	2. Customers receiving service within 10 minutes 3. Rate of compliance with standards for quality cafeteria service
E. Catering Services	1. Customer satisfaction with catering services	2. Customers receiving service within 5 minutes 3. Rate of compliance with standards for quality catering service
F. Vending Services	1. Customer satisfaction with vending services	2. Rate of compliance with quality catering service standards
G. Employee Safety	1. Rate of work related accidents	2. Rate of compliance with safety standards

Source: Reprinted with permission of American Nutri-Tech, Inc., from *Continuous Quality Improvement for Nutrition Care.* Copyright © 1992 by Rita Jackson.

- Assurance—Residents expect staff members to be courteous and knowledgeable, and to convey trust and confidence.
- Tangibles—The environment should be aesthetically pleasing and indicative of good service.
- Empathy—Residents expect to be given individual attention and to be treated with a degree of caring.
- Responsiveness—Residents expect the staff to help them both willingly and promptly.

Patient and family satisfaction interviews and surveys should include one or more questions regarding nutrition assessments, care plans, nutrition intervention, and meals. Also, the dietitian, with assistance from other dietetic personnel, can routinely monitor resident satisfaction with nutritional services.

Various resources can be used to develop indicators of quality care in long-term facilities. For example, standards for nutrition support for residents of long-term care facilities (American Society for Parenteral and Enteral Nutrition 1989) can be used as the basis for indicators and criteria related to assessment, therapeutic plans, specialized feeding, monitoring, termination of therapy, and organization of services related to specialized nutrition support in nursing homes. Also, the American Health Care Association (1992) published a manual containing regulations, forms, procedures, and guidelines for a long-term care survey. As spelled out in this manual, the objectives of dietary services system assessment are the following:

- Review components of the dietary services system that may, if found deficient, have a negative impact on the health and nutritional status of residents.

- Determine whether the facility prepares and serves food in a sanitary manner, limiting the risk of food-borne illness.
- Determine the adequacy of the food preparation system to meet the nutritional needs of residents, including, for example, the adequacy of planned menus, the preparation of modified diets, and the amount of food eaten by residents.
- Determine the quality of life associated with dining, including, for example, preparation of residents for dining, meal service, and the availability of food choices.

CHILD CARE SERVICES AND FACILITIES

As a service to their personnel or the community, many hospitals now provide day-care facilities for children, dependent adults, or the elderly. Dietitians offer consultation services or they oversee nutritional care in these programs. An important aspect of such professional services should be quality assessment and continuous improvement of dietary and nutrition care.

Standards for child day-care were published by The American Dietetic Association (1987c). These standards include

- food service and meal plans consistent with nutrient needs and cultural food patterns
- emotional climate to enhance food acceptance and enjoyment of meal periods
- physical environment to accommodate the age, size, and developmental level of the children, with special provisions to meet needs of the handicapped
- nutrition education to help children, parents, families, and day-care staff make informed choices affecting their health and well-being
- nutrition consultation from a registered dietitian, especially with regard to nutrition assessment and, when necessary, intervention to provide for special nutritional needs
- compliance with local and state regulations related to food quality, food preparation facilities, food safety, and sanitation.

Care of individuals with developmental disabilities requires special attention, particularly when the disability is related to nutritional problems or is associated with high nutritional risk. For example, individuals with cerebral palsy, muscular dystrophy, and Down syndrome have altered nutrient needs and present feeding problems (American Dietetic Association 1992). Inappropriate eating practices, altered body composition, and limited mobility may cause obesity in persons with developmental disabilities. Other nutritional problems are also commonplace among this population.

The American Dietetic Association (1989b) also "encourages and supports nutrition services which are coordinated, interdisciplinary, family centered, and community based for children with special health care needs." Estimates are that 10 percent to 20 percent of children have some type of chronic disease or handicap that challenges their ability to achieve their maximal potential. Since these children often have nutrition-related problems, they require early screening, periodic monitoring, and interdisciplinary interaction for assessment and intervention.

Exhibit 19–2 offers some possible indicators of quality in child day-care settings. These indicators should be modified as needed to meet the specific circumstances for any given situation, such as community-based care or accommodation of those with special needs (Neville 1987; Farthing and Phillips 1987).

HEALTH MAINTENANCE ORGANIZATIONS

The rise of health maintenance organizations (HMOs) in the 1970s provided increased opportunities for dietitians to accelerate their involvement in preventive nutrition services (Feidler 1993). Dietetic activities in HMOs, and the needs of individuals using HMO services, differ from those typically seen in hospital outpatient departments. Thus standards of quality need to be uniquely tailored for use in these settings.

According to The American Dietetic Association (1993), the dietitian's scope of practice includes

- nutrition assessment for the purpose of determining the needs and recommending appropriate nutritional intake to maintain, recover, or improve health
- nutrition counseling and education as components of preventive, curative, and restorative health care
- development, administration, evaluation, and consultation in regard to nutrition care standards.

Exhibit 19–3 contains examples of quality indicators appropriate for HMOs. It may be appropriate to expand the listed

---

Exhibit 19–2 Quality Indicators for Use in Child Day-Care Facilities

- Children/parents satisfied with food services
- Children receiving service within 10 minutes
- Rate of compliance with standards of quality dining service
- Children with foodborne illness (sentinel event)
- Compliance with standards of food safety and infection control
- Meals and snacks conforming to the nutrition guidelines
- Children with adequate nutrient intake
- Children with knowledge of nutrition principles

Source: Reprinted with permission of American Nutri-Tech, Inc., from *Continuous Quality Improvement for Nutrition Care.* Copyright © 1992 by Rita Jackson.

Exhibit 19-3 Quality Indicators for Use in Health Maintenance Organizations

> - Clients with undetected signs of malnutrition (sentinel event)
> - Nutritionally compromised clients with ineffective treatment (sentinel event)
> - Clients involved and satisfied with nutrition intervention
> - Clients with timely nutrition assessments
> - Clients with appropriate nutrition intervention
> - Nutritionally compromised clients with follow-up on a monthly basis
> - Clients with knowledge/compliance with therapeutic diets
> - Clients with improved nutrition status after education
>
> *Source:* Reprinted with permission of American Nutri-Tech, Inc., from *Continuous Quality Improvement for Nutrition Care.* Copyright © 1992 by Rita Jackson.

indicators to include such things as referral to other health care professionals as appropriate; nutrition surveillance; referrals for special services, such as supplemental food assistance or Meals-on-Wheels; in-service education for staff; and outcomes of physician consultations.

QUALITY MANAGEMENT OF HOME CARE

A large segment of the population remains at risk for malnutrition (Finn 1990). More than half of those who receive home care services may need nutrition services of one type or another (Ribovich et al. 1990). Such nutrition services include

- nutrition screening and assessment
- computer analysis of menus or dietary intakes
- evaluation of nutritional problems
- development of nutrition care plans
- patient nutrition counseling and education
- home infusion therapy
- nutrition monitoring
- consultation with nurses, physicians, and other health care professionals or family members
- evaluation of outcomes

Dietetic services are especially important when patients receive home parenteral and enteral nutrition (American Dietetic Association 1989a). It was estimated in 1986 that more than 50,000 individuals were using home enteral nutrition, with this population growing in both numbers and diversity (Hoffman 1989). In order to attain full benefit of home nutrition support, nutritional requirements must be accurately assessed and administration of products must be carefully monitored to ensure compliance with prescribed regimens, as well as the avoidance of potential infections, mechanical difficulties, and metabolic complications.

Standards for home nutrition support (American Society for Enteral and Parenteral Nutrition 1992) delineate standards as they relate to

- organization, including professional roles and responsibilities, policies and procedures, documentation, and care planning
- patient selection
- treatment plans, including formula selection and mode of delivery
- infusion devices, as well as patient and family education
- patient monitoring
- termination of therapy

Consultant dietitians or those employed by home care service agencies should be familiar with these standards and use them as appropriate to monitor and evaluate quality.

Reimbursement for home nutrition services is an important factor in the decision to include a dietitian on the home care team. Nutrition support, particularly enteral and parenteral nutrition, is recognized as a key element in medical care and is appropriately reimbursed by third-party payers (Raymond 1990; Bayer 1992; Crocker 1992). Charges for other nutrition services may be included as consultation or education for professionals whose time or activities are directly covered by Medicare, Medicaid, or private insurers (Raymond 1990).

Quality indicators have been drafted for patients receiving home infusion therapy (Howell 1992; Crocker 1992). The focus of draft indicators includes the type of nutrition support services associated with

- unscheduled inpatient admission
- discontinued infusion therapy
- interruption in infusion therapy
- prevention and surveillance of infection
- reporting adverse drug reactions
- client monitoring and appropriate intervention

Customer service and quality improvement are key elements in home therapy from the standpoint of both accreditation and business success.

Other quality indicators are shown in Exhibit 19-4. These indicators are fairly comprehensive and deal with both process and outcome measures associated with a wide range of nutrition services. Other agencies are at work defining cross-functional quality indicators that include client-centered outcomes and consumer satisfaction (Peters and Eigsti 1991; Shaughnessy et al. 1992).

The dietitian can collaborate with other team members, using multidisciplinary documentation as a tool for professional accountability and quality assurance. As described by Erickson (1992), multidisciplinary documentation uses the problem-solving approach for "improved organization of the record, greater efficiency in documentation, and a cumulative increase in time and energy to deliver home care services." Each standardized care plan outlines needs/problems, outcome objectives with measurable criteria, and therapeutic in-

Exhibit 19-4 Quality Indicators for Home Nutrition Services

- Clients with undetected signs of malnutrition (sentinel event)
- Nutritionally compromised clients with ineffective treatment (sentinel event)
- Clients involved and satisfied with nutrition intervention
- Nutritionally compromised clients with assessments within 7 days of identification of need
- Clients with hospital admissions for nutrition-related problems
- Client with appropriate nutrition intervention
- Nutritionally compromised clients with follow-up on a monthly basis
- Clients with adequate nutrient intake
- Clients satisfied with diet therapy
- Clients with foodborne illness (sentinel event)
- Diets conforming to the RDA's and diet prescriptions
- Clients with accurately implemented diet orders
- Rate of compliance with meal service/assistance protocols
- Rate of compliance with standards of nutrient retention
- Rate of compliance with food safety and infection control standards
- Clients with knowledge/compliance with therapeutic diets
- Clients with improved nutrition status after education

Source: Reprinted with permission of American Nutri-Tech, Inc., from *Continuous Quality Improvement for Nutrition Care.* Copyright © 1992 by Rita Jackson.

terventions specific to each need or problem. As a result of using multidisciplinary documentation,

- quality care is more evident;
- communication and written documentation are enhanced;
- data retrieval is greatly improved;
- reimbursable services are more visible.

The Joint Commission may survey home care services. When a home care program is offered, controlled, or promoted by a hospital, it will generally be surveyed as part of the hospital's accreditation process. In such cases, the same rigors are used to survey both hospital and home care; attentiveness to quality standards and monitoring is essential (Type I of the month 1992).

QUALITY MANAGEMENT FOR HOSPICE PROGRAMS

In 1982 Bohnet (1982) suggested that monetary problems within hospice programs may make revenue generation a focus, diminishing the attention given to quality assurance. To ward off this possibility, Bohnet spearheaded the quality movement in hospice programs through her work at Allegheny General Hospital, Pittsburgh, Pennsylvania. She conducted a small study to evaluate care provided to terminally ill cancer patients before creation of a hospice program. Results of this study were used to define appropriate parameters of hospice care and to develop a few simple quality monitors and assessment tools. The monitors included family evaluations, nursing evaluations, and cost analysis. By reviewing all facets of the hospice program, quality improvement was demonstrated. This success led others to realize that quality management must be built into hospice administration.

The National League for Nursing's *Standards of Excellence for Home Care Organizations* (Mitchell and Storfjell 1989) includes a section on hospice care, as well as a section on dietary services to be provided by a registered dietitian, including

- assessing the nutritional needs of clients as requested
- teaching clients, families, and staff regarding special dietary regimens and nutritional requirements as appropriate
- documenting client care in the clinical record
- attending interdisciplinary case conferences as appropriate
- providing in-service education to program staff.

The Joint Commission may accredit home care and hospice programs. *Quality Assurance in Home Care Organizations and Hospice Programs* (Joint Commission 1990) gives practical guidance in developing, implementing, and evaluating these types of programs. This manual offers specific examples of plans, indicators, and data-collection tracking forms to facilitate quality management.

Some specific quality indicators for nutrition services in hospice care are shown in Exhibit 19-5.

WOMEN, INFANTS, AND CHILDREN PROGRAMS

One out of every three babies born in the United States receives WIC benefits. Today, WIC serves approximately 4.5 million participants per month, at taxpayer expense. A recent report (Prenatal WIC participation 1991) revealed that this is money well spent. Pregnant women who participate in the WIC program have healthier babies who require less Medicaid assistance after birth than those low-income pregnant women who do not participate.

Given their effectiveness in saving scarce health care resources, WIC programs are likely to continue. Such programs are often organized in hospital outpatient departments. In a cost-conscious environment, those who deliver WIC services are likely to be held accountable for the quality of services provided.

The American Dietetic Association (1987a) asserted "that appropriate child nutrition services, including food assistance, food services, nutrition education, nutrition screening, and nutrition assessment and counseling, be available to all children, regardless of economic status." While these services are often incorporated into WIC programs, such programs often

Exhibit 19–5 Quality Indicators for Hospice Care

- Clients with undetected signs of malnutrition (sentinel event)
- Nutritionally compromised clients with ineffective treatment (sentinel event)
- Clients involved and satisfied with nutrition intervention
- Nutritionally compromised clients with assessments within 7 days of identification of need
- Clients with hospital re-admissions for nutrition-related problems
- Clients with appropriate nutrition intervention
- Nutritionally compromised clients with monthly follow-up
- Clients with adequate nutrient intake
- Clients with improved nutrition status after education
- Clients satisfied with diet therapy
- Rate of compliance with meal service and assistance protocols
- Rate of compliance with standards of nutrient retention
- Rate of compliance with food safety and infection control standards
- Clients with knowledge/compliance with therapeutic diets

Source: Reprinted with permission of American Nutri-Tech, Inc., from *Continuous Quality Improvement for Nutrition Care.* Copyright © 1992 by Rita Jackson.

Exhibit 19–6 Quality Indicators for WIC Programs

- Clients with undetected signs of malnutrition (sentinel event)
- Clients receiving nutrition screening within 7 days of referrals or change of condition
- Nutritionally compromised clients with ineffective treatment (sentinel event)
- Clients involved and satisfied with nutrition intervention
- Nutritionally compromised clients with assessments within 7 days of identification of need
- Clients with hospital admissions for nutrition-related problems
- Clients with appropriate nutrition intervention
- Nutritionally compromised clients with follow-up on a monthly basis
- Clients with improved nutrition status after education
- Clients with knowledge/compliance with their diets

Source: Reprinted with permission of American Nutri-Tech, Inc., from *Continuous Quality Improvement for Nutrition Care.* Copyright © 1992 by Rita Jackson.

fail to reach everyone in need of the services. Poor nutrition flourishes when poverty and neglect accompany the lifestyles of the unemployed, adolescent parents, single parents, and even dual-income families who place higher priority on their work than on the health and well-being of their children.

Exhibit 19–6 offers some quality indicators for use in WIC clinics. Additional background information on food assistance, nutrition education, screening, assessment, and counseling is also available (McConnell et al. 1987).

ORGANIZATION FOR QUALITY

Facilities and services whose accreditation and licensing depend on compliance with Joint Commission, OBRA, or state regulations are required to monitor and evaluate quality of services provided. Johnson (1991) suggests the following steps in setting up a quality management program:

- Facility management educates the staff regarding the purpose and goals of a quality evaluation and monitoring program. Responsibility for quality must be incorporated into job descriptions, as well as facility policies and procedures.
- Administrators and managers make quality a top priority. They provide leadership for quality management, encourage staff, and reward quality improvement.
- OBRA regulations require quarterly quality evaluations. Many facilities monitor quality indicators on a monthly basis, using indicators shown in Exhibit 19–7.
- Departments prepare and use checklists or other tools to evaluate different aspects of operations and clinical care. Such lists may be adapted from those shown in this or other quality manuals (Jackson 1992).
- As departments master the process, they serve as models for other departments that are less active in showing quality improvement.
- Quality findings are shared at department meetings both to provide rationale for needed changes and to encourage employees to make a personal investment in the quality process.
- The facility quality committee oversees the entire program, appoints quality teams and team leaders, establishes time frames for reviews, implements changes when recommended, and secures resources to conduct the program.

Exhibit 19–7 Performance Indicators Monitored Monthly

Quality of Care
  Number of residents with pressure sores
  Number of residents with greater than 5% weight loss
  Number of residents on a bladder training program

Quality of Life
  Number of residents physically restrained
  Number of resident complaints
  Number of residents on psychotropic medications

Administration
  Number of vacancies
  Number of staff complaints

Source: Reprinted from Johnson, J., Quality Assurance: The Team Approach, *Provider,* Vol. 17, No. 5, pp. 16–21, with permission of the American Health Care Association, © 1991.

- Action teams monitor quality. They collect information, identify problems, and develop plans to improve quality. They also provide written reports to the quality committee.
- Some facilities appoint a quality-assurance coordinator, often an assistant administrator or director of nurses. This individual facilitates the quality process and serves as a resource, but usually does not actually conduct quality reviews.
- Outcomes and customer (residents, family members, facility staff) satisfaction are important components of quality. These factors must be evaluated as a part of the ongoing quality management program.

Quality frameworks for institutions and community-based programs are very similar, including oversight of both residential services and direct patient care. In long-term care facilities and day-care centers, the dietitian plays a role in monitoring and evaluating a wide range of activities, including food services, nutritional assessment and intervention for residents, sanitation and safety, and disaster plans. Figure 19–1 illustrates the various aspects of quality management in a nursing home (Hart 1985). As shown, in addition to departmental quality concerns, the dietitian may also be a part of multidisciplinary teams or involved in quality circles, infection control, staff development, and resident satisfaction.

Quality management in home care, hospice programs, HMOs, outpatient nutrition centers, and WIC clinics focuses primarily on direct client care. Quality evaluation and improvement programs focus on important aspects of care that are narrower, but just as crucial in maintaining high standards of practice. Although usually not mandated by accrediting agencies, the ten-step process can be used as a valuable tool to guide the establishment and management of a quality improvement program in these settings.

MULTIDISCIPLINARY QUALITY PROGRAMS

Hoff and Saxton (1992) listed several benefits of an interdisciplinary approach to quality management in nursing facilities. These include the following:

- identification of areas for improved resident care
- feedback to staff on performance

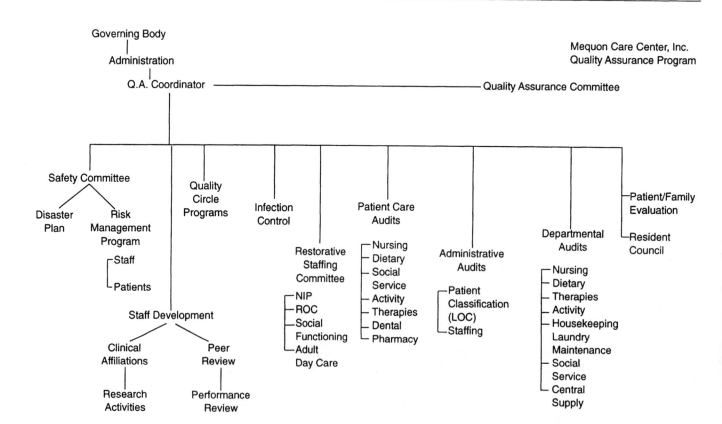

Figure 19–1 Elements of a Quality-Monitoring and Improvement Program for a Non–Hospital Institution Setting. *Source:* Reprinted from Hart, M.A., Quality Assurance Programs in the Long-Term Care Facility, *Quality Assurance: A Complete Guide to Effective Programs*, C.G. Meisenheimer, ed., pp. 279–293, Aspen Publishers, Inc., © 1985.

- compliance with Joint Commission long-term care standards
- provision of a management tool to evaluate programs and services.

The consulting dietitian, dietetic technician, or dietary manager can serve as a vital contributing member of the quality management team. This is particularly true since food service operations have a direct bearing on quality of life and resident satisfaction, which must be assessed under OBRA regulations (Dickes 1991).

Multidisciplinary efforts enhance nutritional care in non-hospital settings. Fielding and Beaton (1992) reported a series of interventions initiated by nurses for nursing home residents who had altered nutrition and inadequate food intakes. Interventions included such things as planned interactions during feeding, increased use of supplements and fluids, oral hygiene, use of appropriate adaptive devices, accommodation of food preferences, verbal cuing, careful positioning during feeding, and monitoring intakes. Weight gains were reported in ten of eleven patients over a five-week period. In the eleventh case, food intake was increased and weight was maintained. Although involvement of the dietitian or other dietetic personnel was not mentioned in this study, it is apparent that such individuals could make a substantive contribution to success of these interventions. Also, when weight gain or weight maintenance is an outcome measure for nutritional care indicators, performance thresholds can rarely be achieved without involvement of both dietary and nursing personnel.

MONITORING AND EVALUATING QUALITY CARE

To initiate the quality management process, a facility or program must form a committee to coordinate the quality process and to take responsibility for quality evaluation and improvement. Each discipline should be represented on the committee. According to Hoff and Saxton (1992), the role of the committee should be to

- identify and appraise all activities related to quality
- promote and assist in the development of standards
- identify concerns affecting multiple services
- set priorities for investigation and resolution
- submit a quarterly report to the administrator and board of directors
- reappraise the quality program annually.

Before launching into an organized quality program, comprehensive in-service training must be provided to all staff members, since all will be involved to some extent in the quality process.

The quality management process begins with a meeting of the quality-coordinating committee to determine authority, responsibility, and accountability for quality within the facility or program. Within dietetic services, a quality plan should be developed to dovetail with the overall quality program. As described in the ten-step process, the dietitian or other responsible dietetic professional should do the following:

1. Assign responsibility. If the dietitian serves on a consulting basis, membership on the quality committee and responsibility for quality management may be defined in the contract. Those employed full time by the facility or organization should define their role in quality management; develop a quality plan; and spell out how dietetic technicians, nutritionists, dietary managers, and other full-time staff members fit into this process.

2. Delineate scope of care. As described earlier, dietary and nutrition services differ between residential care facilities and community-based organizations. These differences will be apparent in the scope of care as defined for the specific situation. Although there will be an overall organization scope of service, the dietitian should spell out those unique nutritional care and dietary services offered.

3. Identify key aspects of care. Here it is essential to enumerate high-volume, high-risk, and problem-prone areas that have greatest impact on outcomes. In other words, spell out those aspects of care important enough to warrant ongoing monitoring. These may include such things as screening for nutritional risk, nutrition counseling, dietary intervention, maintaining or achieving optimal nutrition status, specialized modes of feeding (enteral and parenteral nutrition), and, where appropriate, delivery of food services that meet standards of quality for nutrition and safety.

4. Identify quality indicators. For each important aspect of care, one or two indicators need to be defined. Indicators may be adapted from samples given elsewhere in this manual or other sources, or may be developed using guidelines and forms described in Chapter 10.

5. Establish thresholds for evaluation of each indicator. Thresholds are percentages of cases that trigger in-depth evaluation of a potential problem or opportunity for improvement. As described in Chapter 10, the threshold for nutrition screening might be set at 90 percent. This means that if fewer than 90 percent of residents, clients, or children are screened for nutritional problems, reasons for noncompliance would be studied in-depth to determine why expected levels were not achieved. Thresholds may be set between 0 and 100, taking into account normal variation and historical data that reflect performance in the facility or agency.

6. Collect and organize data. Data may be collected on a monthly or quarterly basis, using forms and checklists

developed or adapted for this purpose. Nutritionists, dietetic technicians, and other dietary personnel may be involved extensively in data collection and summary activities.

7. Evaluate care. The dietitian, or other designated individual, is responsible for reviewing data to determine how they compare with the threshold. A departmental or interdisciplinary (cross-functional) quality committee may also be involved in data analysis and evaluation. If evaluation points to a particular problem, intensive review is needed, as described in Chapter 12. Dietetic technicians, nutritionists, dietary managers, and other personnel may be assigned to problem-solving teams to determine the root of problems. These teams or a quality committee may come up with appropriate action plans.

8. Take action to improve care. The dietitian or other manager institutes needed changes such as training, revised processes, or reorganization. When action plans cross departmental lines, it is essential that all pertinent parties be included in the development and implementation of proposed changes.

9. Assess actions and document improvements. After plans are implemented fully and changes have been integrated into daily activities, a focused review of the same indicator may be conducted to determine whether the changes truly resulted in improved performance of the quality indicator. If no improvement is noted, problem-solving teams should conduct further analyses to be sure the root cause of the problem has been identified and addressed. Additional changes may be outlined in an updated action plan, and further evaluations conducted of the indicator under review.

10. Communicate findings and outcomes. Reports may be submitted to the quality coordinator monthly or quarterly, as specified in the overall quality plan. The reports are approved by the quality-coordinating committee and reviewed by the director or administrator. The quality-coordinating committee may discuss complex issues not resolved at the departmental level and recommend actions to improve performance for certain indicators. A summary report is usually prepared and submitted to the board of directors or trustees, who are ultimately responsible for quality of services provided.

Regardless of the setting, it is important for the dietitian or other qualified professional to integrate nutritional care services into overall health care and to ensure that the quality of care is continually monitored and evaluated. Dietitians can provide leadership for quality-management services, cross-functional quality teams, and patient-focused health care. Nutrition services are a key element in evolving health care delivery systems based on greater efficiency, cost-containment, and accountability. Furthermore, consistent delivery of high-quality clinical nutrition services can position the dietitian for expanded leadership opportunities and greater recognition within the health care environment.

REFERENCES

American Dietetic Association. 1986. *Standards of practice: A practitioner's guide to implementation.* Chicago.

American Dietetic Association. 1987a. Position of the American Dietetic Association: Child nutrition services. *Journal of the American Dietetic Association* 87, no. 2:217.

American Dietetic Association. 1987b. Position of the American Dietetic Association: Nutrition, aging, and the continuum of health care. *Journal of the American Dietetic Association* 87, no. 3:344.

American Dietetic Association. 1987c. Position of the American Dietetic Association: Nutrition standards in day-care programs for children. *Journal of the American Dietetic Association* 87, no. 4:503.

American Dietetic Association. 1989a. Position of the American Dietetic Association: Nutrition monitoring of the home parenteral and enteral patient. *Journal of the American Dietetic Association* 89, no. 2:263–265.

American Dietetic Association. 1989b. Position of the American Dietetic Association: Nutrition services for children with special health care needs. *Journal of the American Dietetic Association* 89, no. 8:1133–1137.

American Dietetic Association. 1992. Position of the American Dietetic Association: Nutrition in comprehensive program planning for persons with developmental disabilities. *Journal of the American Dietetic Association* 92, no. 5:613–615.

American Dietetic Association. 1993. Position of the American Dietetic Association: Nutrition services in health maintenance organizations and other forms of managed care. *Journal of the American Dietetic Association* 93, no.10:1171–1172.

American Health Care Association. 1992. *The long term care survey.* Washington, DC.

American Society for Parenteral and Enteral Nutrition. 1989. Standards for nutrition support for residents of long-term care facilities. *Nutrition in Clinical Practice* 4, no. 4:148–153.

American Society for Parenteral and Enteral Nutrition. 1990. Standards of practice: Nutrition support dietitian. *Nutrition in Clinical Practice* 5, no. 2:74–78.

American Society for Parenteral and Enteral Nutrition. 1992. Standards for home nutrition support. *Nutrition in Clinical Practice* 7:65–69.

American Society of Parenteral and Enteral Nutrition. Board of Directors. 1993. Guidelines for the use of parenteral and enteral nutrition in adult and pediatric patients. *Journal of Parenteral and Enteral Nutrition* 17, no. 4 (Suppl):15A–52A.

Bayer, L.M. 1992. Reimbursement issues in home care. *Support Line: A Newsletter of Dietitians in Nutrition Support* 14, no. 5:4–6.

Bohnet, N.L. 1982. Quality assurance as an ongoing component of hospice care. *Quality Review Bulletin* 8, no. 5:7–11.

Case-McAleer, D., and M. Hopkins 1992. The R. D. power of persuasion—liberalization of the therapeutic diet in the long term care setting. *Journal of the American Dietetic Association* 92 (suppl), no. 9:A95. Abstract.

Crocker, K.S. 1992. Current status of home infusion therapy. *Nutrition in Clinical Practice* 7, no. 6:256–263.

Dickes, S.D. 1991. Food service operations respond to quality of life mandate. *Provider* 17, no. 6:39.

Dimant, J. 1991. From quality assurance to quality management in long term care. *Quality Review Bulletin* 17, no. 7:207–215.

Erickson, G.P. 1992. Multidisciplinary documentation for home care. *Caring* 11, no. 1:22–26.

Farthing, M.A., and M.G. Phillips. 1987. Nutrition standards in day-care programs for children: Technical support paper. *Journal of the American Dietetic Association* 87, no. 4:504–506.

Feidler, K.M. 1993. Managed health care: Understanding the role of the nutrition professional. *Journal of the American Dietetic Association* 93, no. 10:1111–1112.

Fielding, J., and S. Beaton. 1992. Quality assurance generated by professional nursing practice in long-term care. *Journal Nursing Care Quality* 6, no. 2:41–45.

Finn, S.C. 1990. Nutrition in home care. *Caring* 9, no. 10:4–6.

Finn, S.C., et al. 1991. Image and role of the consultant dietitian in long-term care: Results from a survey of three Midwestern states. *Journal of the American Dietetic Association* 91, no. 7:788–792.

Franz, M.J. 1992. Practice guidelines for nutrition care by dietetics practitioners for outpatients with non–insulin-dependent diabetes mellitus: Consensus statement. *Journal of the American Dietetic Association* 97, no. 9:1136–1139.

Gilbride, J.A., and M.D. Simko. 1986. Role functions of dietitians in New York State nursing homes. *Journal of the American Dietetic Association* 86, no. 2:222–227.

Gilmore, S.A., et al. 1993. Standards of practice criteria: Consultant dietitians in health care facilities. *Journal of the American Dietetic Association* 93:305–308.

Gustafson, D.H. 1992. Lessons learned from an early attempt to implement CQI principles in a regulatory system. *Quality Review Bulletin* 18, no. 10:333–339.

Hart, M.A. 1985. Quality assurance programs in the long-term care facility. In *Quality assurance: A complete guide to effective programs*, ed. C.G. Meisenheimer. Gaithersburg, Md: Aspen Publishers, Inc.

Hoff, J.T., and C. Saxton. 1992. Successful QA program requires interdisciplinary approach. *Provider* 17, no. 4:43–44.

Hoffman, K.C. 1989. Psychosocial concerns of home nutrition therapy consumers. *Nutrition in Clinical Practice* 4:51–56.

Howell, W.H. 1992. JCAHO clinical indicators for home infusion therapy. *Support Line: A Newsletter of Dietitians in Nutrition Support* 14, no. 5:1–3.

Jackobs, M.K. 1992. Residents dependent upon enteral formulas and the healing of pressure sores: A quality assurance monitor. *Journal of the American Dietetic Association* 92 (suppl), no. 9:A-66.

Jackson, R. 1992. *Continuous quality improvement for nutrition care*. Amelia Island, Fla: American Nutri-Tech, Inc.

Johnson, J. 1991. Quality assurance: The team approach. *Provider* 17, no. 5:16–22.

Joint Commission on Accreditation of Healthcare Organizations. 1987. *Ambulatory healthcare standards manual, 1988*. Oakbrook Terrace, Ill.

Joint Commission on Accreditation of Healthcare Organizations. 1988a. *Long term care standards manual, 1988*. Oakbrook Terrace, Ill.

Joint Commission on Accreditation of Healthcare Organizations. 1988b. *Standards for the accreditation of home care*. Oakbrook Terrace, Ill.

Joint Commission on Accreditation of Healthcare Organizations 1990. *Quality assurance in home care organizations and hospice programs*. Oakbrook Terrace, Ill.

Kane, R.A. 1981. Assuring quality of care and quality of life in long term care. *Quality Review Bulletin* 7, no. 10:3–10.

McChesney, M. 1991. The right chemistry: Measuring and assuring resident satisfaction. *Contemporary Long Term Care* 14, no. 9:54–56.

McConnell, P.E., et al. 1987. Child nutrition services: Technical support paper. *Journal of the American Dietetic Association* 87, no. 2:218–220.

Mitchell, M.K., and J.L. Storfjell. 1989. *Standards of excellence for home care organizations*. New York: National League for Nursing.

Nelson, K.J., et al. 1993. Prevalence of malnutrition in the elderly admitted to long-term-care facilities. *Journal of the American Dietetic Association* 93:459–461.

Neville, J.N. 1987. Nutrition services in health maintenance organizations and alternative health care delivery systems: Technical support paper. *Journal of the American Dietetic Association* 87, no. 10:1392–1393.

Nutrition Screening Initiative. 1991. *Nutrition screening manual for professionals caring for older Americans*. Washington, DC.

OBRA '87 update. 1991. *Journal of the American Dietetic Association* 91, no. 11:1381.

Peters, D.A., and D. Eigsti. 1991. Utilizing outcomes in home care. *Caring* 10, no. 10:44–51.

Posner, B.M., et al. 1987. Nutrition, aging, and the continuum of health care: Technical support paper. *Journal of the American Dietetic Association* 87, no. 3:345–347.

Prenatal WIC participation increases infant birth weight, reduced Medicaid costs. 1991. *Journal of the American Dietetic Association* 91, no. 1:33.

Raymond, J. 1990. State and federal reimbursement for home nutrition support. *Caring* 9, no. 10:16–18.

Ribovich, E.J., et al. 1990. Nutrition services in home care. *Caring* 9, no. 10:8–14.

Shaughnessy, P.W., et al. 1992. Developing a quality assurance system for home care. *Caring* 11, no. 3:44–48.

Type I of the month: Home care program. 1992. *Briefings on JCAHO* 3, no. 12:1,3–4.

Appendix A

Sample Documents

Appendix A–1. Quality Assurance Plan (University of Chicago Hospitals) **258**
Appendix A–2. Quality Assurance Program Plan (Yale New–Haven Hospital) **279**
Appendix A–3. The Ten-Step Monitoring Program (University of Utah) **286**
Appendix A–4. Nutrition Services Continuous Monitors (Henry Ford Hospital) **291**
Appendix A–5. Quality Assurance Evaluation (Yale New–Haven Hospital) **298**
Appendix A–6. Quality Assurance Plan: Assessment of the Monitoring and Evaluation Process (Olin Teague Veterans' Center) **303**
Appendix A–7. Assessment of Your Departmental Quality Assurance Program (Yale New–Haven Hospital) **306**
Appendix A–8. Clinical Pathway DRG 106 (Iowa Methodist Medical Center) **310**

Appendix A–1

University of Chicago Hospitals Nutrition Services Quality Assurance Plan

I. GOAL

The goal of the Quality Assurance Plan for Nutrition Services is to ensure the provision of quality Nutrition Services to patients through a systematic method of monitoring, documenting, and evaluating patient care services, and resolution of the problems identified.

II. OBJECTIVES

A. To provide effective mechanisms to monitor and evaluate the quality and appropriateness of nutritional care
B. To identify and resolve problems in the delivery of nutritional care
C. To report the results of all QA activities to the QA Department
D. To disseminate pertinent QA information to appropriate persons and committees
E. To evaluate the performance of individuals within the department
F. To maintain compliance with the standards of accrediting, licensing, and regulatory agencies

III. OVERALL RESPONSIBILITY

A. The Director of Nutrition Services has the responsibility for assuring that a Departmental QA plan is developed, implemented, and reviewed at least annually.
B. The Clinical Nutrition Manager is responsible for the coordination of monitoring, evaluation, and problem resolution of departmental quality assurance activities.
C. The clinical dietitians will participate in ongoing collection of data, i.e., Peer Review.
D. The Nutrition Supervisors will monitor and evaluate QA activities pertaining to patient care services provided by the diet technicians and nutrition assistants.
E. The Nutrition Supervisors will monitor and evaluate QA activities pertaining to Satellite Kitchens.
F. The Formula Room supervisor will monitor and evaluate QA activities pertaining to the Nutrition Formula Room.
G. The Department Secretary will be responsible for the typing and distribution of the QA Plan, current indicators, and monthly reports.
H. The UCH Quality Assurance Committee of the Clinical Operations Committee reviews the departmental QA plan at least annually and provides guidance and direction.
I. The UCH Nutrition Committee, composed of representatives from Medicine, Surgery, Pediatrics, Obstetrics, Pharmacy, Nursing, Food Service, and Nutrition reviews the departmental QA plan and results at least annually, and offers guidance and direction.

IV. SCOPE OF NUTRITION SERVICES QA PROGRAM

Nutrition Services shall have an ongoing program for monitoring, documenting, and evaluating the quality and appropriateness of important aspects of patient care services for in-patients on both the adult and pediatric services, for patients in the acute and chronic dialysis centers, and for patients followed by Adult Nutrition Support Service. In addition, QA activities will encompass practices with respect to safety, sanitation, and infection control in the preparation, storage and distribution of formulas and supplement, nourishments, and food for late trays in the Satellite Kitchens.

V. IMPORTANT ASPECTS OF CARE

Nutrition Services has defined the following important aspects of patient care based on the scope of care provided.
A. Nutritional screening for newly admitted patients; assessments for those at risk
B. Development and follow-up of nutritional care plans and goals
C. Periodic assessment of the effects of nutritional therapy
D. Patient and family education/counseling
E. Accuracy and timeliness with respect to preparation and delivery of nutritional formulas, snacks, and late trays

Courtesy of Nutrition Services, University of Chicago Hospitals, Chicago, Illinois.

VI. INDICATORS OF QUALITY

Indicators of the quality of patient care services for important aspects of care are reviewed and/or revised at least annually, and a schedule of monitoring established. See current listing of indicators, thresholds, and monitoring schedule.

VII. METHODS OF MONITORING

Data sources include patient medical records, diet Kardexes, patient menus and snack labels, patient interviews, and sanitation check lists. All supervisory and professional staff are involved in data collection. Results of data collected are compared with established thresholds, evaluated, and action taken to resolve problems. Monitoring is ongoing in order to assess improvement.

Forms used for data collection and evaluation are as follows:

A. Monthly Chart Review Check List
B. Height/Weight Monitoring Sheet
C. Tube Feeding Monitoring Sheet
D. Diet Kardex Card Monitoring Sheet
E. Chart Audit Worksheet (Dialysis Unit)
F. Chart Audit Worksheet (Nutrition Support Service)
G. RD Peer Review Monthly Assignments
H. Supervisors QA Monthly Assignments
I. Menu Accuracy Audit Worksheet
J. Snack/Supplement Audit Worksheet
K. Calorie Count Audit Worksheet
L. Late Tray Audit Worksheet
M. Patient Interview Form (Satisfaction Survey)
N. Nutrition Counseling Review Form
O. Diet/Drug Interaction Monitoring Form
P. Bedside Tables Monitoring Form
Q. Tray Delivery Log Monitoring Form

Exhibit A-1-1 Continuous Quality Improvement Monitoring Plan

Indicator	Discussion	Threshold	Data Source
1. Patients are screened for nutritional risk within 48–72 hr. of admission.	30 charts reviewed monthly; peer review	90%	Medical record
2. Admitting height and weight are documented in medical record for all patients by Nursing.	40 charts reviewed bimonthly by RDs	90%	Medical record
3. Outcome of nutrition intervention is monitored:			
a) Nutritional care goals are met.	30 charts reviewed monthly; peer review by RDs	90%	Medical record
b) Enteral feeding goals are achieved.	15 charts reviewed bi-monthly; peer review by RDs	90%	Medical record; I&O
c) Parenteral feeding goals are achieved (NSS).	5 charts reviewed monthly; peer review by RDs	90%	Medical record; I&O
d) Selected nutritional parameters are within normal limits (dialysis).	5 charts reviewed monthly; peer review by RDs	90%	Medical record
4. Patient tolerance and acceptance of diet is monitored:			
a) Meal rounds are documented weekly.	5 Kardex cards/mo. reviewed by each RD	90%	Diet Kardex cards
b) Food intake is recorded for all Calorie Count patients.	10/mo. reviewed by Nutrition Supervisors	90%	Calorie Count Menus
c) Food preferences are accommodated.	10 pts./mo. interviewed by Nutrition Supervisor	90%	Patient interviews
d) Patient diet card completed.	5 Kardex cards/mo. reviewed by ea. RD	90%	Diet Kardex cards
5. Patients receive diet ordered:			
a) Menus are accurate.	5 menus/tech reviewed monthly by Nutrition Supervisor	90%	Menus
b) Snacks/supplements are correct.	10 snacks/mo. reviewed by Nutrition Supervisor	90%	Kardexes; snack labels Snacks/supplements
6. Timeliness of tray delivery:			
a) Patient trays are delivered on time.	All units rev. 1 wk. by FS/NS Supervisors	90%	Tray delivery log
b) Late trays are delivered within 30 minutes of order.	10 late trays/mo. reviewed by Nutrition Supervisor	90%	Omninote record; late tray delivery log
7. Patients receive appropriate written dietary information:			
a) Patient comprehension of diet counseling is documented.	10 pts. bi-monthly; RD peer review	90%	Medical record; patient interview
b) Instruction re: potential drug/diet interactions for discharge meds is documented.	10 pts. bi-monthly; RD peer review	90%	Medical record

Aspen Publishers, Inc., 1994

Exhibit A-1-2 Continuous Quality Improvement Monitoring Schedule

ASPECT OF CARE	JUL	AUG	SEP	OCT	NOV	DEC	JAN	FEB	MAR	APR	MAY	JUN
Nutrition screening		x	x	x	x	x	x	x	x	x	x	x
Adm. Ht. and Wt.		x		x		x		x		x		x
Nutr. care goals		x		x		x		x		x		x
Enteral fdg. goals			x		x		x		x		x	
Parenteral fdg. goals		x	x	x	x	x	x	x	x	x	x	x
Dialysis parameters		x	x	x	x	x	x	x	x	x	x	x
Meal rounds	x	x	x	x	x	x	x	x	x	x	x	x
Calorie counts		x	x	x	x	x	x	x	x	x	x	x
Food preferences	x	x	x	x	x	x	x	x	x	x	x	x
Accurate menus	x		x		x		x		x		x	
Accurate snacks/suppl		x		x		x		x		x		x
Late trays		x		x		x		x		x		x
Diet comprehension			x		x		x		x		x	
Diet/drug instruction			x		x		x		x		x	
Diet Kardex card done	x	x	x	x	x	x	x	x	x	x	x	x
Bedside tables				x	x	x	x	x	x	x	x	x
Tray delivery times						x	x	x	x	x	x	x

Exhibit A–1–3 Continuous Quality Improvement Monitoring Chart Review Work Sheet

---

Reviewer: _____

Dietitian: _____

Date of Review: _____

Unit: _____

Med. Rec. #: _____

Patient Name (in pencil): _____

Diet Order: _____

Diagnosis: _____

Level of Care: _____

Admission Date: _____

Date of Initial Screening: _____

Yes   No   N/A

SCREENING:

____ ____ ____ Initial screen/assessment is completed and placed in the medical record. (Pol. 220)

____ ____ ____ Time criteria met (48 hrs; 72 hrs if admitted over weekend or holiday). (Pol. 220)

____ ____ ____ If initial screen does not include assessment, the assessment is completed within 2 days after the screen. (Levels 3 and 4)

COMPLETION OF ASSESSMENT/CARE PLAN:

Subjective Data Reported

____ ____ ____ Diet history information, Diet PTA reported for pts on mod diets, with eating problems or 80% IBW.

Objective Data Reported

____ ____ ____ Weight status

____ ____ ____ Lab data—Alb, other pertinent labs

____ ____ ____ Pertinent medications

____ ____ ____ Instructions given

Assessment Data Reported

____ ____ ____ Interpretation of abnormal laboratory values

____ ____ ____ Implications of pertinent medications

____ ____ ____ Determination of present educational needs

____ ____ ____ Assessment of appropriateness of diet order

____ ____ ____ Calculation of patient's calorie/protein needs for patients on Diabetic, Renal Diets, TF, TPN

FOLLOW-UP CARE PROVIDED ____ # of notes: ____

____ ____ ____ Short term goal met
　　　　　　　　If not, why _____
　　　　　　　　Action taken _____

____ ____ ____ Identifies modifications to be made in care plan.

____ ____ ____ Patients understand/receptiveness to instruction or plan.

COMMENTS:

CQI: Chart Rev. (Form A)

Aspen Publishers, Inc., 1994

Exhibit A–1–4 Continuous Quality Improvement Monitoring, Height/Weight Documentation in Medical Record

Nutritional Problem:	Patient's height and weight are needed for nutritional assessment.									
Nutritional Goal:	Nutritional assessment includes evaluation of energy stores.									
Action Plan:	Obtain patient's height and weight at admission.									
Desired Patient Outcome:	Patient's height and weight are recorded in medical record									
	Yes	No	Yes	No	Yes	No	Yes	No	Yes	No
Patient's *height* is documented.										
a) nursing admission form										
b) nutrition assessment form										
c) other										
Patient's *weight* is documented.										
a) nursing admission form										
b) nutrition assessment form										
c) other										

Comments:

CQI: Ht/Wt Monitoring (Form B)

Exhibit A-1-5 Continuous Quality Improvement Monitoring, Enteral Feedings

		1	2	3	4	5
	Unit					
	Pt. Initials					
	Med. Rec. #					
	Diagnosis					
	Adm. date					
	Date of review					
a) Receiving appropriate product						
	Yes					
	No					
b) Receiving volume ordered						
	Yes					
	No					
c) Tube feeding is being advanced toward goal or is at goal						
	Yes					
	No					

Nutritional Problem: Patient cannot meet nutritional needs via oral route.
Nutritional Goal: Provide adequate protein, calories, and other nutrients to meet needs via enteral feeding.
Action Plan: Provide enteral (tube) feeding calculated to meet nutritional needs.
Desired Patient Outcome: Patient receives appropriate amount of enteral feeding to meet nutritional needs.

COMMENTS:

CQI: Enteral Fdg (Form C)

Aspen Publishers, Inc., 1994

Exhibit A–1–6 Continuous Quality Improvement Monitoring, Kardex Cards

Reviewer: _____

	1	2	3	4	5
Patient Initials:					
Unit:					
Room No.:					
Diet Order:					
Admission Date:					
Date of Review:					
DT:					

Desired outcome: Information on patient's dietary problems/needs/preferences is documented in Kardex.

	1	2	3	4	5
	Y N N/A	Y N N/A	Y N N/A	Y N N/A	Y N N/A
Kardex card completed	___ ___ ___	___ ___ ___	___ ___ ___	___ ___ ___	___ ___ ___

Desired outcome: Adequacy of patient food intake is monitored regularly.

	1	2	3	4	5
	Y N N/A	Y N N/A	Y N N/A	Y N N/A	Y N N/A
Meal rounds made	___ ___ ___	___ ___ ___	___ ___ ___	___ ___ ___	___ ___ ___

CQI: DT Kardex (Form D)

Exhibit A–1–7 Continuous Quality Improvement Monitoring Dialysis Unit, Chart Review Work Sheet

---

Reviewer: _____   Patient Initials _____

Dietitian: _____   ESRD 2° to _____

Date of Review: _____   First Dialysis Tx (date) _____

Unit: _____   Initial Assessment (date) _____

Med. Rec. #: _____   Periodic Assessment (date) _____

Yes   No   N/A

SCREENING:

___ ___ ___   Initial assessment completed and placed in medical record.

___ ___ ___   Time criteria met.

___ ___ ___   Periodic assessment completed and put in medical record.

___ ___ ___   Time criteria met.

COMPLETION OF NUTRITION ASSESSMENT:

Subjective Data Reported

___ ___ ___   Diet history information

Objective Data Reported   *Assessment* Data Reported

___ ___ ___   Weight status   ___ ___ ___   Dietary recommendations

___ ___ ___   Lab data   ___ ___ ___   Interpretation of abnormal labs

___ ___ ___   Pertinent medications   ___ ___ ___   Implications of pertinent meds

___ ___ ___   Instructions given   ___ ___ ___   Pt's. understanding of/receptiveness to diet

FOLLOW-UP CARE:

___ ___ ___   Interdialytic weight gains usually acceptable (not >2–3 kg between treatments)

　　　　　　　If not, why _____

　　　　　　　RD action taken _____

___ ___ ___   Serum potassium 3.5–5.5 mEq/L.

　　　　　　　If not, why _____

　　　　　　　RD action taken _____

___ ___ ___   Serum calcium 8.6–11.0 mg/dl.

　　　　　　　If not, why _____

　　　　　　　RD action taken _____

___ ___ ___   Serum phosphorus 3.5–5.5 mg/dl.

　　　　　　　If not, why _____

　　　　　　　RD action taken _____

___ ___ ___   Serum albumin >3.0 g/dl (CAPD) or >3.5 g/dl (HD)

　　　　　　　If not, why _____

　　　　　　　RD action taken _____

CQI: Dialysis (Form E)

Exhibit A–1–8 Continuous Quality Improvement Monitoring, Adult Nutrition Support Service

Reviewer: _____

Nutritional Problem: Patient's nutritional needs cannot be met orally or by enteral route.
Nutritional Goal: Provide adequate nutrients via parenteral route.
Action Plan: Provide PPN/TPN calculated to meet nutritional needs.
Desired outcome: Patient receives adequate nutrition via parenteral feedings.

	1	2	3	4	5
Unit					
Patient Initials					
Med. Rec. #					
Diagnosis					
Consult Date:					
Review Date:					
PPN/TPN Order					

	1 Y N	2 Y N	3 Y N	4 Y N	5 Y N
a) Tolerance to parenteral feeding is documented at least twice weekly (review only current month's notes)	__ __	__ __	__ __	__ __	__ __
b) Caloric goals are met	__ __	__ __	__ __	__ __	__ __

If not, action taken by NSS
1 _____
2 _____
3 _____
4 _____
5 _____

CQI: NSS (Form F)

Exhibit A–1–9 Continuous Quality Improvement Monitoring, RD Assignments

To: _____

From: _____

In re: Continuous Quality Improvement Monitoring

    For _____ (month) _____ QA, please review the following:

    _____ Charts for Nutrition notes by _____ (RD) _____ using Chart Review Work Sheet (Exhibit A–1–3). Select Level 3 or 4 patients.

    _____ Charts for Admitting Heights and Weights on _____ (unit) _____ (Exhibit A–1–4).

    _____ Charts for Tube Feedings on *your floors* (Exhibit A–1–5).

    ___5___ Kardex cards on *your floors* for meal rounds and completion of Kardex card (Exhibit A–1–6).

    Please submit completed monitoring forms *and this sheet* to me by _____

CQI: RD Assignments (Form G)

Exhibit A–1–10 Assignment of Supervisory Responsibility

To: All Supervisors
From: _____
In re: Supervisor Responsibilities— _____ (month) 92

Lead supervisor— _____

Dates of Supervisors' Meetings

    Wed. _____ 2:00 A-04
    Wed. _____ 2:00 A-04
    Wed. _____ 2:00 A-04
    Wed. _____ 2:00 A-04

	LC	EP	CK	AN
Do _____ DT schedules				
Check and approve _____ DT schedules				
Do _____ NA schedules				
Check and approve _____ NA schedules				
Conduct _____ DT meetings (2); turn in attendance and minutes				
Conduct _____ NA meetings (1 per shift); turn in attendance and minutes				
QA—snack audit				
QA—menu audit				
QA—patient interviews				
QA—calorie count audit				
QA—late tray audit				
Follow-up on HR issues				
Miscellaneous:				
Special Audits: Bedside Tables				
Trayline Delivery Times				

CQI: Supervisor's Assignments/FORM H

Exhibit A-1-11 Menu Review Checklist

Menu Review Checklist (5/month)

Technician _____

Date _____

	1	2	3	4	5
A. Menu Selections Comply with Prescribed Diet					
1. Current diet order/information					
2. Correct menu					
3. Stamps, symbols, abbreviations. Correct					
4. Db, renal, low pro meal plans					
5. Diet patterns/guidelines followed					
B. Patient Food Preferences Respected					
1. Patient diet card completed					
2. Preferences/tolerances included					
3. Write-ins/substitutions appropriate					
4. Approved selections are initialed					
5. Neat and legible					
6. "Or You May Prefer" provided					
Type of Diet:					
Patient Name:					
Room Number:					

Please comment on deficiencies: _____
_____
_____

	1	2
C. Diabetic Snacks Are Correct per Diet Order/Meal Plan		
1. Guidelines provided in Kardex		
2. Snack label complete		
3. Snack label correct		
Patient Name: _____		
Room Number: _____		

CQI: Menu Review Checklist (Form I)

Aspen Publishers, Inc., 1994

Exhibit A–1–12 Continuous Quality Improvement Monitoring, Patient Snacks

Reviewer: _____

Unit: _____

Nutritional Problem: Patient's appetite/food intake is insufficient to meet nutritional needs.
Nutritional Goal: Increase intake of calories and protein to ensure that nutritional needs are met.
Action Plan: Provide between-meal snacks/supplements as appropriate.
Desired Outcome: Patient receives snacks/supplements as appropriate to meet nutritional needs.

	1	2	3	4	5
Patient Initials:					
Patient Room No.:					
Date Snacks/Supplements Ordered:					
Date Started:					
Date of Review:					
Snack/Supplement label correct? (Y/N):					
Snack/Supplement matches label (Y/N):					
Patient received correct Snack/Supplement (Y/N):					

Corrective Action Taken (Indicate Date and Initials)

1 _____
2 _____
3 _____
4 _____
5 _____

CQI: Snack Suppl. (Form J)

Exhibit A-1-13 Continuous Quality Improvement Monitoring, 1991/1992, Calorie Intake Records

Reviewer: _____

Unit: _____

Nutritional Problem: Patient's calorie intake appears to be inadequate.
Nutritional Goal: Patient will ingest sufficient calories to meet energy needs.
Action Plan: Place patient on 3-day "Calorie Count" to assess P-F-C-Calorie intake.
Desired Outcome: Patient's food intake is accurately recorded so that calorie intake can be calculated.

	1	2	3	4	5
Patient Initials:					
Patient Room No.:					
Date Calorie Count Started:					
Date of Review:					
All Menus Saved (Y/N):					
Indicate Menus Missing (B-L-D):					

Reason for Missing Menus? 1 _____

2 _____

3 _____

4 _____

5 _____

Action Taken Re: Missing Menus? (Include Date and Initials)

1 _____

2 _____

3 _____

4 _____

5 _____

CQI: Cal. Ct. (Form K)

Aspen Publishers, Inc., 1994

Exhibit A–1–14 Continuous Quality Improvement Monitoring, Timeliness of Late Trays

Reviewer: _____

Unit: _____

Nutritional Problem: Patient needs a meal tray after the regular meal service.
Nutritional Goal: Patient will receive the tray within 30 minutes after order is received (for orders received after tray line ends).
Action Plan: Traypassers will be instructed to check on late tray orders every 15 minutes and to deliver late trays to patients within 30 minutes of order.
Desired Outcome: Patient receives late tray within 30 minutes after order is received.

	1	2	3	4	5
Patient Initials:					
Patient Room No.:					
Date:					
Diet Ordered:					
Time Ordered:					
Time Tray Left Kitchen:					
Time Received by Patient:					

Indicate Reason for Lags
> 30 Minutes:

1. _____
2. _____
3. _____
4. _____
5. _____

CQI: Late Trays (Form L)

Exhibit A–1–15 Continuous Quality Improvement Monitoring, Patient Satisfaction

Please answer the following questions so that we may better serve you.

Patient Name _____

Patient Room Number _____ Date of Admission _____

	Yes	No	Unsure/NA
1. Are you familiar with the present diet you're on?			
2. Have you been interviewed by the dietitian or dietitian technician?	___	___	___
3. Have any specific food preferences (include allergies, special observances/practices) been obtained and accommodated?			
4. Have you been making your own menu selection?			
5. Were you informed in advance when changes in your menu selection were necessary?			
6. Have you been visited during actual meal time by the dietitian or dietitian technician?	___	___	___
7. If your diet included snacks and/or supplements:			
Have they been correct?	___	___	___
Have they been on time?	___	___	___

CQI: Pt. Satisf. (Form M)

Exhibit A–1–16 Continuous Quality Improvement Monitoring, Nutrition Counseling

Reviewer: _____

Unit: _____

Nutritional Problem: Patient needs to modify eating habits.
Nutritional Goal: Patient will understand foods appropriate for his/her medical condition.
Action Plan: Patient will be provided with verbal and written instructions regarding diet.
Desired Outcome: Patient is able to correctly identify three criteria related to understanding of diet (e.g., reason for diet, appropriate food preparation methods, foods allowed/excluded).

Indicate Y, N, or N/A for each criterion

	1	2	3	4	5
Patient Initials:					
Patient Room No.:					
Medical Record No.:					
Date of Review:					
Indicate Diet:					
Interview Results:					
A. Knows Reason for Diet.					
B. Understands Appropriate Food Preparation Methods.					
C. Identifies Foods Permitted/ Excluded on Diet.					
Results of Nutritional Counseling Are Documented in Medical Record.					
Patient Has Been Provided with Written Diet Information, As Appropriate.					

CQI: Nutr. Counseling (Form N)

Exhibit A-1-17 Continuous Quality Improvement Monitoring, Patient Education

---

Patient Education re: Potential Diet/Drug Interactions

Reviewer: _____

Unit: _____

Nutritional Problem: Patient is expected to be discharged on medications that have potential for diet/drug interactions.
Nutritional Goal: Patient will understand dietary adjustments to be made in order to prevent potential diet/drug interactions.
Action Plan: Patients will be provided verbal and written information regarding potential diet/drug interactions.
Desired Outcome: Patient is able to verbalize understanding of dietary action needed to prevent potential diet/drug interactions.

Indicate Y, N, or N/A for each criterion

	1	2	3	4	5
Patient Initials:					
Patient Medical History No.:					
Patient Room No.:					
Date of Review:					
Interview Results:					
• Patient has written materials re: potential diet/drug interaction. (Indicate medication.)					
• Patient is able to verbalize dietary action needed to prevent potential diet/drug interaction.					
Diet/drug counseling and patient comprehension are documented in medical record.					

CQI: Diet/Drug (Form O)

Exhibit A–1–18 Continuous Quality Improvement Monitoring, Special Review: Bedside Tables

Desired Outcome: Patients' Bedside Tables Will Be Cleared in Time for Meal Tray Delivery.

Reviewer _____

Unit _____

Day _____

Meal _____

Room #   Bedside Table Cleared? (Indicate Y, N, or N/A)

CQI: Bedside Tables (Form P)

Exhibit A-1-19 Delivery Schedule

UNIT DELIVERY SCHEDULE								DAY/DATE:			
Breakfast				Lunch				Dinner			
Unit	Sched	Act	Int	Unit	Sched	Act	Int	Unit	Sched	Act	Int
6NE	6:56			CRC	11:25			CRC	4:45		
6NW	7:01			6NE	11:26			C2E	4:57		
D6	7:03			6NW	11:31			INF	4:59		
6SE	7:08			D6	11:32			INT	5:01		
6SW	7:12			6SE	11:38			ICU	5:03		
5NE	7:19			6SW	11:43			C3E	5:05		
5NW	7:25			5NE	11:48			C3W	5:09		
D5	7:27			5NW	11:54			6NE	5:14		
5SE	7:34			D5	11:56			6NW	5:20		
5SW	7:36			5SE	12:02			D6	5:21		
CRC	7:38			5SW	12:06			6SE	5:27		
3NE	7:39			3NE	12:11			6SW	5:32		
3NW	7:43			3NW	12:15			5NE	5:37		
D3	7:46			D3	12:19			5NW	5:43		
3SE	7:47			3SE	12:21			D5	5:45		
3SW	7:48			3SW	12:23			5SE	5:51		
J3	7:49			J3	12:27			5SW	5:56		
W3	7:55			W3	12:37			3NE	6:01		
W4	8:05			W4	12:40			3NW	6:05		
C2E	8:11			C2E	12:42			D3	6:11		
INF	8:13			INF	12:44			3SE	6:13		
INT	8:14			INT	12:49			3SW	6:16		
ICU	8:15			ICU	12:54			J3	6:20		
C3E	8:18			C3E	12:59			W3	6:28		
C3W	8:21			C3W	1:03			W4	6:33		
BR	8:25			BR	1:06			BR	6:36		
D2	8:26			D2	1:07			D2	6:37		
4NE	8:31			4NE	1:12			4NE	6:42		
4NW	8:35			4NW	1:16			4NW	6:46		
D4	8:36			D4	1:17			D4	6:47		
4SE	8:41			4SE	1:22			4SE	6:52		
4SW	8:44			4SW	1:25			4SW	6:55		

CQI: Delivery Audit (Form Q)

Aspen Publishers, Inc., 1994

Appendix A-2

Department of Food and Nutritional Services Quality Assurance Program

1. Assignment of Responsibility
   The Associate Director of Clinical Nutrition is responsible for monitoring and evaluating the quality and appropriateness of patient nutrition care at YNHH. This individual is accountable to the Director of Food and Nutritional Services, Hospital Administration, the Medical Board's Nutrition Committee, and the Medical Director Quality Assurance for the purpose of assuring that clinically, patients receive quality nutrition care as part of their overall medical management. The Assistant Director of Clinical Nutrition is responsible for implementing departmental quality assurance activities using a peer review process with the professional clinical nutrition staff.

2. Scope of Care
   Because of the size and complexity of YNHH, a tertiary care facility, the scope of inpatient nutrition care is on clinically managing nutritionally high-risk patients (those with known or suspected Protein/Calorie Malnutrition—PCM) and subsequently on those patients where diet plays an important role in medical treatment. By intervening in the care of patients with PCM as a priority, the potential exists for improving the outcome and quality of medical therapy by preventing increased morbidity and mortality due to inadequate or inappropriate nutrition care. Therefore, the major clinical activities provided by dietitians encompass the following areas:
   - **Direct Patient Care**
     — Screening, assessing, and implementing nutrition care plans on a priority basis without a formal consult according to defined and approved hospital standards of practice.
     — Responding to formal consults for nutrition assessment and counseling.
     — Adjusting or correcting diet orders to support optimal patient medical treatment.
     — Providing nutrition counseling as needed.
     — Communicating verbally and in writing, nutrition care plans and recommendations via informal consults, patient care rounds, and chart documentation.
   - **Indirect Patient Care**
     — Educating health care providers on the nutritional aspects of patient care.
     — Participating in hospital and departmental quality assurance (and quality improvement) activities.

3. Aspects of Care
   To provide quality nutrition care that supports the medical treatment of hospitalized patients, the following are considered the most important clinical nutrition aspects of care:
   a. Ensuring that those patients at the greatest nutritional risk (i.e., those with PCM) are identified and assessed.
   b. Ensuring that clinical nutrition standards of practice are followed to provide appropriate and timely patient care.
   c. Ensuring that appropriate nutrition care is actually delivered.

4.–5. Indicators/Triggers
   The clinical indicators used are measurable variables that relate to the structure, process, and outcome of patient care. They are as follows:
   a. Aspect of Care—Identifying and assessing patients with PCM. Indicators (Structure-Based)/Triggers:
      i. Patients with a LOS >3 days are screened for nutritional risk/<95%.
      ii. Patients with known or suspected PCM are assessed:
         – Those on Parenteral or Enteral Nutrition/<98%.
         – Those consuming oral diets/<95%.
      iii. Baseline objective parameters are available for nutrition assessment:
         – Height/Weight— <90%.
         – Albumin (when PCM is known or suspected)— <90%.
      iv. Serial objective parameters are available for nutritional follow-up:
         – Wt— <90%.
         – Alb— <90%.
   b. Aspect of Care—Clinical nutrition standards of practice are followed. Indicators (Process-Based)/Triggers:
      i. Upon initial screening all appropriate risk factors are identified/<90%.
      ii. Upon initial screening patients are appropriately classified (prioritized)/<95%.
      iii. When PCM is suspected or identified, the type and extent of malnutrition are documented in patient medical records/<98%.
      iv. The type and extent of PCM are correctly identified/<98%.

Courtesy of the Department of Food and Nutritional Services, Yale–New Haven Hospitals, New Haven, Connecticut.

v. Calorie and protein requirements are correctly identified based on the type and extent of PCM or other Dx/<98%.
vi. Initial dietitian (RD) nutrition care plans and recommendations are:
  – Clearly identified/<100%.
  – Appropriate/<95%.
  – Timely/<95%.
vii. Subsequent RD nutrition care plans and recommendations are:
  – Clearly identified/<100%.
  – Appropriate/<95%.
  – Timely/<90%.
c. Aspect of Care—Appropriate nutrition care is delivered. Indicators (Outcome-Based)/Triggers:
  i. Initial diet orders are appropriate given patient's medical condition and treatment/<75%.
  ii. Subsequent diet orders are appropriate/<90%.
  iii. Physicians implement dietitian recommendations/<80%.
  iv. Nurses implement dietitian recommendations/<80%.
  v. Patients clinically demonstrate an improved or stable nutritional status or receive appropriate nutrition support given medical condition (Exception: Patients in whom it is too soon to evaluate change in status)/<90%.
6. Data Collection and Organization
Data collection is conducted quarterly on all indicators, as the indicators/thresholds are used not only to monitor the quality of nutrition care but also to objectively monitor and evaluate dietitian work performance. Data are collected and organized using two monitoring tools to measure the indicators of important aspects of care.

— A Patient Acuity Survey is conducted on *all* hospitalized patients to determine whether those patients at highest nutritional risk are identified and assessed. (See Exhibit A–2–1.)
— The High Risk Patients Evaluation is conducted on a prospective, random sample of assessed patients (30 charts/quarter) using a peer review process to determine adherence to standards of practice and the delivery of appropriate nutrition care. (See Exhibit A–2–2.)

Collected data are then compared with thresholds.
7. Evaluation of Care
Based on the data collected, indicators that do not meet thresholds are identified as issues that require analysis and recommendations. Unusual patterns or trends may also be noticed and reported on, as well as positive indicators when they support conclusions for issues identified.
8. Action Plans
For each issue identified, an action plan will be developed, and reviewed and approved by professional clinical nutrition staff and the Medical Board's Nutrition Committee. These plans will then be implemented.
9. Action Plan Assessment
Subsequent clinical reviews will determine the effectiveness of developed action plans and whether further action is necessary to demonstrate improvement.
10. Communication of Relevant Information
Quarterly, an in-depth QA report will be written by the department with a QA summary face sheet bulleting major issues. This report will be distributed to professional clinical nutrition staff, the department director, and the medical director QA. Members of the Nutrition Committee and Administration will receive the summary face sheet.

Exhibit A–2–1 Clinical Nutrition Patient Acuity Survey

Reviewer: _____

Quadrant(s) Reviewed: _____

Date: _____

# of Patients in Kardex: _____   # of Patients with a LOS > 3 days: _____

# of Patients Kardex Classified LOS > 3 days: _____   # of Patients Kardex Classified LOS < 3 days: _____

# of Patients with a LOS > 3 days on Service not Kardex Classified: _____

# of Patients not Kardex Classified with a LOS > 3 days but on Service < 3 days: _____

DO NOT COUNT KCs < 3 DAYS WHEN DOING THE TALLY, UNLESS THEY ARE ALREADY ASSESSED, TOO.

Stick Tally

	Classified	Assessed	Total Classified	Total Assessed
# Class I				
# Class II				
# Class IIIA				
# Class IIIB				
# Class IIIC				
# Class IIID				
# Class IV				

1. All pertinent information (Age, Dx/Tx, Diet, Met/Mech Prob., Labs, Meds, Ht/Wt) is documented on cards of patients.

    Yes _____ No _____

    If no, why not? _____
    _____
    _____
    _____
    _____
    _____

Exhibit A-2-2 Patient Acuity Survey

Reviewer: _____ Quadrant: _____ Date: _____

1) PATIENT CLASSIFICATION INFORMATION

A) Number of patients on NAR: _____
B) Number of patients with a LOS >3 days: _____
C) Number of patients screened: _____
D) Number of patients with a LOS >3 days not screened: _____
E) Number of patients with a LOS >3 days but on service <3 days not screened: _____
F) Classification Breakdown at the time of the survey (Stick Tally, then total)

Class	Number
I	
II	
IIIA	
IIIB	
IIIC w/o PCM	
IIIC w/PCM	
IIID	
IV	

II) TIMELINESS OF INITIAL NOTE OR 1ST FOLLOW-UP. Stick tally results. (Classification at time of adm. or initial note)

Was the initial note or 1st F/U timely according to standards of practice?	CLASS I	CLASS II	CLASS IIIA	CLASS IIIB	CLASS IIIC (with PCM)	CLASS IIIC (without PCM)	CLASS IIID	CLASS IV
YES								
No/but F/U was done								
NO								
F/U not required yet based upon standards of practice								

Exhibit A–2–3 Department of Food and Nutritional Services: High-Risk Patients Evaluation

---

Reviewer(s): _____

Baseline Information

Patient Initials: _____ Age _____ Sex _____ Service _____ Rm# _____ Unit#_____

Principal Diagnosis: _____

Metabolic/Mechanical Problems: _____

_____ Current Diet: _____

Today's Date _____ Admission Date _____ Attending Physician_____

EVALUATION AND OUTCOME OF NUTRITIONAL CARE

A. Availability of Baseline Nutrition Parameters

   1. Was an admitting/initial weight available?

      Yes ____ No ____ Where? _____

   2. Was an albumin available?

      Yes ____ No ____ Not Applicable _____

B. Availability of Subsequent Nutrition Parameters

   1. Were weights available for patient follow-up?

      Consistently ____ Sporadically ____ Not at All ____

      Too Soon to Evaluate ____ N/A ____

   2. Were serial albumin or other pertinent labs available for patient follow-up?

      Consistently ____ Sporadically ____ Not at All ____

      Too Soon to Evaluate ____ N/A ____

C. Initial Assessment—Date Completed: _____ By: _____ Patient Class: _____

   1. Are the risk factors appropriately identified?

      Yes ____ No ____ Not applicable ____

      If no, explain _____

   2. Was the patient classified correctly?

      Yes ____ No ____ Not applicable ____

      If no, explain. _____

   3. Were the nutritional care plan and recommendations:

      a. Clearly identified? Yes ____ No ____

      b. Appropriate, given data available and medical condition?

         Totally ____ Partially ____ Not at all ____

         If partially or not at all, explain. _____

      c. Timely, given standards of practice?

         Yes ____ No ____

         If no, explain. _____

Exhibit A–2–3 Continued

D. Metabolic Profile—Date Completed: _____ By: _____

    1. Was a Metabolic Profile completed?

        Yes ____

        No (baseline parameters not available): ____

        Not Applicable: ____

        Not done, but should have been: ____

        Explain: _____

        If MP was not done, skip to Section F.

    2. Was the type of malnutrition correctly identified?

        Yes ____ No ____ Not applicable ____

        If no, explain. _____

    3. Are the protein/calorie recommendations appropriate given the patient's condition and nutritional status?

        Yes ____ No ____ Not applicable ____

        If no, explain. _____

E. Documented Follow-up

    1. Were nutrition care plan and/or recommendations

        a. Clearly identified?

            Consistently _____ Sporadically _____

            Not at all _____ Too soon to evaluate _____

        b. Appropriate?

            Consistently _____ Sporadically _____

            Not at all _____ Too soon to evaluate _____

        c. Timely?

            Consistently _____ Sporadically _____

            Not at all _____ Too soon to evaluate _____

            Comments: _____

F. Diet Orders

    1. Was the initial diet order appropriate?

        Yes ____ No ____

        What was it? _____ If no, explain. _____

    2. Were subsequent diet orders appropriate?

        Consistently ____ Sporadically ____

        Not at all ____ Not applicable ____

        Why? _____

Aspen Publishers, Inc., 1994

H. Physician/Nurse Follow-Through

Were Dietitian's recommendations followed by Physician/Nursing Staff?

	Consistently	Sporadically	Not at All	Too Soon to Evaluate	N/A
Physicians					
Albumin ordered					
Other labs ordered					
Diet orders changed					
TF initiated					
HAL initiated					
TF form. or rate changed					
HAL form. or rate changed					
Other _____					
Nurses					
Follow-up weights obtained					
Calorie counts done					
Other _____					

COMMENTS: _____

_____

_____

H. At the time of this review, what was the patient's overall nutritional condition?

1. Improved.

    Explain. _____

    _____

2. Deteriorated (check reason).

    a. Deteriorated despite adequate nutrition support.

    b. Deteriorated due to inadequate nutrition support while there was aggressive (not palliative) medical management.

    c. Deteriorated due to terminal prognosis.

3. Unchanged (check reason).

    a. Stable.

    b. No subjective/objective parameters to evaluate _____ days since admission and/or _____ days since dietitian request.

    c. Too soon to evaluate.

4. Condition too unstable to objectively evaluate. Explain.

    _____

    _____

Appendix A-3

The Ten-Step Monitoring Program Nutrition Care Service

Step 1: Assigned Responsibility.

The assigned responsibility for monitoring and evaluation activities and the ultimate responsibility for everything within the Nutrition Care Service Department rests with the Department Director. The Department Director is involved in and in charge of the direction of the NCS department. The Director has delegated specific responsibilities to the department's Quality Assurance Committee.

The Quality Assurance Committee design indicators, data collections systems, frequency of review, evaluation methodologies, etc.

The NCS Department Director approves the plan, the NCS Clinical/Administrative staff agrees on the indicators, and all QA findings are discussed with entire staff at routinely scheduled staff meetings.

Step 2: Scope of Care.

I. *Evaluation and assessment of patient's nutritional needs.*
   A. Identification of patients requiring nutritional assessment.
   B. Timeliness of nutritional intervention.
   C. Proper performance of nutritional assessments.
II. *Development of nutritional plans and goals.*
   A. Appropriateness of diet order versus diagnosis.
   B. Appropriateness of enteral/parenteral nutrition.
   C. Patients should be NPO or on unsupplemented clear liquids for no longer than 7 days.
III. *Periodic assessment of the effects of nutritional therapy.*
   A. Nutritional adequacy of patient menu selections.
   B. Accuracy of the implementation of diet order.
   C. Patient consumption of food served.
   D. Patient consumption of nourishment provided.
   E. Accuracy of administration of enteral nutrition.
   F. Change in patient's status.
   G. Patient satisfaction.
IV. Education and consultation of patient and/or family.
   A. Appropriateness of patient and/or family nutritional counseling.
   B. Comprehension of diet instructions or consultations.
   C. Change in patient's status.
V. Quality control of food services.
   A. Meal schedule.
   B. Accuracy of the tray.
   C. Quality of food and presentation.
VI. Environmental assessment.
   A. Sanitation.
   B. Infection Control.
   C. Safety.

Step 3, Step 4, Step 5, Step 6 (See Attached Copy Quality Assurance Department).

Step 7: Quarterly and Annual Review Results are Discussed with Qualified NCS Clinical and Administrative Staff. This peer review allows the area in question to be undertaken when the threshold of evaluation is reached. The annual and quarterly reviews, as well as peer review, assist/document analysis trends and check the appropriateness of indicators.

Indicators recently deleted by evaluation of care process include:

1. Timeliness of calorie count implementation.
2. Identification of patients on unsupplemented NPO or clear liquid diets to assure appropriate nutritional intervention.

Indicators which will require continued monitoring in the program as defined by using evaluation of care process include:

1. Identification of patients requiring nutritional intervention by questionnaire and/or screen form according to Nutrition Care Service policy #6-31. Continue to monitor 5 day initial screen.
2. Accuracy of calorie counts.
3. Accuracy of tray assembly not critical with respect to diet order but important to patient's satisfaction.
4. Accuracy of implementation of diet order.
5. Proper performance of nutritional assessment for consultation/care plan.
6. Accuracy of administration of enteral and TPN.
7. Patient surveys which indicate a need for improvement in service provided.
8. Accuracy of implementation of nourishments critical to diet order.

Step 8: Take Actions To Solve Identified Problems.

In monitoring some aspects of care/indicators, the Quality Assurance Committee and qualified clinical nutritionists concluded that there were no problems in delivery of care of spe-

---

Courtesy of the University of Utah Health Science Center, Salt Lake City, Utah.

cific indicators (see above for details). If opportunities for improvement were not found after sufficient time (six months or one year) then indicator was deleted or randomly monitored. This was dependent upon its scope and aspect of care. The indicators, thresholds for evaluation, data collection method, and evaluation procedures were re-evaluated to determine their utility in assessing that aspect of care (see quarterly reviews, annual reviews, and QA program).

Corrective action is initiated whenever an area of needed improvement is identified. These actions reflect an understanding of personnel's knowledge, behavior, and established hospital system(s). All the facts are gathered at staff meetings and QA committee meetings before action is taken.

Action taken to solve identified problems include:

Inservices, counseling, revisions/development of policies and procedures, development of departmental standards, development of new indicators, and deletion or revision of established indicators.

Step 9: Assess the Actions and Document Improvement.

Indicators are reviewed on quarterly basis and/or annual basis. Indicators whose thresholds have been met are deleted. New indicators to measure appropriateness or administration of quality care are added after collaborating with QA committee and qualified NCS staff.

If problems are identified, the following steps are taken by Quality Assurance committee.

1. Review problem areas during staff meeting.
2. QA results are discussed at staff meeting. Minutes report any deficiencies.

If problem continues to "fall out" then review of data is done.

1. Individuals are counseled if necessary.
2. A change in department standards, policy, procedure, data collection form(s) is made if necessary. Inservice is given as necessary.
3. Review and discuss problem areas during staff meeting.

Then indicator is reviewed again and determination of consequences of outcome is determined. This process continues until the action taken resolves the problem and thresholds are met. Assess the impact on problem or areas of improvement.

Step 10: Communicate Relevant Information to the Organization-Wide Quality Assurance Program.

The QA monitoring and evaluation findings are reported to University of Utah Hospital Quality Assurance Department every quarter and annually. Quality assurance problem/resolution flowchart is included in quarterly and annual reports.

Exhibit A-3-1 Monitoring and Evaluation of Health Care

QUALITY ASSURANCE DEPARTMENT

Department: Nutrition Care Service
Division:
Date:

MONITORING AND EVALUATION OF HEALTH CARE

Scope and Aspect of Care	Indicator	Criteria	Threshold	Data Collection
I. High volume Problem prone High risk	I. Identification of patients requiring nutritional intervention by questionnaire and/or screen form according to Nutrition Care Policy #6-31	I. a. By the 5th day of admission to the floor all patients will be screened by clinical nutritionist. 1. Adults: wt. loss of 10% or more of usual body weight within the last 3 months or 80% of IBW for height. Children: ht. or wt. < 5%. 10% wt. loss over last 3 months. Ht. for wt < 5% 2. Albumin < 3.0 gm/dl 3. Patient with diagnosis of malnutrition b. All patients will be rescreened every 10 days following initial screen as long as patient remains hospitalized.	95%	Sample Size: Minimum 2 random current patient by division forms per unit per month Frequency: Random Responsible Person: Administrative Nutritionist and Clinical Nutritionists Data Source: Current patient list by division
II. High volume High risk Problem prone	II. Accuracy of calorie counts	II. a. Tech calculation of calorie intake within 100 calories or 90% of clinical nutritionist calculations. b. Tech calculations of protein intake within 10 grams protein or 90% of clinical nutritionist calculations.	95%	Sample Size: Random checks of five calorie counts per dietitian Frequency: Monthly Responsible Person: Clinical Nutritionist Data Source: Calorie count sheet
III. High volume High risk Problem prone	III. Accuracy of tray assembly not critical with respect to diet order, but important to patient satisfaction	III. Tray assembly accuracy will be measured against: a. omission (items missed) b. substitutions (unavailable items) Not critical with respect to diet order.	90%	Sample Size: Random checks of five trays per clinical nutritionist per month Frequency: Monthly Responsible Person: Clinical nutritionist Data Source: Diet order accuracy sheet
IV. High volume High risk Problem prone	IV. Accuracy of implementation of diet order	IV. Tray accuracy will be measured against diet order and parameters set by approved diet manual.	95%	Sample Size: Random checks of five trays per clinical nutritionist per month Frequency: Monthly Responsible Person: Clinical nutritionist Data Source: Diet order accuracy sheet
V. High risk Problem prone	V. Proper performance of nutritional assessment for consultation/care plan (continued)	V. a. Assessment of nutritional status should include 3 or more of the following: 1. Evaluation of anthropometric indices: height, weight, % of IBW, growth history (when appropriate), weight history, tricep skin fold (when appropriate), midarm muscle circumference (when appropriate). 2. Evaluation of pertinent biochemical indices of nutritional status. 3. Evaluation of mechanical, physiological, psychological, economic or social factors which may interfere with ingestion, digestion, absorption, or metabolism of nutrients.	94%	Sample Size: Random two charts minimum per clinical nutritionist Frequency: Quarterly Responsible Person: Nutrition Care Service Quality Assurance Committee Data Source: Medical Record

Exhibit A–3–1 continued

Scope and Aspect of Care	Indicator	Criteria	Threshold	Data Collection

V. High risk Problem prone	V. Proper performance of nutritional assessment for consultation/care plan (continued)	V.
		4. Evaluation of response to nutritional therapy; averaging calorie counts or other nutrients and suggesting modifications in nutrient delivery based upon disease state and nutritional status.
		5. Evaluation of diet history.
		6. Evaluation of drug nutrient interaction (when appropriate).
		7. Evaluation of response to nutritional therapy based on tolerance to diet.
		8. Quantitative and qualitative evaluation of nutrient needs based on department standards.
		9. Evaluation of educational need of patient (see appropriateness of patient caregiver's nutrition counseling/instruction).
		b. Assessment of appropriateness of patient/caregivers nutritional counseling/instruction. All criteria must be met if diet instruction is given.
		  1. Evaluation of patient's/caregiver's understanding of nutritional counseling/instruction.
		  2. Evaluation of "actual" compliance to diet therapy or counselor's perception of patient's ability/willingness to comply with diet therapy.
		  3. Provision of written educational material when appropriate.
		  4. Provision of "alternative methods" if patient demonstrates an inability to comprehend/understand nutritional counseling/instruction, i.e., contact caregivers for nutritional counseling.
		c. Nutrition Care Plan is based on at least one of the following:
		  1. Instruction, review or provision of nutritional information to patient and/or caregiver.
		  2. Monitoring of nutritional needs.
		  3. Statement of methods of goal/accomplishments.
		d. Complete implementation of nutritional care plan. Implementation will include documentation of all components of established nutrition care plan.
		e. Documentation of nutritional assessment and care plan in SOAP format.
		f. Documentation of nutritional assessment and care plan with physician's written request.
			94%	Sample Size: Random two charts minimum per clinical nutritionist
				Frequency: Quarterly
				Responsible Person: Nutrition Care Quality Assurance Committee
				Data Source: Medical Records

Exhibit A-3-1 continued

Scope and Aspect of Care	Indicator	Criteria	Threshold	Data Collection
V. High risk Problem prone	V. Proper performance of nutritional assessment for consultation/care plan (continued)	V. g. Documentation of nutritional assessment and care plan within 24 hours after dietitian receives written order. Exception: Bone Marrow Transplant assessment documented within 72 hours after written request received.	94%	Sample Size: Random two charts minimum per clinical nutritionist Frequency: Quarterly Responsible Person: Nutrition Care Service Quality Assurance Committee Data Source: Medical Records
VI. High risk Problem prone	VI. Accuracy of administration of enteral and TPN (i.e., tube feedings and TPN)	VI. Nursing notes (intake and output) should be monitored to assess appropriate administration.	90%	Sample Size: Random checks—five nursing notes vs. physicians' order per clinical nutritionist Frequency: Monthly Responsible Person: Clinical Nutritionist Data Source: Enteral/parenteral flow sheet
VII. High volume High risk Problem prone	VII. Patient surveys which indicate a need for improvement in service provided	VII. Patient surveys should indicate that nutritional services are good.	90%	Sample Size: Minimum 50 per month Frequency: Monthly Responsible Person: Nutrition care staff Data Source: Patient satisfaction surveys
VIII. High risk Problem prone	VIII. Accuracy of implementation of nourishment critical to diet order	VIII. Nourishment accuracy measured against diet order, i.e., diabetics, post gastrectomy, dumping syndrome and physician/dietitian ordered nourishments. a. Nourishment Kardex accuracy measured against diet order slips. b. Patient received nourishment as ordered.	95%	Sample Size: Random checks of five nourishments per technician Frequency: Monthly Responsible Person: NCS technician Data Source: Nourishment check sheet

Aspen Publishers, Inc., 1994

Appendix A-4

Nutrition Services Continuous Monitors

Monitor	Criteria	Data Sources/ References	Frequency*	Responsibility
Clinical RD Peer Review	For each RD 85% of peer review criteria met.	Medical Records Policy 711.00 Form 711.01 Form 711.02 Form 711.03 Policy 806.00	20 charts/RD/year and as needed	Clinical RDs, Associate Director, Department of Dietetics
DI Peer Review	For each DI on staff responsibility 85% of peer review criteria met.			
Clinical RD/DT/DI Review of Timeliness/Appropriateness of Nutrition Screening	90% of newly admitted patients will be screened for risk of malnutrition within 72 hours (3 business days).	Policy 854.00 Form 854.01 Form 854.02	2x/year All DIs on S.R. approx. 40 patients/review/RD or DT @ 80 pts./review/DI	Coordinator, Nutrition Services or designee
Clinical RD/DI Review of Timeliness & Appropriateness of Nutritional Intervention	75% of patients at risk for malnutrition will receive appropriate nutritional intervention within 96 hours (4 business days) of admission as evidenced by charting.‡	Policy 854.00 Form 854.01 Form 854.02	2x/year All DIs on S.R. approx. 40 patients/review/RD/DI @ 80 pts./review/DI	Coordinator, Nutrition Services or designee
Clinical RD Enteral Feeding Review	80% of patients (not followed by NSS) receiving Enteral Tube Feedings will receive appropriate nutritional intervention within 72 hours of the placement of the enteral feeding tube.	Policy 856.00 Medical Records Patient Assessment forms. RD Documentation on form 856.01	2x/year Patients on enteral feeding for 2-week period	Coordinator, Nutrition Services or designee

Courtesy of the Henry Ford Hospital, Detroit, Michigan.

Monitor	Criteria	Data Sources/References	Frequency*	Responsibility
Clinical RD, Dietetic Technician, Dietetic Intern Review of Food—Nutrient Interaction Counseling	65% of patients needing FDI counseling, as designated by pharmacy list will receive written instructions and documentation in medical record and discharge summary form.	Policy 839.00 Policy 855.00 Form 855.01	2x/year (15–20 records/review)	Coordinator, Nutrition Services or designee
NPO Status Review	90% of all patients with a diet order of NPO for ≥ 5 days will be receiving or have documented plans for an alternate form of nutritional support, i.e., tube feeding, TPN, PPN.	Medical Records 24° Diet logs, RD Kardex, Kitchen Kardex, Policies 806.00, 861.00 Forms 861.01; 861.02; 861.03	2x/year Review of 100% of patients with a diet order of NPO	Coordinator, Nutrition Services or designee
Clear Liquid Status Review	100% of all patients receiving a clear liquid diet for > than 3 days will automatically receive nutritional supplements.	Medical records, 24° Diet Logs, RD Kardex, Kitchen Kardex, Policy 860.00 Forms 860.01; 860.02	2x/year Review of 100% of patients with Clear Liquid diet order	Coordinator, Nutrition Services or designee
H.S. Nourishment Delivery Review	90% of all patients without an ADA diet order will receive H.S. nourishments as ordered or with appropriate substitutions. 100% of all patients with an ADA diet order will receive H.S. nourishments as ordered or with an appropriate substitution.	H.S. nourishment lists 24° Diet Log, Kitchen Kardex, Policy 858.00 Form 858.01	2x/year Review of 100% of patients with ADA diet order on 10 units. Review of 100% patients without ADA diet order	Coordinator, Nutrition Services or designee
HT/WT Availability Review	80% of patients admitted will have ht/wt listed on their wall chart or in their medical record.	Medical Records Patient wall charts Policy 859.00 Form 859.01	2x/year Review of 100% of patients on 10 pt. units	Coordinator, Nutrition services or designee

Appendix A   293

Monitor	Criteria	Data Sources/ References	Frequency*	Responsibility
Charted Nutrition Care Tally Total Interventions Total care level 3 & 4 patients charted in the General Memo	Nutrition screening. Nutrition assessment. Nutrition history. Nutrition counseling/diet review. Food preferences and tolerances. Base and diet plan. Nutrient intake analysis. Comprehensive nutrition counseling. Food–drug instructions. Mailed diet instruction. (expected charting = minimum 50 notes/month per Nutrition Team). Expected ≥ 50% of charting on care level 3s & 4s.‡	Policy 817.00 Forms 817.01 817.02 Medical Records Policy 804.00 805.00 806.00 807.00 812.00	Monthly	Clinical RDs
Dietetic Technician Nutritional Assessment Review	90% of patients reviewed will have IBW calculated. Appropriate care level assigned, pertinent lab data listed, assessment of Kcal/pro needs, a nutritional care plan and food preferences and allergies noted.	Policy 848.00 848.04	1x/year for DTs (10 records/review)	Supervising RDs
Dietetic Technician Meal Rounds Review	90% of patients visited will have assessment of Tray Accuracy, Food Quality, and Food Intake. Nutrition Screening and Food Drug Counseling will be given 100% of time as appropriate.	Policy 848.00 Form 848.03	1x/year for DTs 20 patient visit observations/ review	Supervising RDs
Dietetic Technician Nutrient Intake Analysis Review	90% of nutrient intake analyses reviewed will show accurate calculation and documentation in medical record.	Policy 848.00 Form 848.02	1x/year for DTs (1 intake analysis/review)	Supervising RDs
Dietetic Technician Menu Checking Review	90% of patient menus will be accurate for the established criteria and in compliance with diet order and diet manual.	Policy 848.00 Form 848.05 Policy 810.00 Standard Menu Writing	1x/year for DTs (20 patient menus/review)	Supervising RDs
Dietetic Technician Nutrition Note Review	90% of nutrition notes will be written according to established criteria.	Policy 848.00 Form 848.01 Policy 807.00 Recording in the patient's medical record	1x/year for DTs (general memo note & 1 discharge teaching menu/review)	Supervising RDs

Monitor	Criteria	Data Sources/ References	Frequency*	Responsibility
Dietetic Technician Monthly Status Reports	25 charts read per week. 25 patients screened per week. 5 patients rescreened per week. Average of 10 patients per day seen on meal rounds.	Policy 817.00 Form 817.01 Form 817.02	Monthly	Dietetic Technicians with follow-up from supervising RDs
Diet Clerk Review of Diet Prescription Processing	100% of diet orders from nursing will be accurately transferred to the Kardex, to the menus and to the nourishment sheets.	Policy 841.00 Form 841.01	1x/year for DCs (30 diet changes/review)	Supervising RDs
Diet Clerk Review of Nourishment Sheets	95% of nourishments listed in Kardex will be accurately written on the nourishment sheets.	Policy 841.00 Form 841.02	1x/year for DCs (20 nourishment orders/review)	Supervising RDs
Diet Clerk Review of Hot and Cold Specials	95% of standardly tallied items will be counted accurately. 100% of written specials will be tallied accurately.	Policy 841.00 Form 841.03	1x/year for DCs (3 standard tallied items & 1 unit written special/review)	Supervising RDs
Diet Clerk Review of Census	100% of patients will be counted on daily floor census.	Policy 841.00 Form 841.03	1x/year for DCs (2 units/review)	Supervising RDs
Accuracy of Patient Trays	97% quarterly assessment. 96% patient survey index Q/A tray.	Patient Survey (Policy 408.00) QA tray assessment (Policy 409.00) Quarterly Accuracy Assessments (Policy 852.00) Form 63-68 Policy 404.00	Monthly 20–24/month Quarterly	FSS/Manager FSS/Manager RD/Tech

Monitor	Criteria	Data Sources/References	Frequency*	Responsibility
Sanitation Checklist (includes equipment compliance)	100% compliance with action plan noted for repairs not meeting standards.	Sanitation checklists	6 areas 1/month	FSS
Food Temperatures	100% within safe zone for hot foods. 15 index** for QA assessment per Oregon study.	Policy 306.00 Daily temp. log sheets QA tray assessments	Daily 20–24/month	FSS
Patient Satisfaction Survey and Tray Assessment	Return rate: 50% Adjusted Satisfaction Score: 96.5% HFHCC Satisfaction Score: 76% Food & Beverage Temperature: 95% (15 index**) Quality and Taste of Food: 94% (15 index**) Food and Tray Appearance: 98% (Food = 15 index**) (Tray = 5 index**) Variety: 96% Tray Accuracy: 95% (15 index**) Courtesy of Staff: 99% Overall food/service quality: 98% (85 index**)	Patient Survey (Policy 408.00) QA tray assessment (Policy 409.00) Policy 307.00 403.00 404.00	Survey = 1/month QA tray assessments = 20–24/month	FSS/RD Manager
Tray Delivery within 12 minutes of assembly	5 Index**	Q/A tray assessment (Policy 409.00)		Manager/FSS
Infection Control	Employees follow established infection control procedures. Meet 100% compliance on daily basis.	Area specific inservice as needed. QA tray assessment	Departmental inservice 1/year Daily evaluation	FSS/Manager Ed. Coordinator

Monitor	Criteria	Data Sources/References	Frequency*	Responsibility
Safety	Employees' injuries are evaluated for follow-up and/or preventive measures. Employees follow dept. safety procedures.	Unusual Occurrence forms Safety Audits Supervisor–Manager observations/Employee reports	Dept. inservice 1/year Daily assessment Evaluated 1x/year on P/A	Ed. Coordinator FSS/Manager Ed. Coordinator

* As needed when problem identified.
**Exceptions: Obstetric Unit;
                Cardiology Unit.
† Exceptions: Patients who are followed by Nutrition Support Service for nutritional care.
                Patients who have become at 1 nutritional risk (96° after risk increased).
                Patients who are not medically stable and nutritional needs cannot be determined within 96° of admission.
                Patients who are in surgery/recovery (96° after stabilized).
                Patients who were transferred to various units (96° after being on IPD Unit).
                Patients where death is imminent and documented as such.

Aspen Publishers, Inc., 1994

QUALITY ASSURANCE (Policy 007.00)

JAN	FEB	MAR	APR	MAY	JUNE	JULY	AUG	SEPT	OCT	NOV	DEC
RD Peer Review X 2	RD Peer Review X 2	NO Peer Review	RD Peer Review X 2	NO Peer Review	DI Peer Review X 3	DI Peer Review X 3	NO Peer Review	RD Peer Review X 2	RD Peer Review X 1	RD Peer Review X 1	NO Peer Review
Martha (2/1)	Cheryl (3/1)	Julie (5/1) Toni (6/1)	Cathy (5/1) Steve (6/1)	Janice (6/1) Mini (6/1)	Mary (7/1) Jennifer (7/1)	Bevelyn (8/1)	Christina (9/1)	Robin (10/1) Sheila (10/1)	Mary Lu (12/1) Beverly (12/1)	Mary (1/1) Lisa (1/1)	Leslie (1/1)
↕ FDI		↕							↕ FDI		
			↕ HS Nourishments			↕ HS Nourishments					
			↕ Enteral Feeding						↕ Enteral Feeding		
↕ NPO							↕ NPO				
			↕ Clear Liquid						↕ Clear Liquid		
↕ Height and Weight						↕ Height and Weight					
↕ Timeliness			↕ Timeliness			↕ Timeliness			↕ Timeliness		
↕ Special Audits As Needed											

Appendix A–5

Clinical Nutrition Quality Assurance Evaluation

Department/Facility _____ Date _____

Issues Identified	Conclusions/Recommendations	Action/Follow-up
A. Positive outcomes		
1. Patients with LOS > 3 days are screened for nutritional risk (99%).	Results remain consistent with last year's average of 98%. This screen is the initial step in the clinical nutrition care process.	Continue to evaluate via acuity screen and patient nutrition care monitor (PNCM).
2. Clinical issues pertaining to process indicative of positive interaction (correctly identified malnutrition, appropriate initial, follow-up care plans, Kcalorie/Protein recommendations, and Kardex classification; follow-up recommendations clearly identified and dietitian followed up on recommendations).	7 of 13 clinical process indicators > threshold, and this is indicative of positive nutrition interaction.	Actual process of screening and assessment is being totally revised to streamline functions and improve accuracy. A new screening form was developed and approved by Medical Records Committee, as well as charting standards of practice. Implementation is planned for _____. Review all clinical process indicators via staff inservice and QA meetings. Continue to evaluate via Patient Nutrition Care Monitor.
3. Initial diet order is appropriate (84%).	This reflects improvement in diet order appropriateness; the initial diet order has exceeded the threshold value.	Continue to evaluate via PNCM.
B. Areas in need of improvement		
1. Patients with known or suspected PCM are assessed (IIIAs) in a timely manner (72%).	Results are below annual average of 89% and last quarter average of 86%. This monitor reflects timeliness only not that these patients were never assessed.	Continue to monitor via acuity screening tool. Develop a separate indicator on whether or not patients were actually assessed.
2. Availability of admission weight/albumin and follow-up weight/albumin continues to be a problem.	Figures still remain far below threshold values for all four areas.	Develop protocols for albumin/weight pending formation of hospital interdisciplinary QA committee.
3. Clinical issues pertaining to process (risk factors appropriately identified, nutrition assessment appropriate, type/extent of PCM documented and timely follow-up).	In most cases, completeness of listing mechanical metabolic problems was the problem (i.e., patient with femur fracture in traction x 21 days not noted as a risk factor). In two patients, preliminary nutrition assessment was incomplete, and in two patients estimated protein/calorie recommendations were not appropriate. Completeness and accuracy need to be addressed.	As noted in A2 screening and assessment documentation is being revised. Continue to evaluate further via PNCM.

Courtesy of Yale–New Haven Hospital, New Haven, Connecticut.

Issues Identified	Conclusions/Recommendations	Action/Follow-up
4. Outcome issues below threshold include subsequent diet orders, physician's follow-up on dietitian recommendations, and nurse's follow-up on dietitian recommendations.	In majority of patients, the enteral/parenteral feedings that were ordered by physicians were not meeting estimated needs. Two PO diet orders were not correct. As previously discussed in QA reports, follow-up weight and follow-up albumin were not obtained by nurse and physician respectively.	Will continue to monitor TF/PN orders. Dietitians to continue to adjust diet orders contraindicated by physician. Develop protocols for albumin/weight pending information of hospital interdisciplinary QA committee.
5. Patients receive appropriate nutritional care (88%).	Improved from previous quarter.	Continue to monitor via PNCM.

Exhibit A–5–1 Quarterly QA Data Collection Summary Sheet

Date: _____
Patient Population:
Med, Surg, Peds, Gyn, Psych, PCUs

Aspect of Care—Patients at Highest Nutritional Risk Are Screened and Assessed.

			RESULTS			
Indicators (Structure)	Threshold	Previous Annual Average	1st Qtr. % (#)	2nd Qtr. % (#)	3rd Quarter % (#)	Annual Ave. % (#)
1. Patients with a LOS > 3 days are screened.	95%	98%	98% (354/361)	97% (358/369)	99% (306/308)	
2. Patients receiving PN or EN support are assessed (Class IVs).	98%	98%	N/A	99% (72/73)	94% (45/48)	
3. Patients with known or suspected PCM are assessed (Class IIIAs).	95%	89%	N/A	86% (42/49)	72% (48/67)	
4. Baseline objective parameters are available for nutrition assessment:						
a. height/weight.	90%	71%	69% (24/35)	66% (23/35)	56% (18/32)	
b. albumin (when PCM is known or suspected).	90%	61%	60% (21/35)	57% (20/35)	52% (16/31)	
5. Serial objective parameters are available for follow-up:						
a. weight.	90%	63%	74% (20/27)	71% (22/31)	42% (13/31)	
b. albumin.	90%	62%	76% (16/21)	50% (11/22)	52% (13/25)	
		(Based on 138 charts)	(Based on 35 charts)	(Based on 35 charts)	(Based on 32 charts)	

Aspen Publishers, Inc., 1994

Exhibit A-5-2 Quarterly QA Data Collection Summary Sheet

Date: _____
Patient Population:
Med, Surg, Peds, Gyn, Psych, PCUs

Aspect of Care—Clinical Nutrition Standards of Practice Are Followed.

Indicators (Process)	Threshold	Previous Annual Average	1st Qtr. % (#)	2nd Qtr. % (#)	3rd Qtr. % (#)	4th Qtr. % (#)	Annual Ave. % (#)
1. Kardex classified (screened) correctly.	95%	N/A	94% (33/35)	97% (34/25)	97% (31/32)		
2. RFs are appropriately identified.	90%	85%	88% (29/33)	66% (23/35)	83% (25/30)		
3. Patients are appropriately classified.	95%	96%	97% (32/33)	89% (31/35)	93% (28/30)		
4. Nutrition assessment appropriate.	90%	N/A	84% (27/32)	82% (28/34)	81% (26/32)		
5. Initial dietitian Nutrition Care Plan and Recommendations are:							
a. clearly identified.	100%	93%	85% (28/33)	100% (35/35)	97% (31/32)		
b. appropriate.	95%	95%	97% (32/33)	77% (27/35)	97% (31/32)		
6. Type and extent of PCM are:							
a. documented.	98%	96%	96% (23/24)	95% (19/20)	77% (10/13)		
b. correctly identified.	98%	94%	91% (21/23)	100% (19/19)	100% (10/10)		
7. Calorie & Protein recommendations are appropriate given type and extent of PCM or other diagnoses.	98%	97%	96% (25/26)	95% (18/19)	100% (10/10)		
8. Dietitian follow-up on own recommendations	90%	N/A	67% (8/12)	96% (24/25)	94% (15/16)		
9. Follow-up Dietitian Nutrition Care Plan and Recommendations are:							
a. clearly identified.	100%	98%	100% (18/18)	88% (22/25)	100% (17/17)		
b. appropriate.	95%	96%	92% (17/19)	100% (25/25)	100% (17/17)		
c. timely.	90%	NA	NA	77% (20/26)	76% (13/17)		

Exhibit A–5–3 Quarterly QA Data Collection Summary Sheet

Date: _____
Patient Population:
Med, Surg, Peds, Gyn, Psych, PCUs

Aspect of Care—Appropriate Nutrition Care Is Delivered.

RESULTS

Indicators (Outcome)	Threshold	Previous Annual Average	1st Qtr. % (#)	2nd Qtr. % (#)	3rd Quarter % (#)	Annual Ave. % (#)
1. Initial diet orders are appropriate.	75%	79%	83% (24/35)	77% (27/35)	84% (27/32)	
2. Subsequent diet orders are appropriate.	90%	77%	85% (22/26)	75% (18/24)	78% (18/23)	
3. Physicians follow-up on dietitian recommendations.	80%	67%	68% (32/47)	72% (36/50)	61% (22/36)	
4. Nurses follow-up on dietitian recommendations.	80%	72%	66% (23/35)	79% (27/34)	57% (16/28)	
5. Quality of nutrition care is appropriate (i.e., patient improved, remained stable, or was given appropriate support from all team members).*	90%	78%	73% (16/22)	82% (18/22)	88% (14/16)	

*Does not include patients admitted too soon for objective decision or whether objective information is lacking.

7/92—QASUM3

Appendix A-6

Quality Assurance Plan: Assessment of the Monitoring and Evaluation Process for FY 92

A summary of the significant findings of monitoring and evaluation activities during FY 92 is presented below:

Key Functions	Summary of Findings/Actions	Outcome/Impact on Patient Care/Services
1. Follow-up nutrition education		
a. Provision of informational handout	a. Initial 82% compliance; provider specific information indicated problem to be interpretation of information in medical record rather than lack of documentation. Staff reviewed terminology at June 22, 1992 staff meeting. Indicator met at 100 percent.	a. All clinical dietitians interpret nutrition education provided to patients data in the same manner.
b. Verbalization of understanding by patient.	b. Initial 89% compliance first quarter; 100% fourth quarter. Staff was reminded to document verbalization of understanding by patient.	b. Quality of care provided to patient may not have changed; however, the quality of medical record documentation did improve.
c. Nutrition education provided within seven working days of assessed need or prior to discharge.	c. Consistently exceeded threshold (94% first quarter; 100% fourth quarter). Staff encouraged to continue to strive for 100% compliance.	c. Dietitians became more aware of the need for documentation of provision of education. Clinical staff also made greater efforts to avoid a patient slipping by without proper education or documentation.
d. Nutrition education is provided within seven (7) working days of assessed need or prior to discharge.	d. Education provided was consistently above threshold (94% first quarter; 100% fourth quarter). Staff encouraged to strive for 100% compliance.	d. Dietitians became more aware of need for documentation of provision of education. They also made greater efforts to avoid a patient slipping by without proper education or documentation.
2. Appropriate and timely communication, feeding and documentation is completed for those patients ordered Nutrient Intake Analysis (NIA).		
a. Consult received by Clinical Dietetic Technician via consult option of DHCP Program.	a. Initially Dietetics was notified of an NIA in a variety of ways, including via Additional Order of DHCP; consults to dietitian or verbal communication to Clinical Dietetic Technician (CDT). After a review of proper procedure through the supervisor of the unit secretaries, and sending provider specific information for follow-up, consults for NIA were appropriately received. Dietitians were reminded in staff meetings to send a consult to CDT as well so that documentation exists. Improved results.	a. If consult is received correctly, the initiation of the NIA will result in more timely implementation of the NIA and timely recording of data in patient record.

Courtesy of Olin E. Teague Veterans' Center, Temple, Texas.

Key Functions	Summary of Findings/Actions	Outcome/Impact on Patient Care/Services
b. NIA ordered by 3 p.m. initiated by breakfast meal of following day.	b. One CDT receives all consults for NIAs via DHCP (current computer program does not allow consult to go to 2 people). Procedure works well unless this CDT is off or on annual leave. Relief CDT did not always remember to check consults of other CDT in her absence. Provider specific information forwarded to CDT's supervisor for review of proper procedure. Relief CDT needs to check consults daily to ensure that NIAs get initiated on time. This did not improve over time.	b. This procedure for consults works well except when CDTs off on annual leave. Problem with CDTs did not surface until the second half of the fiscal year. Delay in implementation of NIA results in delay of record information in medical records of patient. This may result in delay of decision on nutrition support. This will be continued into next fiscal year to ensure that this has been correct.
c. Results of NIA recorded in medical record by CDT within 48 hours of order.	c. If NIA is not initiated on time, it cannot be recorded on time in most cases. On one occasion, CDT unable to obtain data needed for recording until it was time for her to go off duty. CDT was asked to communicate this to supervisor in the future, so someone else can be assigned the task.	c. CDTs responsible for initiating and recording NIA information in medical records are both more aware of the importance of timely documentation in medical record because of this monitor.
d. All NIA trays are checked by Health Technician (HT) or designated employee, with intake correctly recorded on menu.	d. Despite communication with HTs and Food Service Worker Foremen and initiation of "Report of Missing NIA Tray," procedure as currently established does not appear to work well. Procedure will be revised and initiated in FY 93. Also, some new or part-time employees did not recognize NIA trays on several occasions and disposed of them. Education & Staff Development Dietitian has been asked to help with this training of new employees. Recording remained high throughout year but not at threshold.	d. Recording of NIA trays by HT remain consistently high at or about 90% but not at threshold set at 100%. NIA data are incomplete for patient's medical record. Change in procedure may help resolve this. ESDD dietitian will provide help with training of new employees on procedures for NIA trays. Monitor to continue into FY94. Thresholds may be changed as this is a problem prone but not high-risk monitor.
3. Sentinel Events: Imminent Danger to Patient or Employee	Ten incidents occurred from January to August. All incidents were reported to Nursing Service and at Dietetic Service's morning meeting. They were also discussed at the following meetings: Employee, QA, Supervisors, and Safety/Sanitation. Medical Media is contacted when a foreign object is found on a tray and two instant photographs are taken, one for Nursing Service, the other for our file.	With the use of an instant picture, timeliness of feedback to the wards has greatly improved. Our rate of incidence dropped by 40% from 25 incidents to 10.

Key Functions	Summary of Findings/Actions	Outcome/Impact on Patient Care/Services
4. Focus Review Problems identified in Patient Satisfaction Surveys.	Trending over the year indicates some fluctuation from quarter to quarter in perception of the food served. All areas improved from first quarter to fourth with the most substantial changes being seen in food temperatures (12%), variety (9.7%), and overall quality (8.8%). Of the six indicators three exceed threshold, two fall one percent below, and two appetizing appearance and courteous employees, although good, fall slightly below the threshold.	Implementation of special meals in addition to a variety of different ideas and a great deal of effort on staff members' part has had a positive impact on patient satisfaction during the year. Other innovative concepts being considered for the next fiscal year may already help to improve patient satisfaction and maintain our overall upward trend.

5. Departmental Minutes
  a. Minutes captured pertinent agenda information in a way that the information was clear and understandable and does not exceed two pages, unless there is an unusually long agenda.
  b. Preparation of minutes are computed in a timely manner and does not exceed four hours total preparation time per set of minutes.
	a. Initial review indicated that both indicators were substantially out of compliance. Some of this was contributed to newer clerical staff not being familiar with department operations. The information and format required when QA items are discussed was also a major contributing factor. Inservice class with the Center's QA Coordinator provided a better, more time-efficient method of transcribing minutes, which reduced the time of minute preparation and number of pages produced. For four of the monthly meetings, the standard of the number of pages minutes should not exceed was changed from two to three.	a/b. Modifications to the format for meeting minutes involving format resulted in a reduction in the amount of transcription time and the number of pages. The length of time clerical spends on minutes has improved since the first recording period.

6. "JCAHO" Mock Review	The department is in substantial compliance as a whole. The review identified the areas of staffing of the department, enforcement of compliance to written policies, safe/sanitary storage of food, and required work. These problems have been communicated to the appropriate individuals for corrective action.	Staffing is substandard as a result of hiring lags and freezes; request to fill vacancies has been submitted. As this review was conducted in fourth quarter, impact recommendations cannot be evaluated at this time.

DISCUSSION OF PROGRAM EFFECTIVENESS.
1. The monitor on Follow-Up Nutrition Education brought to light the differences in interpretation of information in medical records. It also resulted in improved dietitians' documentation in the medical record and a greater effort made to make sure that patients received the necessary education.
2. Having identified the problems of why NIAs were not being received, unit personnel were then trained on the proper procedures for communicating the request. Clinical Dietetic Technicians are aware of the importance of timely initiation and documentation of NIAs. Also, NIA tray identification will be included routinely in new employee training.
3. Overall, the QA program was successful. The staff was more accepting of the QA program as they have become more oriented to the problem-solving process. They showed very little resistance to the QA process as they have seen how the program enhances patient care. The Interdisciplinary Committee's QA process also appeared to work well as the members worked as a team to problem solve patient care issues.

Appendix A–7

Yale–New Haven Hospital Quality Assurance Program Evaluation

Assessment of Your Departmental Quality Assurance Program

Department: Food and Nutrition Services (Dietary)

Please respond to the following questions and return this form to the Quality Improvement Department. If you have any questions about the information requested, please call the QI Manager. (Use separate sheets of paper to answer the questions below, if necessary.)

I. An ongoing QA Program designed to objectively and systematically evaluate the quality and appropriateness of patient care services, pursue opportunities to improve care and resolve identified problems is implemented.

   a. Define the scope of care or services provided by your department. The scope of services includes all the major clinical functions for inpatients and outpatients (e.g., conditions/diagnosis treated, treatment or activities performed).

   Inpatient—At YNHH, 35–40% of the patient population with a LOS > 3 days have or are at high risk for protein/calorie malnutrition (PCM). Therefore, the scope of inpatient nutrition care is on clinically managing these patients and subsequently on those patients where diet plays an important role in medical treatment. By intervening in the care of patients with PCM as a priority, the potential exists for improving the outcome and quality of medical therapy by preventing increased morbidity and mortality due to inadequate or inappropriate nutrition care. The major clinical activities provided by registered dietitians encompass the following areas:

   —Screening, assessing, and implementing nutrition care plans on a priority basis without a formal consult according to defined and approved hospital standards of practice.
   —Responding to formal consults for nutrition assessment and counseling.
   —Adjusting diet orders or recommending enteral/parenteral feeds to support optimal patient medical treatment.
   —Providing nutrition counseling as needed.
   —Communicating verbally and in writing, nutrition care plans and recommendations via informal consults, patient care rounds, and chart documentation.
   —Educating health care providers on the nutritional aspects of patient care.
   —Participating in hospital and department quality assessment (and quality control) activities.

   Outpatient—The focus of outpatient nutritional care is on nutrition (diet) counseling and/or nutrition education for individuals and groups to help prevent or treat any disease or condition in which dietary intervention may play a role. YNHH physician referred outpatients and community-based physician referrals are serviced by registered dietitians in the following clinics:

   —YNHH Nutrition Clinic
   —Renal Stones and Bones Clinic
   —Yale Vascular Center
   —Genetics Clinic
   —High Risk Obstetrics Clinic
   —Pedi Nephrology Clinic
   —Pedi Endocrine Clinic
   —Adult Endocrine Clinic
   —Pedi Virology Clinic
   —ENT Clinic
   —RT Clinic
   —Eating Disorders Clinic
   —Guilford Shoreline Nutrition Center

   The following services are provided:

   —Individual and group dietary counseling.
   —Community service via nutrition education to consumer groups and schools.
   —Nutrition consulting to Business and Industry, and other health care agencies.
   —Nutrition education to consumers and employees via wellness fairs and newsletters.
   —Fielding consumer, and electronic and print media requests for nutrition information.
   —Development and sale of nutrition educational material.
   —Participation in hospital and department quality assessment activities.

   b. For your scope of service, the important aspects of care in the major clinical functions have been identified (e.g., activities that are high-volume, high-risk or problem prone). Please list these below:

   Inpatient—To provide quality nutrition care that supports the medical treatment of hospitalized patients, the following are considered the most important clinical nutrition aspects of care:

a. Ensuring that those patients at the greatest nutritional risk (i.e., those with PCM) are identified and assessed.

b. Ensuring that clinical nutrition standards of practice are followed to provide appropriate and timely patient care.

c. Ensuring that appropriate nutrition care is actually delivered.

Outpatient—Currently under development.

c. List the indicators used to measure these important aspects of care monitored over the past 12 months with the time frames for how often the data was evaluated.

The clinical indicators used are measurable variables that relate to the structure, process, and outcome of patient care. All indicators are evaluated quarterly (time frame). They are as follows:

a. **Aspect of Care**—Identifying and assessing patients with PCM. Indicators (Structure-Based)/Thresholds:

 i. Patients with a LOS > 3 days are screened for nutritional risk/95%.

 ii. Patients with known or suspected PCM are assessed:

 —Those on Parenteral or Enteral Nutrition/98%.
 —Those consuming oral diets/95%.

 iii. Baseline objective parameters are available for nutrition assessment:

 —Height/weight—90%.
 —Albumin (when PCM is known or suspected)—90%.

 iv. Serial objective parameters are available for nutrition follow-up:

 —Weight—90%.
 —Albumin—90%.

b. **Aspect of Care**—Clinical nutrition standards of practice are followed. Indicators (Process-Based)/Thresholds:

 i. Upon initial screening all appropriate risk factors are identified/90%.

 ii. Upon initial screening patients are appropriately classified (prioritized)/95%.

 iii. When PCM is suspected or identified, the type and extent of malnutrition are documented in patient medical records/98%.

 iv. The type and extent of PCM are correctly identified/98%.

 v. Calorie and protein requirements are correctly identified based on the type and extent of PCM or other diagnosis/98%.

 vi. Initial dietitian nutrition care plans and recommendations are:

 —Clearly identified/100%.
 —Appropriate/95%.
 —Timely/95%.

 vii. Subsequent dietitian nutrition care plans and recommendations are:

 —Clearly identified/100%.
 —Appropriate/95%.
 —Timely/95%.

c. **Aspect of Care**—Appropriate nutrition care is delivered. Indicator (Outcome-Based)/Thresholds:

 i. Initial diet orders are appropriate given patient's medical condition and treatment/75%.

 ii. Subsequent diet orders are appropriate/90%.

 iii. Physicians implement dietitians' recommendations/80%.

 iv. Nurses implement dietitians' recommendations/80%.

 v. Patients clinically demonstrate an improved or stable nutritional status or receive appropriate nutrition support given medical condition (Exception: Patients in whom it is too soon to evaluate change in status)/90%.

d. Did you monitor the important aspects of care delivered by your department? (Do your indicators correspond to the items you listed in b. above?)

 _x_ Yes ___ No

e. In implementing indicators, did you use objective, measurable criteria that reflect current practice and clinical experience while assessing your quality assurance information? Did you have a threshold for evaluation set-up for each indicator, if appropriate?

Criteria	_x_ Yes	___ No
Threshold	_x_ Yes	___ No

Courtesy of Yale–New Haven Hospital, New Haven, Connecticut.

f. Did you evaluate your indicators according to your specified time frames? (Compare data collected to pre-established criteria.)

   _X_ Yes ___ No

g. Did you analyze the data further when criteria were not met?

   _X_ Yes ___ No

h. Did you draw conclusions at each evaluation period?

   _X_ Yes ___ No

i. Did you revise your indicators or select a different aspect of care when data consistently met standards?

   Revise _X_ Yes and _X_ No
   Select New Aspect _N/A_ Yes _N/A_ No

II. The quality of patient care is improved and identified problems are resolved through actions taken, as appropriate.

a. Were problems or opportunities to improve care identified through your monitoring or other QA activities over the past 12 months? List these problems or opportunities for improvement.*

   1. Patients on oral diets with known or suspected malnutrition are not all assessed at the time of review.

   2. Availability of admission weight/albumin and follow-up weight/albumin is a problem.

   3. Nutritional risk factors are not consistently appropriately identified by the dietitian upon screening.

   4. Initial/subsequent diet orders are not consistently appropriate.

   5. Physicians/nurses follow-up on dietary recommendations is not consistent.

   6. Patients do not consistently receive appropriate nutrition care despite dietitian recommendations being consistently appropriate.

b. What action was taken in relation to the opportunities identified?

   1. Dietitian services were first realigned without an ↑ in FTEs to ↓ pt. load on high pt. acuity services and ↑ pt. load or provide consult only services to low pt. acuity services. This substantially increased the number of patients assessed. However, it was also discovered that the monitoring tool was not sensitive enough to standards of practice. This tool was revised and implemented in the first quarter of Fiscal '92.

   2. Discussed at Medical Board's Nutrition Committee. There is *wide* variation in the availability of these parameters from Patient Care Unit to Patient Care Unit. At present this problem remains unresolved as different protocols will need to be set up dependent upon individual medical and nursing departments (i.e., Medicine, Surgery, Psych., Peds, Ob/Gyn).

   3. Clarification of risk factors and standards of practice including charting were the topic of weekly QA meetings throughout the Summer of '91. The result was internal revisions to the YNHH Nutritional Classification and Assessment Manual.

   4. The more complex the diet order (i.e., multiple restrictions, enteral or parenteral feeds) the less likely it was to be appropriate. Based on Nutrition Committee and Medical QA Committee recommendations, the initial nutrition assessment form was revised to more clearly indicate nutritional recommendations and diet order adjustments, and enteral and parenteral feeding recommendations. The revised form was approved by the Nutrition and Medical Records Committees and implemented in March '91.

   5. See numbers 3 and 4 above.

   6. This problem remains unresolved. Refer to number 2 above.

c. Were problems followed up and monitored to determine if they were resolved or reduced to an acceptable level?

   _X_ Yes ___ No

d. What problems were not resolved? Why?

   All are still problems. For #1, resolution depends on results of revised monitoring tool. For #3, inservice continues on a monthly basis. For #'s 2, 4, 5, and 6, an interdisciplinary group is needed to address these issues.

e. Please identify any process/outcome issues that require interdisciplinary problem solving.

   Same as above, problems 3, 4, 5 and 6.

f. Did you document conclusions, recommendations, action and follow-up monitoring results?

   _X_ Yes ___ No

---

*These were areas where thresholds were not met.

g. Did you report the results of your monitoring activities to your appropriate staff and to your administrator?

   _X_ Yes ___ No

III. The main goal of quality assessment is improvement in patient care.

   a. What improvements in patient care have resulted directly from the activities of your quality assurance program in the past 12 months?

      1. ↑'ed number of high risk patients on oral diets who are nutritionally assessed (76% → 89%).

      2. Now exceed threshold (95%) for patients being appropriately screened for nutritional risk (96%).

      3. Now exceed threshold (95%) for appropriateness of dietitian recommendations for patient nutritional care (96%).

   b. Did you use the findings of quality assurance activities to evaluate staff performance, modify policies and procedures, change systems, and so on? Please list these.

      The first and second aspects of care relating dietitian performance are incorporated into their criteria-based performance goals. Also, as mentioned previously, dietitian services were realigned, and standards of practice were revised and clarified.

   c. Were your educational programs for staff or trainees, at least in part, based on the findings from monitoring and evaluation activities? Please correlate QA findings with educational programs given by your department.

      QA Finding   *Education Program*

      Please refer to IIb and IId.

   d. Please review your present indicators and criteria to determine their clinical importance, effectiveness in identifying problems and improving patient care.

      1. Problems or opportunities for improvement were identified.

         _X_ Yes   ___ No

      2. The problems identified impact patient care.

         _X_ Yes   ___ No

      3. Indicators were based on high-volume, high-risk, problem prone areas.

         _X_ Yes   ___ No

   e. Please list your indicators with their corresponding major aspects of care.

      Indicators   *Aspect*

      Same   Same

Completed by:

Administrative
Review Signature: _____ Date _____

Appendix A-8

Clinical Pathway DRG 106: Coronary Artery Bypass Graft (CABG)

DRG 106   Target LOS: 7 Days Post Operative
Clinical Pathway—CABG (Stable Angina)

	Cath Day Admit	*Pre Operative Day*
Diagnostics	CBC, SMA-7, UA, Coagulation Studies (PT, PTT, Platelets), EKG, CXR	Type and Crossmatch—3 units Bun, Creat—drawn day after cath and results available prior to surgery

Treatments	Identify and document: diabetes, drug allergies, height, weight (actual)
	Precath Hibiclens scrub prep and clip
	Vital signs
	  Precath:
	    Baseline prior to cath (BP, pulse, resp, temp, peripheral pulses)
	  Postcath:
	    Q15min x 4—recovery room (BP, pulse, resp, peripheral pulses)
	    Q30min x 2
	    Q1° x 4 (include temp on last set)
	    Q8° Shift
	Measure output—postcath
		Hibiclens shower day before surgery
		Clip in surgery

Medications	As ordered per physician	As ordered per anesthesia
Consults	Contact surgical consult postcath (Recovery Room to contact surgeon office)	Anesthesia (called by OR)
Nutrition	NPO after midnight Diet as ordered	Diet as ordered NPO after midnight
Activity	Up ad lib 6–8 hours post cath	Up ad lib
Teaching	Cardiac educator conducts with family present; booklet, video, verbal teaching Give educational binder to patient	Cardiac educator conducts; booklet, video, TCDB, IS-baseline readings
Social		

Courtesy of Iowa Methodist Medical Center.

Aspen Publishers, Inc., 1994

	Operative Day	*Post Operative Day 1*
Diagnostics	HCT and K-Q6° x 2 (postop), glucose, ABG's—1/2 hour after on vent, PT, PTT, platelets, CPK—12° postop repeat in 24° if CPK-MB > 5%, CXR, EKG	CXR, EKG HCT, K in am

Treatments	Admit to CCU
	NG as indicated
	Ventilator (changes per Resp as indicated)
	Wright Studies prior to extubation
	Extubate 8–12° postop per protocol
	IS Q2° after extubation
	DC NG 1° after extubation
	Cardiac monitor
	Arterial line, CVP, Swan-Ganz
	Temp Q2°
	I/O Q1°
	Foley
	Pulse oximeter
	Chest tubes
	Pacer wires
	Vital signs
	   Q15 min. until stable (BP, pulse, resp)
	   then Q1° when stable
	   Q2° temp
		CCU → Transfer Younker 7 late in day if following criteria are met and adequate staffing exists:
		   Off vasopressors
		   Hemodynamically stable
		   Up in chair without incident
		   Extubated
		   O$_2$ by cannula
		O$_2$ per cannula post-extubation
		DC pulse oximeter when on O$_2$-cannula
		IV's/saline lock for antibiotics (Transfer with saline lock)
		Chest tubes
		DC CVP, arterial line, foley
		Vital signs Q2°
		DC dressing at 24°
		Dressing care prn if drainage
		I/Q Q shift
		Pacer wires
		IS Q2° (post-extubation)
		Weight when OOB

Medications	Antibiotics (Ancef) Q8° x 3–6 doses Pain meds as indicated Antacids per NG tube Beta blocker/Digoxin as indicated Inotropes/Vasodilators as indicated Blood or blood products as indicated K, magnesium replacements	DC antibiotics if 3–6 doses completed PO Analgesia Diuretics as indicated Digoxin/Beta blockers Ascriptin Cardiac Rehab Phase I
Consults		
Nutrition	NPO	Liquids—advance as tolerated to Level I cardiac diet
Activity	Bedrest	Dangle Up in chair 2–3 times Chair for meals
Teaching		Orientation to Y7—call light, bed controls, meal times, activities, rehab, visiting hours, explanation of spirometry every 2–4 hours, cough and deep breaths every 4 hours
Social		

	Post Operative Day 2	Post Operative Day 3	Post Operative Day 4
Diagnostics			
Treatments	DC chest tube Chest physiotherapy as indicated O₂ cannula Saline lock IS Q2—4° (Increase as warranted) DC Pacer wires Vital signs Q4 (while awake) Dressing care prn if drainage I/O—Q shift Daily weight Legs elevated on 2 pillows in bed 30 min TID	O₂ cannula DC saline lock IS Q2—4° Vital signs Q4° (while awake) Daily weight Cough, deep breathing Dressing care prn if drainage I/O (if on diuretics or weight is > 5 lbs of admission weight) Legs elevated on 2 pillows in bed 30 min TID	O₂ cannula IS Q2—4° Vital signs Q4° (while awake) Daily weight Cough, deep breathing Dressing care prn if drainage I/O (if on diuretics or weight is > 5 lbs of admission weight) Legs elevated on 2 pillows in bed 30 min TID
Medications	DC antibiotics if 3–6 doses completed Analgesics as indicated Diuretics as indicated Ascriptin Beta blockers/Digoxin Bronchodilators as indicated	Analgesics as indicated Diuretics as indicated Ascriptin Beta blockers/Digoxin Bronchodilators as indicated	Analgesics as indicated Diuretics as indicated Ascriptin Beta blockers/Digoxin Bronchodilators as indicated
Consults			Cardiac Rehab Phase II
Nutrition	2000 cc fluid restriction Level I cardiac diet	2000 cc fluid restriction Level I cardiac diet	2000 cc fluid restriction Level I cardiac diet
Activity	Attend daily Rehab classes Chair for meals Cardiac Rehab Phase I as indicated Walk on PM's x 2—(nursing)	Up ad lib, assistance as necessary Cardiac rehab Phase I BID	Up ad lib, assistance as necessary Cardiac rehab Phase I BID Shower if pacer wire DC'd Begin paperwork for post-discharge cardiac rehab
Teaching	Reinforce cough and deep breathing IS Q4°	Medication teaching verbal and written teaching cards—Aspirin, Darvocet, Tenormin, diuretics optional Review disease process and surgery procedure with client and family	Daily education classes, i.e.: diet, risk factors, stress or watch channel 10 video—"Change of Heart—Taking It in Stride" Give Risk factor sheet and stress sheet if appropriate Begin reviewing CABG discharge instruction sheet (wound care, lifting restrictions, depression, medications, activities, diet, returning to work, smoking risks)
Social	Review patients in care conference regarding individual risk factors High-risk profile Referral to Social Service as indicated		

Aspen Publishers, Inc., 1994

	Post Operative Day 5	Post Operative Day 6	Post Operative Day 7
Diagnostics	EKG, CXR (PA and lat), HCT, Chem—7—(as indicated one day before discharge)		
Treatments	DC O_2 cannula IS Q2—4° Vital signs Q4° (while awake) Daily weight Cough, deep breathing Dressing care prn if drainage I/O (if on diuretics or weight is > 5 lbs of admission weight) Legs elevated on 2 pillows in bed 30 min TID	IS Q2—4° Vital signs Q6° (while awake) Daily weight Cough, deep breathing Dressing care prn if drainage I/O (if on diuretics or weight is > 5 lbs of admission weight) Remove Staples Legs elevated on 2 pillows in bed 30 min TID	IS Q2—4° Vital signs Q6° (while awake) Daily weight Cough, deep breathing Dressing care prn if drainage I/O (if on diuretics or weight is > 5 lbs of admission weight) Legs elevated on 2 pillows in bed 30 min TID
Medications	Analgesics as indicated Diuretics as indicated Ascriptin Beta blockers/ Digoxin Bronchodilators as indicated	Analgesics as indicated Diuretics as indicated Ascriptin Beta blockers/Digoxin Bronchodilators as indicated	Analgesics as indicated Diuretics as indicated Ascriptin Beta blockers/Digoxin Bronchodilators as indicated
Consults			
Nutrition	2000 cc fluid restriction Level I cardiac diet	2000 cc fluid restriction Level I cardiac diet	2000 cc fluid restriction Level I cardiac diet
Activity	Up ad lib, assistance as necessary Cardiac rehab Phase I BID Shower if pacer wire DC'd	Up ad lib, assistance as necessary Cardiac rehab Phase I BID Shower if pacer wire DC'd	Up ad lib, assistance as necessary Cardiac rehab Phase I BID Shower if pacer wire DC'd
Teaching	Identify individual risk factors and instruct in modification of risk factors	Review previous teaching with patient and document understanding Review prescriptions and follow up doctor appointments	Review discharge instruction sheet and assess level of understanding
Social			

Discharge Criteria: CABG Patient

Diagnostics	Potassium within normal limits or at baseline Renal function within normal limits or at baseline Hemoglobin ≥ 8 or patient asymptomatic CXR completed 2 days prior to discharge; pleural effusion less than 1/4 of pleural space
Treatments	No major fluctuations in blood pressure 24° prior to discharge Temperature < 100°F Heart rhythm stable within 24° prior to discharge Within 5 pounds of preoperative weight Chest wound healing showing no sign of infection Adequate bowel and bladder functioning
Nutrition	Adequate caloric intake
Activity	Independent ambulation
Teaching	Discharge instructions completed regarding written plan from Cardiac Rehab, referral for Phase II Rehab, medications, diet restrictions, risk factor modification and follow-up doctor's appointment
Social	Patient and family ready for discharge and social circumstances satisfactory

Exhibit A–8–1 CABG Monitoring Work Sheet

Return to the UM Office for Filing
Iowa Methodist Medical Center—Confidential
DRG 106—CABG Monitoring Worksheet

Patient's Name _____ Total LOS _____

MR # _____ Admitted _____ Discharged _____

Key Monitors	Yes	No	Variance Code
1. If patient needed a CABG, was the cardiac surgeon contacted by the recovery room?			
2. Was there a difference in date of surgery and date of catheterization? If so, number of days _____			
3. Was the actual height and weight done on admission?			
4. Were 3 units PRBCs typed and crossmatched pre op CABG?			
5. Did the OR notify Anesthesia of impending CABG?			
6. If the patient met established criteria within 24 hours, was he/she transferred to Y7?			
7. What was the number of hours from CCU admission to extubation? _____ Was this within 8–12 hours per protocol?			
8. What was the number of hours from CCU that patient activity was initiated? _____ Was this within protocol?			
9. On day 2, was the chest tube removed?			
10. Were all abnormal labs/x-rays addressed by the physician?			
11. Did nursing documentation show that legs were elevated on 2 pillows 30 minutes tid from postop day 2 until D/C?			
12. Did discharge occur on the date expected? Were outcomes met? If not, why?			
13. Did any clinical complications occur? If so, what?			

Expected Outcomes	Date Achieved	Difference from D/C Date (days)
1. Patient shall have potassium within normal limits or at baseline.		
2. Patient shall have renal function within normal limits or at baseline.		
3. Patient shall have hemoglobin ≥ 8 or be asymptomatic.		
4. Chest X-ray completed 2 days prior to discharge; pleural effusion less than 1/4 of pleural space.		
5. Patient shall show no major fluctuations in blood pressure 24 hours prior to discharge.		
6. Patient shall have temperature < 100°F.		
7. The patient's heart rhythm shall be stable within 24 hours prior to discharge.		
8. The patient shall be within 5 pounds of his/her preoperative weight.		
9. The patient chest shall be healing showing no sign of infection.		
10. The patient shall demonstrate adequate bowel and bladder functioning.		
11. The patient shall demonstrate adequate caloric intake.		
12. The patient shall demonstrate independent ambulation.		
13. The patient shall have instructions completed: • Referral for phase II Rehab • Regarding a written Cardiac Rehab Plan • Medications • Diet restrictions • Risk factor modification • Follow-up doctor's appointment		
14. The patient and family shall demonstrate readiness for discharge and satisfactory social circumstances.		

Exhibit A-8-2 Variance Codes and Descriptions

Return to UM Office for Filing—Confidential
Iowa Methodist Medical Center
DRG 106—CABG

System 300		Practitioner 400		Patient 500	
Variance Code	Variance Description	Variance Code	Variance Description	Variance Code	Variance Description
301	Admissions	401	Abnormal Test—No F/U	501	Allergies
302	Dietary	402	Anesthesiologist Delay	502	Altered V/S
303	EKG	403	Assessment	503	Bleed/Hematoma
304	Equipment Breakdown	404	Consent	504	Clinical Complication
305	Equipment Unavailable	405	Consultation	505	Family Delay
306	ICF Bed Unavailable	406	Education	506	Instruction
307	Laboratory	407	Health & Physical	507	Insufficient O_2 Saturation
308	OR Change	408	IV/Lines	508	Late
309	Other	409	MD Delay Procedure	509	Money for Prescription
310	Other	410	MD Discharge Delay	510	More Extensive
311	Patient Bumped	411	Medication	511	N/V
312	Pharmacy	412	Order Delay	512	No Home Support
313	Prior Case	413	Other	513	No show
314	Procedure Area	414	Other	514	NPO
315	Radiology	415	Other	515	Other
316	Schedule	416	Other	516	Pain
317	SNF Bed Unavailable	417	RN-Delay Discharge	517	Psych/Social/Legal
318	Transport	418	RN-Delay Procedure	518	Transportation
319	Variance	419	Social Service	519	Unable to Void
320	Waiting for Bed	420	Unit Staff Delay	520	Wants to See MD

Variance Analysis for TURPs
First Six Months, 1990

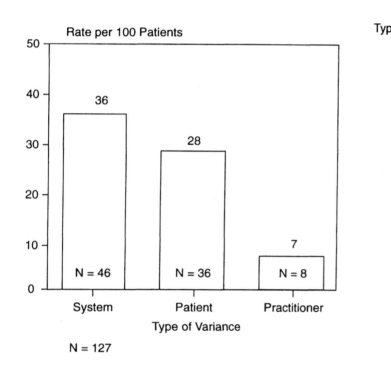
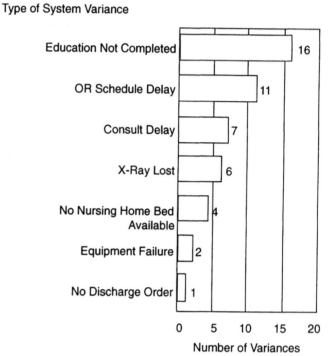

Figure A–8–1 Variance Analysis for Transurethral Resection of Prostrate

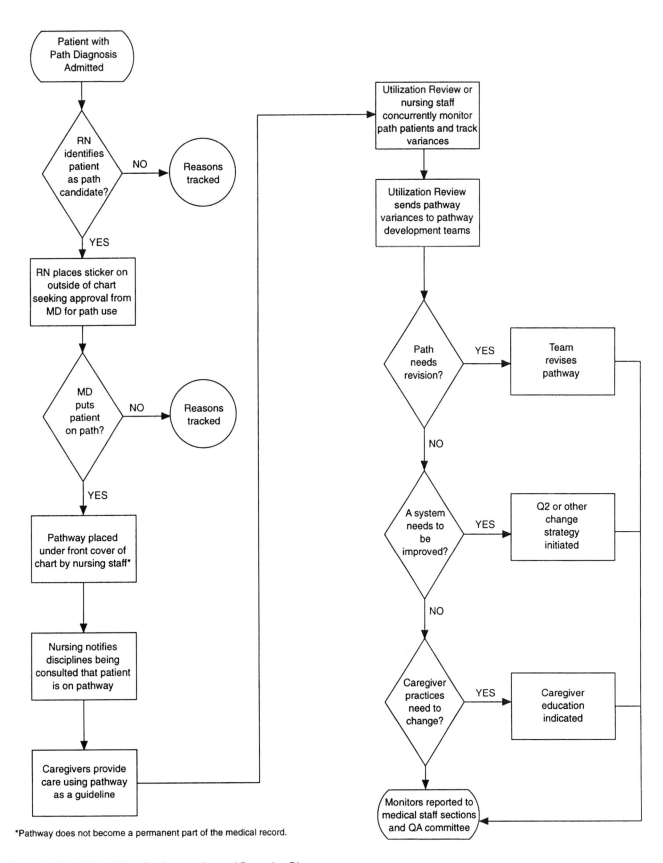

Figure A–8–2 Clinical Path Implementation and Reporting Plan

Exhibit A-8-3 Preliminary Measuring and Monitoring Plan

Iowa Methodist Medical Center
Clinical Pathway Implementation
Preliminary Measuring and Monitoring Plan

Date Element	Monitoring Frequency	Responsibility	Reported To
OUTCOMES DATA			
External Analysis • Regional/National benchmark comparison	Annually	Utilization Management	Clinical Pathways Committee Medical Staff Sections QA Committee
Internal Analysis • LOS • Complication rates • Charges (by department)	Quarterly	Utilization Management	Clinical Pathways Committee Medical Staff Sections QA Committee
PROCESS DATA			
Clinical Outcomes • Expected discharge indicators • Variance analysis	Concurrently/Quarterly	Utilization Management	Clinical Pathways Committee Medical Staff Sections QA Committee
Key Process Outcomes • (see attached)	Concurrently/Quarterly	Utilization Management	Clinical Pathways Committee Medical Staff Sections QA Committee

Aspen Publishers, Inc., 1994

Appendix B

American Dietetic Association Clinical Indicator Project Summary of Results

Source: Reprinted from Queen, P.M., Caldwell, M., and Balogun, L., Clinical Indicators for Oncology, Cardiovascular, and Surgical Patients: Report of the A.D.A. Council on Practice Quality Assurance Committee, *Journal of the American Dietetic Association*, Vol. 93, pp. 338–344, with permission of the American Dietetic Association, © 1993.

ADA CLINICAL INDICATOR PROJECT SUMMARY OF RESULTS

Indicators	% Met Indicator	Type of Indicator	Frequency of Data Collection	Explanation of Results	Recommendations
CARDIOVASCULAR CARE *Population: Patients with Cardiovascular Disease (DRG codes 103–109, 112, 115–118, 121–127, 129, 132–145) *Variables: Adult, 18 years of age and older, acute care setting * Exclusions: Pregnant and lactating women with cardiovascular disease					
RECOMMENDED INDICATORS: THOSE THAT SURVIVED THE FIELD-TEST PROCESS					
CV1. Patients at high risk for impaired nutrition status receive nutritional intervention within 5 days of admission or within 5 consecutive days of NPO, IV dextrose and/or clear liquid status, subcategorized by patients with multi-system organ failure, patients with sepsis.	88%	sentinel event; process	quarterly	Considerable comments that the test sites were unable to obtain weight and albumin information; however, 88% of the medical records met the criteria and IBW, albumin or recent voluntary weight loss was found.	Task force suggested that this indicator be recommended to the Joint Commission.
CV2. Patients/caregivers demonstrated/describe appropriate food choices and/or feeding regimens consistent with therapeutic diet, subcategorized by patients who are discharged within 4 days of admission patients with unavailable significant other.	28%	comparative rate; outcome	quarterly	The task force felt that the reasons only 28% met the indicator were 1) information on comprehension of the diet was not documented, 2) the field testers were looking for very specific information (i.e., describe actual food choices) and not more general information regarding comprehension of diet, and 3) many patients attended group education classes and comprehension was not assessed and documented.	Even though test sites did not document comprehension, the task force suggested that this indicator be recommended to the Joint Commission.

Appendix B 321

Indicators	% Met Indicator	Type of Indicator	Frequency of Data Collection	Explanation of Results	Recommendations
INDICATORS TESTED BUT DID NOT SURVIVE THE FIELD-TEST PROCESS					
After initial screening and weekly follow-up patients identified at nutrition risk have a nutrition care plan within 2–4 days.	45%	comparative rate; process	quarterly	Only 45% of the medical records met the indicator due to how it was worded.	The task force suggested that this indicator not be recommended to the Joint Commission. They did believe that this indicator is extremely important in monitoring patient care but that it needs revision and subsequent field testing. The recommended wording change for the indicator statement is: *After initial screening and subsequent follow-up screening, patients identified at nutrition risk have a nutrition care plan within 4 days.*
Patients assessed with inadequate knowledge about diet have a discharge plan.	31%	comparative rate; process	quarterly	The task force felt that the reason only 31% met the indicator was lack of documentation of formal assessment of knowledge.	The task force suggested that this indicator not be recommended to the Joint Commission. However, they believed the indicator highlights a critical area of patient monitoring, but it needs further revision and subsequent field testing. The recommended wording change for the indicator statement is: *Patients who require additional knowledge about diet have a discharge plan.*

ADA CLINICAL INDICATOR PROJECT SUMMARY OF RESULTS

Indicators	% Met Indicator	Type of Indicator	Frequency of Data Collection	Explanation of Results	Recommendations
ONCOLOGY Population: Oncology patients ICD-9-CM Diagnosis 1 codes between 140.0–208.9 or V58.0, V58.1 or Diagnosis 2 codes between 14.0 & 208.9 *Variables: 18 years of age or older; acute-care setting.					
RECOMMENDED INDICATORS: THOSE THAT SURVIVED THE FIELD-TEST PROCESS					
0.1. Patient is NPO or on Clear Liquid Diet without nutrition support for greater than 5 days, subcategorized by gastrointestinal tract surgical patients.	44% (negative indicator)	sentinel event; process	monthly	Several test sites had policies addressing NPO orders. Patients who exceeded the 5 day time-frame had medical justification.	The task force suggested that this indicator be recommended to the Joint Commission.
0.2. All patients at moderate or high risk are identified by screening and assessed within 72 hours of hospitalization, subcategorized by moderate risk, high risk, bone marrow transplant, surgery patients.	20%	comparative rate; process	monthly	Although each of 5 test sites had a different screening/assessment mechanism established, each had different time limits: One was 3 working days, one was within 4 work days, one related to screening within 72 hours. Others had different systems to prioritize patient status.	The importance of the indicator should not be diminished due to the low percentage of patients meeting the indicator. Each hospital/clinical center needs to apply the indicator to its own population and adjust it appropriately. A specific, realistic but clinically useful time frame should be established in implementing this indicator if the stated "within 72 hours of hospitalization" does not comply with a given clinical site.

Indicators	% Met Indicator	Type of Indicator	Frequency of Data Collection	Explanation of Results	Recommendations
0.3. Moderate or high risk patients/caregivers demonstrate knowledge/skills required to implement nutrition care plan for discharge, subcategorized by moderate nutritional risk and high nutritional risk patients.	R1—8%* R2—15%	comparative rate; outcome	monthly	Although the indicator had low results, the problem was the lack of documentation of the data elements.	The test sites felt this indicator was important for monitoring care, improving patient care and patient outcome. The task force suggested that this indicator be recommended to the Joint Commission.
INDICATORS TESTED BUT DID NOT SURVIVE THE FIELD-TEST PROCESS					
Patients with weight loss of greater than 10% usual body weight continue to lose weight during current hospitalization.	(results not usable)	sentinel event; outcome	monthly	Indicator was not met; 3 of 5 test sites had too few patients to evaluate this indicator. The main problem was the lack of documented weights during hospitalization. Documentation of continued weight or usual body weight was the hardest data to find.	Although there was consensus that this indicator highlighted a problematic or critically important area of patient care, the majority of test sites would not use this indicator to monitor patient care. The task force suggested that this indicator not be recommended for use by the Joint Commission and that it should be rewritten.
Patients receive TPN when it is documented that the gut is or will be impaired and/or dysfunctional for more than 10 days.	34%	sentinel event; process	quarterly	Only 53 patients were in the indicator population, the smallest of any indicator. Problems with documentation and understanding the intent of the indicator were identified.	Since the indicator does not highlight a problematic or critical area of care and 4/5 test sites felt it was not appropriate for evaluating care, the task force suggested that this indicator be deleted.

ADA CLINICAL INDICATOR PROJECT SUMMARY OF RESULTS

Indicators	% Met Indicator	Type of Indicator	Frequency of Data Collection	Explanation of Results	Recommendations
Patient screened and assessed at moderate or high risk receive no nutritional intervention within 3 days of recommendation for intervention.	(results not usable)	sentinel event; process	monthly	4/5 test sites did not meet the indicator. Problems occurred with documentation and understanding intent of the indicator.	The task force questioned the understanding/clarity/interpretation of this indicator although 4 out of 5 sites indicated they would use this indicator to monitor and evaluate care. The task force recommended this indicator not be used. Revisions suggested were to change the indicator to read positive such as: *"Patients screened and assessed at moderate or high risk have nutritional intervention within 3 days."*

SURGERY

Population: Surgical patients (All Surgical DRG codes except 112)
Variables: 18 years of age or older; acute care setting

RECOMMENDED INDICATORS: THOSE THAT SURVIVED THE FIELD-TEST PROCESS

Indicators	% Met Indicator	Type of Indicator	Frequency of Data Collection	Explanation of Results	Recommendations
S.1. Patients who are NPO or on clear liquid diets without nutrition support for 7 or more consecutive hospital days.	10% (negative indicator)	comparative rate; process	prospective review with quarterly summary	Since this was a negative indicator, only 10% were not fed within 7 days.	The task force felt this indicator was important for monitoring and improving care; that it was a problematic area of patient care. This indicator was recommended to the Joint Commission.
S.2. Patients with weight loss greater than 10% of admission weight at discharge subcategorized by major burn, pulmonary edema, congestive heart failure patients.	2% (negative indicator)	comparative rate; outcome	retrospective review quarterly	Only 2% of the sample lost more than 10% of body weight. There was some difficulty in getting weights on patients.	The emphasis needs to be on getting weight documented on patients. This indicator was recommended to the Joint Commission.

Appendix B 325

Indicators	% Met Indicator	Type of Indicator	Frequency of Data Collection	Explanation of Results	Recommendations
S.3. Calorie and protein and/or volume goals for patients receiving tube feeding or parenteral nutrition are documented in the medical record.	100%	comparative rate; process	quarterly	No problem identified.	Although the indicator measures process only, and may not always imply positive outcome or that nutrient goals were achieved, the task force recommended this indicator to the Joint Commission.
S.4. Postoperative patients initiated on tube feeding and/or parenteral nutrition receive at least 1000 calories/day by day four, subcategorized by tube feeding, total parenteral nutrition (TPN), peripheral, parenteral nutrition (PPN) concurrent combination tube feeding and TPN, concurrent combination tube feeding and PPN.	67%	comparative rate; process	ongoing, prospective collection with quarterly summary	No problems identified except that prospective data collection would be more efficient.	The task force suggested this indicator to be recommended to the Joint Commission.
INDICATORS TESTED BUT DID NOT SURVIVE THE FIELD-TEST PROCESS					
Patients receive nutrition intervention when there is an admitting diagnosis of malnutrition or admission albumin less than 3.0 g/dl, or unintentional weight loss of greater than 5% usual body weight within 1 month prior to admission, subcategorized by major burn, solid organ transplant, trauma patients.	51%	comparative rate; process	quarterly	Documentation of weights, weight history and serum albumin were problematic. However, the test sites felt this to be the most important indicator for patient outcome.	Since the data elements are hard to find, the field testers felt that other ways of defining nutrition risk may be more appropriate. The task forces did not recommend this indicator to the Joint Commission.

ADA CLINICAL INDICATOR PROJECT SUMMARY OF RESULTS

Indicators	% Met Indicator	Type of Indicator	Frequency of Data Collection	Explanation of Results	Recommendations
Patients on TPN with 2 blood glucose levels greater than 250 mg % on consecutive days who do not receive treatment within 1 day of the second laboratory report, subcategorized by patients with diabetes, receiving steroids, on peritoneal dialysis.	0% (negative indicator)	sentinel event; process	quarterly	Not a problem area.	The task forces suggested that this indicator be included in guidelines of practice and not be recommended to the Joint Commission.

*R1, R2 = Rater 1 and Rater 2. Each test site had at least 2 dietitians evaluating medical records. This indicator showed a difference between raters in the overall percent of medical records meeting the indicator.

Appendix C

Indicators from University Consortium

Courtesy of University Hospital Consortium, Oak Brook, Illinois.

University Hospital Consortium Quality Indicators
Department: Nutrition

Aspect of Care	Indicator	Threshold	High Risk	S/P/O	Discipline Responsible
Nutritional Assessment (Initial and Follow-up)	Assessment note is: —dated	< 100%		P	
	—signed by dietitian with full name or title.	< 100%		P	
	Subjective assessment includes: —current intake —appetite —typical eating patterns —food allergies/intolerances —eating problems —activity level —previous diet.				
	Objective assessment includes: —age, sex, height, weight and previous weight —ideal body weight and/or % of usual or ideal body weight —basal energy expenditure (BEE) —calorie counts —laboratory values when pertinent —medications —diet order.	< 100%		P	
	Documented assessment includes: —a problem list —diagnosis and related nutritional implications —energy/caloric/protein requirements —adequacy of the present nutrition prescription.	< 100%		P	
	Nutrient analysis includes: —selected nutrient/energy intake (carbohydrate, protein, fat and kilocalories) —detailed calculation or extensive computerized nutrient/energy analysis.	< 100%		P	
	Plan of care includes (if applicable): —pertinent laboratory values —frequency of weights —diet change —diet instructions —follow-up instructions.	< 100%		P	

Aspen Publishers, Inc., 1994

Aspect of Care	Indicator	Threshold	High Risk	S/P/O	Discipline Responsible
Nutritional Intervention for High-Risk Patients	Evaluation notes address: —date of note —new problems —calorie count analysis —need for further follow-up.	<100%		P	
	Adequate intake is recorded by calorie counts.	<95%		P	
	Improved calorie intake.	<75%		O	
	Improved protein intake.	<75%		O	
	Improved serum proteins.	<40%		O	
	Improvement in nutritional status as evidenced by weight stabilization.	<50%		O	
	Patient is screened within 48 hours of admission.	<90%		P	
	Nutritional assessment and care are documented in the medical record within 5 days of admission.	<90%		P	
	The presence of potential nutritional problems are documented: —on initial assessment —at seven-day intervals.	<90%		P	
	Patients on modified diets are screened within 48 hours of modified diet orders.	<90%		P	

	Registered dietitians will assess the following high-risk patients:
	—admissions to intensive care or burn units
	—patients receiving parenteral or enteral nutrition
	—patients diagnosed with:
	    acute renal failure
	    endstage renal disease
	    chronic renal failure
	    liver failure
	    cirrhosis
	    cardiac surgery
	    congestive heart failure
	    thoracic surgery
	    COPD
		<100%		S	D, PH

Aspect of Care	Indicator	Threshold	High Risk	S/P/O	Discipline Responsible
	diabetes				
	short gut syndrome				
	GI surgery				
	pancreatitis				
	anorexia				
	bulemia				
	morbid obesity				
	—all transplant patients with abnormal laboratory values				
	albumin < 3.5				
	glucose > 200				
	cholesterol > 240				
	BUN > 40				
	creatinine > 3.0				
	amylase > 200				
	—weight below 80% of ideal body weight.				
Medium-Risk	Registered dietitians will assess the following medium-risk patients:	< 100%		S	
	—antepartum				
	—hypertensive				
	—patients with cholesterol levels of 200–240.				
Low-Risk	Registered dietitians' diet aides will assess low-risk patients:	< 100%		S	
	—other patients per protocol.				
Nutritional Care for Patients Who Are NPO	Patients who are NPO > 5 days are assessed by NPO day 3.	100%		S	
	Assessment includes:	100%		P	
	—date of intervention				
	—diagnosis				
	—clinical data				
	signs/symptoms of nutritional deficiencies				
	response to NPO status				
	—height				
	—weight				
	current				
	ideal				

Aspect of Care	Indicator	Threshold	High Risk	S/P/O	Discipline Responsible

	—biochemical data
	    total lymphocyte count
	    pre-albumin
	    albumin
	    transferrin
	    BUN/creatinine
	    serum glucose
	    cholesterol/triglyceride
	—dietary history
	    estimate of usual intake
	    food allergies or intolerances
	—a statement of estimated calorie/protein needs using Harris-Benedict Equation.
	—a statement of adequacy of present nutrition regimen and length of time patient has been NPO.				

	The nutrition care plan documentation includes: —appropriate recommendations and route of nutrition support (i.e., oral, enteral, or parenteral hyperalimentation).	< 95%		P	
Patients on Clear Liquid Diets	Patients receiving unsupplemented clear liquid diets for greater than three (3) days have an assessment of nutritional status.	< 100%		P	
Patients Receiving Parenteral Nutrition Support	Nutrition consults are ordered for patients on central parenteral nutrition.	< 100%		P	
	Assessment of patient receiving TPN is performed within 2 days after initiating TPN or receiving a consult.	< 100%		P	

	Assessment includes:
	—clinical data
	    signs/symptoms of nutritional deficiencies
	—Biochemical data
	    total lymphocyte count (TLC)
	    pre-albumin
	    albumin
	    transferrin
	    retinol binding protein
	    cellular immunity skin test
	    other indicators
	    height				

Aspect of Care	Indicator	Threshold	High Risk	S/P/O	Discipline Responsible

	weight
	  current
	  ideal
	body mass index = weight/height$_2$.				

	Product ordered was administered within 24 hours of the order.	< 100%		P	
	Laboratory studies are ordered at the initiation of parenteral nutritional therapy.	< 100%		P	P, D, N
	A follow-up nutrition support note will be done per protocol.	< 100%		P	
	During the course of parenteral nutrition therapy the patient will not experience the following: —blood sugar < 100 mg% —blood sugar > 250 mg% —serum sodium > 155 mmol/l —serum potassium > 6.0 mg% —serum magnesium < 1.4 mEq/l —serum magnesium > 3.0 mEq/l —serum phosphorus < 2.0 mg% —serum phosphorus > 6.0 mg%.	< 100%	HR	O	P, D, N, PH

Patients Receiving Enteral Support	Assessment includes:
	—Clinical data
	  reason for diagnosis inadequate
	  nutrients by oral route
	—Biochemical data initiation
	  total lymphocyte count (TLC)
	  pre-albumin
	  albumin
	  height
	  weight
	    current
	    ideal.
		< 100%		P	

	Intolerance of tube feedings: —nausea —vomiting	< 100%	HR	P	N, D

Aspen Publishers, Inc., 1994

Appendix C   333

Aspect of Care	Indicator	Threshold	High Risk	S/P/O	Discipline Responsible
	—cramps —abdominal distention, fullness —diarrhea —constipation.				

	Continued monitoring includes:
	—weight 3 times/week
	—laboratory data
	    blood glucose
	    urine glucose
	    other pertinent data
	—BEE or REE
	—tube placement
	—method of delivery.
		<95%	P		N, D, P

	Complications identified and addressed:
	—gastrointestinal
	—mechanical
	    clogged tube
	    broken tube
	    displaced tube
	—aspiration
	—metabolic
	    hyperglycemia
	    hypo; hypernotremia
	    other.
		>0%
		>0%
		
		>0%
		>0%
			P
			O
			
			O
			O		

Patient/Family Education	Patient/family education for discharge addresses the following: —mixing formula —sanitation —amount/duration of feedings —intervals between feedings —cost of feedings —complications —patient/family ability to demonstrate safe feeding techniques —patient/family ability to perform care of tube (nasal or enterostomy).	<100%	P		D, N, DP

334 TOTAL QUALITY MANAGEMENT FOR HOSPITAL NUTRITION SERVICES

Aspect of Care	Indicator	Threshold	High Risk	S/P/O	Discipline Responsible
Discharge Planning for Patients	Discharge planning will include:				DP, N, D
	—follow-up appointment	< 100%		P	
	—emergency phone number	< 100%		P	
	—patient and/or caregiver perform demonstration of feeding technique (enteral/parenteral)	< 100%		P	
	—proper equipment is ordered.	< 100%		P	
Nutritional Assessment and Monitoring of Renal Dialysis Patients	Nutritional assessment is:				
	—dated	< 100%	HR	P	
	—includes problem list.				
	Subjective data addresses:	< 100%	HR	P	
	—appetite				
	—meal intake				
	—eating pattern				
	—eating problems				
	—weight history (new patient)				
	—activity level (new patient)				
	—previous diet.				
	Objective data addresses:	< 100%	HR	P	N, L, D, P
	—age/sex				
	—diet order				
	—height/weight/IBW/EBW				
	—laboratory values				
	—medications				
	—diet instructions.				
	Assessment includes:	< 100%		P	
	—evaluation of diet order				
	—fluid weight				
	—K cal/protein requirements				
	—previous diet instructions				
	—compliance with diet				
	—nutritional problems.				

Aspen Publishers, Inc., 1994

Aspect of Care	Indicator	Threshold	High Risk	S/P/O	Discipline Responsible
	Plan addresses: —request for diet change —monitoring of weight/labs —monitoring of intake (calorie count) —reinforcing/instructing on diet.	< 100%		P	
	Follow-up assessment includes: —evaluation of appetite/intake —monitoring weight/labs —reinforcing diet —statement of new nutrition-related problems.	< 100%		P	
Nutritional Assessment and Monitoring of Patients Receiving Hemodialysis	Patients on hemodialysis are assessed per protocol of beginning therapy.	< 100%	HR	P	
	Initial assessment includes: —previous diet history —weight history —nutrient needs assessment (protein, potassium, sodium, phosphorus, fluid).	< 100%	HR	P	
	Follow-up assessment: —review of lab values and trends —renal diet reinforced.	< 100%	HR	P	
	Assessments and interventions are documented per protocol for stable patients.	< 100%	HR	P	
	Ongoing nutritional assessment and intervention is documented per protocol for unstable patients.	< 100%	HR	P	
	Unstable parameters are: —BUN < 40 or > 90 —K < 3.5 or > 5.8 —Phos < 2.4 or 6.0 —Alb < 3.5. —unintentional weight gain or weight loss of > 3 kg. —interdialytic weight gains exceed threshold on more than 3 consecutive clinic visits.	> 0%		O	N, L, D, P

Aspect of Care	Indicator	Threshold	High Risk	S/P/O	Discipline Responsible
Nutritional Assessment and Monitoring of Patients Receiving CAPD	Assessment includes: —previous diet history —weight history —assessment of nutrient needs (protein, sodium, phosphorus, potassium, fluid, calories) —calories contributed by CAPD.	< 100%		P	N, P
	Patients on CAPD are counseled per protocol.	< 100%		P	N, P
	Patients on CAPD are reviewed per protocol if the following conditions exist: —BUN < 40 or > 90 —K < 3.0 or > 5.8 —Phos < 2.4 or > 6.0	< 100%		O	
	Follow-up review addresses the following: —weight changes —changes in diet order —changes in medical condition —identification of specific problem areas.	< 100%		O	
Timeliness of Nutritional Interventions in Critical Care Patients	Nutritional assessment and support is initiated within 3 days of admission to the unit.	< 100%	HR	P	N, P, D
	Recommendations are made by a registered dietitian.	< 100%	HR	P	
	Physician response to recommendations occur within 2 days.	< 100%		P	
	Physician orders reflect dietary recommendation.	< 100%		P	
Nutritional Assessment of the Adult Patient Diagnosed with Cancer	Assessment is performed within 2 working days.	< 100%		P	

	Assessment includes:
	—clinical data
	    primary organ site affected, presence of metastases, clinical symptoms
	    treatment modality and potential side effects
	    effects of malignancy on nutrient ingestion, tolerance and/or utilization
	—biochemical data
	    serum albumin/pre-albumin
	    transferrin
		< 100%		P	

Aspect of Care	Indicator	Threshold	High Risk	S/P/O	Discipline Responsible
	blood glucose				
	BUN/creatinine				
	cholesterol/triglyceride.				
Nutritional Assessment of the Adult Burn Patient	Assessment is performed within 2 working days.	< 100%		P	N, D
	Assessment includes:	< 100%		P	
	—clinical data				
	signs/symptoms of nutritional deficiencies				
	—biochemical data				
	total lymphocyte count (TLC)				
	pre-albumin				
	albumin				
	transferrin				
	retinol binding protein				
	cellular immunity skin test				
	other indicators (i.e., weight/height).				
Nutritional Assessment of High Risk Obstetric Patients	High risk OB patients are automatically referred to the registered dietitian:	< 100%	HR	P	N, D
	—age 18 or less				
	—diagnosis of IDDM				
	—diagnosis of gestational DM				
	—multiple pregnancies				
	—diagnosis of eating related disorders such as anorexia, bulemia, etc.				
	—others identified by M.D.				
Nutritional Assessment of Adult Patients with Hypercholesterolemia	Assessment will occur within 2 working days.	< 100%		P	N, P, D
	Assessment includes:	< 100%		P	
	—clinical data evaluated to determine cause of hypercholesterolemia				
	—primary causes				
	definite hyperlipidemia				
	coronary heart disease				
	—secondary causes				
	diabetes mellitus				
	hypothyroidism				
	nephrotic syndrome				
	renal failure				
	—positive risk factors				
	height				

Aspect of Care	Indicator	Threshold	High Risk	S/P/O	Discipline Responsible

	weight
	  ideal
	  usual
	—biochemical data
	  serum cholesterol
	  < 200 mg/dl—desirable
	  200–239 mg/dl—borderline
	  > 240 mg/dl—high risk
	  HDL & LDL values.
		< 100%		O	

Nutritional Assessment of the Adult Patient Diagnosed with Type I and Type II Diabetes	Assessment will occur within 2 working days.	< 100%		P	N, D

	Assessment includes:
	—clinical data
	  medications and/or insulin regimen
	  home glucose monitoring
	—biochemical data
	  BUN/creatinine
	  cholesterol
	  triglyceride
	  glucose
	  glycohemoglobin.
		< 100%		P	

Discharge Instruction Related to Food/Drug Interactions	Patients will be instructed regarding interactions.	< 100%	HR	P	D, N, PH
	Prescription drugs will be labeled with potential interactions.	< 100%	HR	P	
	Patients on Warfarin—discharge information includes: —avoid excess consumption of leafy green vegetables, cabbage, broccoli, cauliflower, spinach, liver, egg yolk. —avoid large doses of vitamin A, E, and C supplements. —avoid consumption of alcohol with the drug.	< 100%		P	
	Documentation of the patient's understanding of the instructions is confirmed by: —patient signature on discharge form —patient signature on the standard of care form —documentation of understanding.	< 100%		P	

Aspen Publishers, Inc., 1994

Aspect of Care	Indicator	Threshold	High Risk	S/P/O	Discipline Responsible
Orientation and In-service Training for Staff	All new staff attend staff hospital orientation/guest relations program within one week of start date.	< 100%	HV	S	D, IC
	New staff attend staff departmental orientation program on or before start date.	100%		S	
	Retained staff receive 90-day orientation training program.	100%		S	
	All staff attend monthly in-service programs.	90%		S	
Performance Appraisals	Staff will receive performance appraisals at end of probationary period and annually.	100%		S	
Departmental Staff Meetings	Departmental professional staff, supervisor staff, and support staff meetings are held monthly.	90%		S	
Continuing Education Requirements for Dietitians	Dietitians maintain registration credentials.	100%		S	
Sanitation	Temperatures are maintained within required ranges for the following: —dish machines —food —refrigerators.	< 100% < 100% < 100%	HR	S	
	In-service education on sanitation and infection control is attended annually and documented.	< 100%		S	
Safety	Wound inspection of food service support staff occurs per protocol.	< 100%	HR	S	
	Incident reports.	0%	HR	O	
	Number of employee accidents.	0%	HR	O	
	Number of needles/syringes discovered on trays returned to dish room.	0%	HR	O	
	In-service education programs on safety procedures and emergency plans are attended annually and documented.	100%			D, IC

Aspect of Care	Indicator	Threshold	High Risk	S/P/O	Discipline Responsible
Tray Accuracy	Patient trays will have no incorrect items or missing items.	100%		P	
Diet Accuracy	Physician orders are accurately transcribed —modified diets —non-modified diets	100%		P	
Enteral Tube Feeding Accuracy	Enteral tube feeding orders are accurately transcribed from the physician orders.	100%	HR	P	

Appendix D

Sample Monitoring Forms

Appendix D–1. Checklist for Evaluating Medical Record Notes 342
Appendix D–2. Productivity Codes and Form To Conduct Time Study as a Basis for Establishing Productivity Levels 343
Appendix D–3. Quality Assessment and Improvement Work Sheet 345
Appendix D–4. Quality Assurance Improvement Monitor Development Form 347
Appendix D–5. QA Criteria Form and Data Retrieval Form 349
Appendix D–6. Patient Nutrition Care Monitors 351
Appendix D–7. Quality Assurance Annual Appraisal 355

Appendix D–1

Checklist for Evaluating Medical Record Notes

Criteria	Sample 1	2	3	4	5	Comments
Format conforms to institutional policy						
Location is appropriate						
Patient's name appears on every page						
All entries signed and dated						
All entries are permanent; rules for correcting notes are followed						
Screening and assessment information given if patient hospitalized more than 48 hours						
Notes are brief, clear, and precise						
Only standard abbreviations used						
Accurate calculations; precise details are given when necessary						
Notes are legible						
Facts are given; generalizations and impressions avoided						
Patient education needs recorded						
Evaluation of care documented						
No extraneous remarks and criticisms						
Recommendations straightforward and definitive						
Dietitian's recommendations accepted; reflected in physician orders						
Countersignatures affixed as appropriate						
Referral and discharge needs indicated						
Y—Meets criteria; N—Does not meet criteria						

Source: Reprinted from *Handbook for Clinical Nutrition Services Management*, by M.R. Schiller, J.A. Gilbride, and J.O. Maillet, p. 273, Aspen Publishers, Inc., © 1991.

Appendix D-2

Productivity Codes and Form To Conduct Time Study as a Basis for Establishing Productivity Levels

CLINICAL PRODUCTIVITY CODES

DPC (Direct patient care)
- procedures conducted in patient's presence
- diet hx
- counseling
- diet instruction/patient classes

IPC (Indirect patient care)
- procedures not conducted in patient's presence
- arrange day
- chart review
- charting
- NCPs
- conferences (MD, RN, DT)
- rounds
- Literature review
- NIA

QA
- taste testing
- prep patient classes
- prep for instruction
- phone
- screening

SS
- Support Services
- menu changes/menu writing
- Nx orders
- Kardex notes
- charges
- missing trays, food

P
- projects

Supv
- supervision
- DTs, DAs, DUCs
- scheduling

Tch
- teaching of students—includes any preparation time

CB
- coffee break

L
- Lunch

Adm
- *administrative*
- sick leave
- annual leave
- conferences
- prep for talk
- activities, professional

Work Undone
- **Scr** — screen
- **NIA** — intakes
- **A** — assess
- **Cht** — chart
- **Fu** — follow-up
- **Kn** — Kardex notes

Source: Courtesy of Susan DeHoog, Seattle, Washington.

FORM FOR TIME STUDY

NAME: _____ DATE: _____
POSITION: _____ DAY OF THE WEEK: _____

CODE	1	2	3	4	5	6	7	8	9	10	WORK UNDONE
DPC											
IPC											
SS											
P											
SUPV											
TCH											
CB											
L											
ADMIN											
Total Census Unit									Hi Risk	Mod Risk	
Total Census Unit									Hi Risk	Mod Risk	
Number of Screens Completed											
NUMBER OF INITIAL ASSESSMENTS								Total:			
NUMBER OF FOLLOW-UPS*								Total:			

HOURS/DAY

*Does not receive a chart entry (i.e., data collection, chart review, etc.).

Aspen Publishers, Inc., 1994

Appendix D-3

Quality Assessment and Improvement Work Sheet

Department _____

I. **Assign Responsibility:**

   Director:

   Assistant Director(s)/Supervisors:

   Staff:

   Department Quality Assurance Representatives:

II. **Delineate Scope of Care/Service:**

   Who are the recipients of your care/service?

   What are your primary clinical activities?

   What major diagnoses do you treat?

   What are the high-volume diagnostic/treatments/procedures you provide?

   What is your staff mix (skill level of various positions, educational requirements, etc.)?

III. **Identify Important Aspects of Care/Service:**

   Must identify at least two quality indicators in each of the following categories:
   High Volume:

   High Risk:

   Problem Prone:

IV. **Identify Indicators:**

   Each aspect of care must be monitored at some point during the year. An indicator may monitor more than one aspect of care.

   How are indicators determined, i.e., who, when?

   What criteria do you use to determine indicators?
   1. The indicator reasonably relates to at least one important aspect of care selected.
   2. The indicator measures a patient outcome, or measures a structure or process of patient care that is reasonably expected to be related to a patient outcome.

Source: Courtesy of Medical Center Hospital, Chillicothe, Ohio.

3. The indicator is reasonably related to the care provided by the discipline or department/service/unit and/or function using the indicator.
4. The rationale of reasoning behind the selection of the indicator is documented or verbalized; that is, how the structure, process, or outcome measured by the indicator affects patient outcomes.
5. The indicator has demonstrated utility to the department/service/unit and/or function in identifying opportunities to improve quality of care provided.

Do you use criteria? If yes, how do you determine what criteria to use?

V. Establish Thresholds for Evaluation:

How are thresholds for evaluation determined?

Who sets them?

VI. Collect and Organize Data:

How are data collected (who, when, etc.)?

Who organizes the data?

What data sources will you use?

Will you do concurrent or retrospective review?

How will data be displayed?

Will you use sampling or 100% review?

VII. Initiate Evaluation:

When will you evaluate the data?

Who will analyze the data?

What will you be evaluating the data for?

Data will be evaluated for trends, patterns, problems, and/or opportunities to improve care.

Appendix D–4

Quality Assurance Improvement Monitor Development Form

Part A

Administrative/Clinical Service: DIETETIC SERVICE   Monitor/Title/NO: _____
                                                                                    (as listed in QA plan)

ASPECT OF CARE: (Identify as high-risk, high-volume, problem prone.)

INDICATORS: (At least two indicators are required; each must have a designated threshold. Identify as structure, process, or outcome.)

THRESHOLDS: (If threshold is met or exceeded, further in-depth assessment is required.)

DATA COLLECTION AND ORGANIZATION:

   a. Person responsible for data collection: _____

   b. Concurrent or retrospective review: _____

   c. Data source: _____

   d. Sample size: _____

   e. Frequency of data collection: _____

   f. Frequency of reporting: _____

   g. Reported to (person/committee): _____

_____   _____
SERVICE/SECTION CHIEF   **DATE**

Source: Courtesy of Olin E. Teague Veterans' Center, Temple, Texas.

Part B

Date: _____

Report Period: _____

Monitor Title/NO: _____

FINDINGS: (Summary of raw data)

CONCLUSION(S):

RECOMMENDATION(S):

ACTION(S):

EFFECTIVENESS OF ACTION: (Did the action improve the care or service? Is the improvement documented and maintained?)

TRENDS:

	Threshold	1st Quarter	2nd Quarter	3rd Quarter	4th Quarter
Indicator 1:					
Indicator 2:					

_____ _____
SERVICE CHIEF/COMMITTEE CHAIRPERSON   DATE

_____ _____
QUALITY ASSURANCE COORDINATOR   DATE

Appendix D-5

QA Criteria Form and Data Retrieval Form

QA CRITERIA FORM

THRESHOLD OF COMPLIANCE __100__ %
INDICATOR: Nutrition Counseling is performed according to department standards
SOURCE: Nutrition Services Policy/Procedure Manual
METHOD: Retrospective Chart Review

CRITERIA	THRESHOLD	EXCEPTION	COMPLIANCE
1. Written documentation by the Clinical Dietitian or Clinical Assistant within 48 hours of receipt of written order.	100%	Order written in Medical Record but not forwarded by nursing unit.	The Clinical Dietitian or Clinical Assistant has documented that either the order was received and full consult pending, or the full consult is charted within 48 hours of receipt of written order.
2. SOAP charting is used	100%	1. Brief note written that order received and consult pending 2. Instruction ordered same day as discharge. 3. Follow-up notes	All initial assessments or documentation is written in SOAP format in the interdisciplinary progress notes in the patient's Medical Record.
3. Assessment of patient/significant other understanding of instructions	100%	1. Patient not able to communicate 2. Family not available	Documentation in medical record indicates an objective assessment of the patient's/significant other's understanding of the diet instruction.
4. Follow-up measures given to patient/significant other	100%	None	Written Documentation includes follow-up measures were given to patient for further questions or instructions if needed.
5. Discharge Planning Flow sheet completed	100%	1. Not in chart to complete and so indicated in SOAP note	Discharge Plan flow sheet includes date and type of instructions given to patient with signature.

Source: Courtesy of Rhonda Billman, Clinical Manager, Nutrition Services, Southwest General Hospital, Middleburg Heights, Ohio.

DATA RETRIEVAL FORM

UNIT: _____										
DATE: _____						1) + Met Criteria		3) O Variation		
RESPONSIBLE RECORDER: _____						2) – Met Exception		4) J Justified Variation		

Nutrition Counseling Is Performed According to Department Standards Medical Record #	Documentation within 48 hours	SOAP Format	Assessment of Patient's/Significant Other's Understanding	Follow-Up Measures Given	Discharge Planning Flow Sheet Complete					
TOTAL #/% COMPLIANCE										**AVERAGE #/% COMPLIANCE**
TOTAL #/% OF NONCOMPLIANCE										**AVERAGE #/% NONCOMPLIANCE**

Aspen Publishers, Inc., 1994

Appendix D–6

Patient Nutrition Care Monitors

Reviewer(s): _____

Patient Initials _____ **Age** _____ **Sex** _____ **Service** _____ **Rm#** _____ **Unit#** _____

Principal Diagnosis _____

Metabolic/Mechanical Problems _____

_____ Current Diet: _____

Today's Date _____ Adm. Date _____ Attending MD _____

EVALUATION AND OUTCOME OF NUTRITIONAL CARE

A. Availability of Baseline Nutrition Parameters

   1. Was an admitting/initial weight available?

      Yes ____ No ____ Where? _____

   2. Was an albumin available?

      Yes ____ No ____ Not applicable ____

B. Availability of Subsequent Nutrition Parameters

   1. Were weights available for patient follow-up?

      Consistently ____ Sporadically ____ Not at all ____

      Too Soon to Evaluate ____ N/A ____

   2. Were serial albumin or other pertinent labs available for patient follow-up?

      Consistently ____ Sporadically ____ Not at all ____

      Too soon to evaluate ____ N/A ____

C. Was Patient Initially Kardex Classified Correctly? Yes ____ No ____ If no, explain _____

_____

D. Initial Assessment—Date Completed: _____ By _____ Patient Class _____

   1. Are the risk factors appropriately identified?

      Yes ____ No ____ Not applicable ____

   2. Was the patient classified correctly at the time of the initial assessment?

      Yes ____ No ____ Not applicable ____

      If no, explain _____

Source: Courtesy of Yale–New Haven Hospital, New Haven, Connecticut.

3. Was the nutrition assessment appropriate?

   ____ Yes (Assessment of diet order vs. estimated needs, and nutritional status is acceptable.)

   ____ No   If no, explain _____

   _____

4. Were the nutritional care plan and recommendations:

   a. Clearly identified? Yes ____ No ____

   b. Appropriate, given data available and medical condition?

   Totally ____   Partially ____   Not at all ____

   If partially or not at all, explain _____

   _____

E. Metabolic Profile—Date Completed: _____ By: _____

   1. Was a Metabolic Profile completed?

      Yes ____

      No (baseline parameters not available): ____

      Not applicable: ____

      Not done, but should have been: ____

      Explain: _____

      _____

      If MP was not done, skip to Section F.

   2. Was the type of malnutrition correctly identified?

      Yes ____   No ____   Not applicable ____

      If no, explain _____

   3. Are the protein/calorie recommendations appropriate given the patient's condition and nutritional status?

      Yes ____   No ____   Not applicable ____

      If no, explain _____

F. Documented Follow-up/Number of Follow-up Numbers _____ (Tally for each follow-up reviewed)

   If no follow-up done, complete #1 only and skip to Section G.

   1. Was the F/U note timely?

      Yes ____   No ____   Too soon to evaluate ____

   2. Did RD follow up on previous nutrition recommendations?

      Yes ____   No ____   Not applicable ____

      If no, explain _____

   3. Were nutrition care plan and/or recommendations:

      a. Clearly identified?

Totally _____   Partially _____

Not at all _____   Too soon to evaluate _____

b. Appropriate?

Totally _____   Partially _____

Not at all _____   Too soon to evaluate _____

G. Diet Orders

1. Was the initial diet order appropriate?

   Yes ____ No ____

   What was it? _____   If no, explain _____
   _____

2. Were subsequent diet orders appropriate?

   Consistently ____   Sporadically ____

   Not at all ____   Not applicable ____

   Why? _____
   _____

H. MD/RN Follow-Through

Were RD's recommendations followed by MD/RN staff?

	Consistently	Sporadically	Not at All	Too Soon to Evaluate	N/A
MDs					
Albumin ordered					
Other labs ordered					
Diet orders changed					
TF initiated					
HAL initiated					
TF form. or rate changed					
HAL form or rate changed					
Other _____					
RNs					
Follow-up weights obtained					
Calorie counts done					
Other _____					

COMMENTS: _____
_____
_____

I. Quality of nutrition care/support received (at time of this review)

    ____ 1.   Appropriate. Why?

        ____ a. Nutritionally improved (as indicated by positive clinical improvement or changes in medical/nutritional parameters).

        ____ b. Nutritionally stable.

        ____ c. Medically and nutritionally compromised, but receiving appropriate nutrition care.

    ____ 2.   Inappropriate. Why?

        ____ a. Nutritionally/medically compromised due to excessive nutrition support.

        ____ b. Nutritionally/medically compromised due to inadequate nutrition support.

    ____ 3.   Nutrition care ordered/received seems appropriate but no clinical parameters available to assess nutritional status or change in status.

    ____ 4.   Too soon to evaluate.

Aspen Publishers, Inc., 1994

Appendix D–7

Quality Assurance Annual Appraisal

Department: _____ **Year:** _____

Person Completing Appraisal: _____

Please answer the following questions as they relate to your department/service's quality assurance program during the year.

1. Does your program conform to the JCAHO 10-step monitoring and evaluation process? YES ____ NO ____
2. Does the Scope of Services, as outlined in your plan, still apply at this time? YES ____ NO ____
3. Have you added or deleted any services/programs during the past 12 months? YES ____ NO ____

   If yes, specify: _____
   _____

4. Are you satisfied with the important aspects of care as stated in the plan? YES ____ NO ____

   If no, specify changes: _____
5. Are the indicators written in a way which identifies the data you want? YES ____ NO ____
6. Have you identified problems or opportunities to improve patient care during the past year? YES ____ NO ____

   If yes, please specify: Problems: _____
   _____

   Opportunities: _____
   _____

7. Have the action plans been effective? YES ____ NO ____
8. Have you effected patient care as a result of the monitoring and evaluation process? YES ____ NO ____

   If yes, please specify: _____
   _____
   _____

9. Have you documented the communication of the QA activities in your departmental minutes, etc.?

   YES ____ NO ____

10. Have you communicated your QA findings monthly to the hospital-wide QA Program on time?

    YES ____ NO ____

11. Do you think your QA program has been effective during the past year? YES ____ NO ____

Source: Courtesy of Rhonda Billman, Clinical Manager, Nutrition Services, Southwest General Hospital, Middleburg Heights, Ohio.

12. Are you planning to revise your program? YES ____ NO ____
    If yes, what changes do you anticipate? _____
    _____
    _____

13. Has your medical director, if applicable, been involved monthly with your program?
    YES ____ NO ____ NA ____ If no, how do you plan to involve him/her? _____
    _____
    _____

14. Does your plan for the year correlate to what you actually did? YES ____ NO ____
15. Were all identified problems resolved? YES ____ NO ____ If no, please specify: _____
    _____

Appendix E

Table of Random Numbers

Appendix E–1

Table of Random Numbers

31 75 15 72 60	68 98 00 53 39	15 47 04 83 55	88 65 12 25 96	03 15 21 91 21
88 49 29 93 82	14 45 40 45 04	20 09 49 89 77	74 84 39 34 13	22 10 97 85 08
30 93 44 77 44	07 48 18 38 28	73 78 80 65 33	28 59 72 04 05	94 20 52 03 80
22 88 84 88 93	27 49 99 87 48	60 53 04 51 28	74 02 28 46 17	82 03 71 02 68
78 21 21 69 93	35 90 29 13 86	44 37 21 54 86	65 74 11 40 14	87 48 13 72 20
41 84 98 45 47	46 85 05 23 26	34 67 75 83 00	74 91 06 43 45	19 32 58 15 49
46 35 23 30 49	69 24 89 34 60	45 30 50 75 21	61 31 83 18 55	14 41 37 09 51
11 08 79 62 94	14 01 33 17 92	59 74 76 72 77	76 50 33 45 13	39 66 37 75 44
52 70 10 83 37	56 30 38 73 15	16 52 06 96 76	11 65 49 98 93	02 18 16 81 61
57 27 53 68 98	81 30 44 85 85	68 65 22 73 76	92 85 25 58 66	88 44 80 35 84
20 85 77 31 56	70 28 42 43 26	79 37 59 52 20	01 15 96 32 67	10 62 24 83 91
15 63 38 49 24	90 41 59 36 14	33 52 12 66 65	55 82 34 76 41	86 22 53 17 04
92 69 44 82 97	39 90 40 21 15	59 58 94 90 67	66 82 14 15 75	49 76 70 40 37
77 61 31 90 19	88 15 20 00 80	20 55 49 14 09	96 27 74 82 57	50 81 69 76 16
38 68 83 24 86	45 13 46 35 45	59 40 47 20 59	43 94 75 16 80	43 85 25 96 93

Source: Reprinted from *Quality Assurance for Patient Care: Nursing Perspectives,* by M.G. Mayers, R.B. Norby, and A.B. Watson, pp. 278–279, Appleton & Lange, 1977.

25 16 30 18 89	70 01 41 50 21	41 29 06 73 12	71 85 71 59 57	68 97 11 14 03	
65 25 10 76 29	37 23 93 32 95	05 87 00 11 19	92 78 42 63 40	18 47 76 56 22	
36 81 54 36 25	18 63 73 75 09	82 44 49 90 05	04 92 17 37 01	14 70 79 39 97	
64 39 71 16 92	05 32 78 21 62	20 24 78 17 59	45 19 72 53 32	83 74 52 25 67	
04 51 52 56 24	95 09 66 79 46	48 46 08 55 58	15 19 11 87 82	16 93 03 33 61	
15 88 09 22 61	17 29 28 81 90	61 78 14 88 98	92 52 52 12 83	88 58 16 00 94	
71 92 60 08 19	59 14 40 02 24	30 57 09 01 94	18 32 90 69 99	26 85 71 92 34	
64 42 52 81 08	16 55 41 60 16	00 04 28 32 29	10 33 33 61 68	65 61 79 48 34	
79 78 22 39 24	49 44 03 04 32	81 07 73 15 43	95 21 66 48 65	13 65 85 10 81	
36 33 77 45 38	44 55 36 46 72	90 96 04 18 49	93 86 54 46 08	93 17 63 48 51	
05 24 92 93 29	19 71 59 40 82	14 73 88 66 67	43 70 86 63 54	93 69 22 55 27	
56 46 39 93 80	38 79 38 57 74	19 05 61 39 39	46 06 22 76 47	66 14 66 32 10	
96 29 63 31 21	54 19 63 41 08	75 81 48 59 86	71 17 11 51 02	28 99 26 31 66	
98 38 03 62 69	60 01 40 72 01	62 44 84 63 85	42 17 58 83 50	46 18 24 91 26	
52 56 76 43 50	16 31 55 39 69	80 39 58 11 14	54 35 86 45 78	47 26 91 57 47	
78 49 89 08 30	25 95 59 92 36	43 28 69 10 64	99 96 99 51 44	64 42 47 73 77	
49 55 32 42 41	08 15 08 95 35	08 70 39 10 41	77 32 38 10 79	45 12 79 36 86	
32 15 10 70 75	83 15 51 02 52	73 10 08 86 18	23 89 18 74 18	45 41 72 02 68	
11 31 45 03 63	26 86 02 77 99	49 41 68 35 34	19 18 70 80 59	76 67 70 21 10	
12 36 47 12 10	87 05 25 02 41	90 78 59 78 89	81 39 95 81 30	64 43 90 56 14	
09 18 82 00 97	32 82 53 95 27	04 22 08 63 04	83 38 98 73 74	64 27 85 80 44	
90 04 58 54 97	51 98 15 06 54	94 93 88 19 97	91 87 07 61 50	68 47 66 46 59	
73 18 95 02 07	47 67 72 62 69	62 29 06 44 64	27 12 46 70 18	41 36 18 27 60	
75 76 87 64 90	20 97 18 17 49	90 42 91 22 72	95 37 50 58 71	93 82 34 41 78	
54 01 64 40 56	66 28 13 10 03	00 68 22 73 98	20 71 45 32 95	07 70 61 78 13	
08 35 86 99 10	78 54 24 27 85	13 66 15 88 73	04 61 89 75 53	31 22 30 84 20	
28 30 60 32 64	81 33 31 05 91	40 51 00 78 93	32 60 46 04 75	94 11 90 18 40	
53 84 08 62 33	81 59 41 36 28	51 21 59 02 90	28 46 66 87 95	77 76 22 07 91	
91 75 75 37 41	61 61 36 22 69	50 26 39 02 12	55 78 17 65 14	83 48 34 70 55	
89 41 59 26 94	00 39 75 83 91	12 60 71 76 46	48 94 97 23 06	94 54 13 74 08	
77 51 30 38 20	86 83 42 99 01	68 41 48 27 74	51 90 81 39 80	72 89 35 55 07	
19 50 23 71 74	69 97 92 02 88	55 21 02 97 73	74 28 77 52 51	65 34 46 74 15	
21 81 85 93 13	93 27 88 17 57	05 68 67 31 56	07 08 28 50 46	31 85 33 84 52	
51 47 46 64 99	68 10 72 36 21	94 04 99 13 45	42 83 60 91 91	08 00 74 54 49	
99 55 96 83 31	62 53 52 41 70	69 77 71 28 30	74 81 97 81 42	43 86 07 28 34	
60 31 14 28 24	37 30 14 26 78	45 99 04 32 42	17 37 45 20 03	70 70 77 02 14	
49 73 97 14 84	92 00 39 80 86	76 66 87 32 09	59 20 21 19 73	02 90 23 32 50	
78 62 65 156 94	16 45 39 46 14	39 01 49 70 66	83 01 20 98 32	25 57 17 76 28	
66 69 21 39 86	99 83 70 05 82	81 23 24 49 87	09 50 49 64 12	90 17 37 95 68	
44 07 12 80 91	07 36 29 77 03	76 44 74 25 37	98 52 49 78 31	65 70 40 95 14	

Aspen Publishers, Inc., 1994

41 46 88 51 49	49 55 41 79 94	14 92 43 96 50	95 29 40 05 56	70 48 10 69 05
94 55 93 75 59	49 67 85 31 19	70 31 20 56 82	66 98 63 40 99	74 47 42 07 40
41 61 57 03 60	64 11 45 86 60	90 85 06 46 18	80 62 05 17 90	11 43 63 80 72
50 27 39 31 13	41 79 48 68 61	24 78 18 96 83	55 41 18 56 67	77 53 59 98 92
41 39 68 05 04	90 67 00 82 89	40 90 20 50 69	95 08 30 67 83	28 10 25 78 16
25 80 72 42 60	71 52 97 89 20	72 68 20 73 85	90 72 65 71 66	98 88 40 85 83
06 17 09 79 65	88 30 29 80 41	21 44 34 18 08	68 98 48 36 20	89 74 79 88 82
60 80 85 44 44	74 41 28 11 05	01 17 62 88 38	36 42 11 64 89	18 05 95 10 61
80 94 04 48 93	10 40 83 62 22	80 58 27 19 44	92 63 84 03 33	67 05 41 60 67
19 51 69 01 20	46 75 97 16 43	13 17 75 52 92	21 03 68 28 08	77 50 19 74 27
49 38 65 44 80	23 60 42 35 54	21 78 54 11 01	91 17 81 01 74	29 42 09 04 38
06 31 28 89 40	15 99 56 93 21	47 45 86 48 09	98 18 98 18 51	29 65 18 42 15
60 94 20 03 07	11 89 79 26 74	40 40 56 80 32	96 71 75 42 44	10 70 14 13 93
92 32 99 89 32	78 28 44 63 47	71 20 99 20 61	39 44 89 31 346	25 72 20 85 64
77 93 66 35 74	31 38 45 19 24	85 56 12 96 71	58 13 71 78 20	22 75 13 65 18
91 30 70 69 91	19 07 22 42 10	36 69 95 37 28	28 82 53 57 93	28 97 66 62 52
38 43 49 46 88	84 47 31 36 22	62 12 69 84 08	12 84 38 25 90	09 81 59 31 46
48 90 81 58 77	54 74 52 45 91	35 70 00 47 54	83 82 45 26 92	54 13 05 51 60
06 91 34 51 97	42 67 27 86 01	11 88 30 95 28	63 01 19 89 01	14 97 44 03 44
10 45 51 60 19	14 21 03 37 12	91 34 23 78 21	88 32 58 08 51	43 66 77 08 83
12 88 39 73 43	65 02 76 11 84	04 28 50 13 92	17 97 41 50 77	90 71 22 67 69
21 77 83 09 76	38 80 73 69 61	31 64 94 20 96	63 28 10 20 23	08 81 64 74 49
19 52 35 95 15	65 12 25 96 59	86 28 36 82 58	69 57 21 37 98	16 43 59 15 29
67 24 55 26 70	35 58 31 65 63	79 24 68 66 86	76 46 33 42 22	26 65 59 08 02
60 58 44 73 77	07 50 03 79 92	45 13 42 65 29	26 76 08 36 37	41 32 64 43 44
53 85 34 13 77	36 06 69 48 50	58 83 87 38 59	49 36 47 33 31	96 24 04 36 42
24 63 73 87 36	74 38 48 93 42	52 62 30 79 92	12 36 91 86 01	03 74 28 38 73
83 08 01 24 51	38 99 22 28 15	07 75 95 17 77	97 37 72 75 85	51 97 23 78 67
16 44 42 43 34	36 15 19 90 73	27 49 37 09 39	85 13 03 25 52	54 84 65 47 59
60 79 01 81 57	57 17 86 57 62	11 16 17 85 76	45 81 95 29 79	65 13 00 48 60
94 01 54 68 74	32 44 44 82 77	59 82 09 61 63	64 65 42 58 43	41 14 54 28 20
74 10 88 82 22	88 57 07 40 15	25 70 49 10 35	01 75 51 47 50	48 96 83 86 03
62 88 08 78 73	95 16 05 92 21	22 30 49 03 14	72 87 71 73 34	39 28 30 41 49
11 74 81 21 02	80 58 04 18 67	17 71 05 96 21	06 55 40 78 50	73 95 07 95 52
17 94 40 56 00	60 47 80 33 43	25 85 25 89 05	57 21 63 96 18	49 85 69 93 26
66 06 74 27 92	95 04 35 26 80	46 78 05 64 87	09 97 15 94 81	37 00 62 21 86
54 24 49 10 30	45 54 77 08 18	59 84 99 61 69	61 45 92 16 47	87 41 71 71 98
30 94 55 75 89	31 73 25 72 60	47 67 00 76 54	46 37 62 53 66	94 74 64 95 80
69 17 03 74 03	86 99 59 03 07	94 30 47 18 03	26 82 50 55 11	12 45 99 13 14
08 34 58 89 75	35 84 18 57 71	08 10 55 99 87	87 11 22 14 76	14 71 37 11 81

27	76	74	35	84		85	30	18	89	77		29	49	06	97	14		73	03	54	12	07		74	69	90	93	10
13	02	51	43	38		54	06	61	52	43		47	72	46	67	33		47	43	14	39	05		31	04	85	66	99
80	21	73	62	92		98	52	52	43	35		24	43	22	48	96		43	27	75	88	74		11	46	61	60	82
10	87	56	20	04		90	39	16	11	05		57	41	10	63	68		53	85	63	07	43		08	67	08	47	41
54	12	75	73	26		26	62	91	90	87		24	47	28	87	79		30	54	02	788	86		61	73	27	54	54
33	71	34	80	07		93	58	47	28	69		51	92	66	47	21		58	30	32	98	22		93	17	49	39	72
85	27	48	68	93		11	30	32	92	70		28	83	43	41	37		73	51	59	04	00		71	14	84	36	43
84	13	38	96	40		44	03	55	21	66		73	85	27	00	91		61	22	26	05	61		62	32	71	84	23
56	73	21	62	34		17	39	59	61	31		10	12	39	16	22		85	49	65	75	60		81	60	41	88	80
65	13	85	68	06		87	64	88	52	61		34	31	36	58	61		45	87	52	10	69		85	64	44	72	77
38	00	10	21	76		81	71	91	17	11		71	60	29	29	37		74	21	96	40	49		65	58	44	96	98
37	40	29	63	97		01	30	47	75	86		56	27	11	00	86		47	32	46	26	05		40	03	03	74	38
97	12	54	03	48		87	08	33	14	17		21	81	53	92	50		75	23	76	20	47		15	50	12	95	78
21	82	64	11	34		47	14	33	40	72		64	63	88	59	02		49	13	90	64	41		03	85	65	45	52
73	13	54	27	42		95	71	90	90	35		85	79	47	42	96		08	78	98	81	56		64	69	11	92	02
07	63	87	79	29		03	06	11	80	72		96	20	74	41	56		23	82	19	95	38		04	71	36	69	94
60	52	88	34	41		07	95	41	98	14		59	17	52	06	95		05	53	35	21	39		61	21	20	64	55
83	59	63	56	55		06	95	89	29	83		05	12	80	97	19		77	43	35	37	83		92	30	15	04	98
10	85	06	27	46		99	59	91	05	07		13	49	90	63	19		53	07	57	18	39		06	41	01	93	62
39	82	09	89	52		43	62	26	31	47		64	42	18	08	14		43	80	00	93	51		31	02	47	31	67
59	58	00	64	78		75	56	97	88	00		88	83	55	44	86		23	76	80	61	56		04	11	10	84	08
38	50	80	73	41		23	79	34	87	63		90	82	29	70	22		17	71	90	42	07		95	95	44	99	53
30	69	27	06	68		94	68	81	61	27		56	19	68	00	91		82	06	76	34	00		05	46	26	92	00
65	44	39	56	59		18	28	82	74	37		49	63	22	40	41		08	33	76	56	76		96	29	00	08	36
27	26	75	02	64		13	19	27	22	94		07	47	74	46	06		17	98	54	89	11		97	34	13	03	58

Aspen Publishers, Inc., 1994

Appendix E–2

Table of Random Digits

85967	73152	14511	85285	36009	95892	36962	67835	63314	50162
07483	51453	11649	86349	76431	81594	95848	36738	25014	15460
96283	01898	61414	83525	04231	13604	75339	11730	85423	60698
49174	12074	98551	37895	93547	24769	09404	76548	05393	96770
97336	39941	21225	93629	19574	71565	33413	56087	40875	13351
90474	41469	16812	81542	81652	45554	27931	93994	22375	00953
28599	64109	09497	76235	41383	31555	12639	00619	22909	29563
25254	16210	89717	65997	82667	74624	36348	44018	64732	93589
28785	02760	24359	99410	77319	73408	58993	61098	04393	48245
84725	86576	86944	93296	10081	82454	76810	52975	10324	15457
41059	66456	47679	66810	15941	84602	14493	65515	19251	41642
67434	41045	82830	47617	36932	46728	71183	36345	41404	81110
72766	68816	37643	19959	57550	49620	98480	25640	67257	18671
92079	46784	66125	94932	64451	29275	57669	66658	30818	58353
29187	40350	62533	73603	34075	16451	42885	03448	37390	96328
74220	17612	65522	80607	19184	64164	66962	82310	18163	63495
03786	02407	06098	92917	40434	60602	82175	04470	78754	90775
75085	55558	15520	27038	25471	76107	90832	10819	56797	33751
09161	33015	19155	11715	00551	24909	31894	37774	37953	78837
75707	48992	64998	87080	39333	00767	45637	12538	67439	94914
21333	48660	31288	00086	79889	75532	28704	62844	92337	99695
65626	50061	42539	14812	48895	11196	34335	60492	70650	51108
84380	07389	87891	76255	89604	41372	10837	66992	93183	56920
46479	32072	80083	63868	70930	89654	05359	47196	12452	38234
59847	97197	55147	76639	76971	55928	36441	95141	42333	67483
31416	11231	27904	57383	31852	69137	96667	14315	01007	31929
82066	83436	67914	21465	99605	83114	97885	74440	99622	87912
01850	42782	39202	18582	46214	99228	79541	78298	75404	63648
32315	89276	89582	87138	16165	15984	21466	63830	30475	74729
59388	42703	55198	80380	67067	97155	34160	85019	03527	78140
58089	27632	50987	91373	07736	20436	96130	73483	85332	24384
61705	57285	30392	23660	75841	21931	04295	00875	09114	32101
18914	98982	60199	99275	41967	35208	30357	76772	92656	62318
11965	94089	34803	48941	69709	16784	44642	89761	66864	62803
85251	48111	80936	81781	93248	67877	16498	31924	51315	79921
66121	96986	84844	93873	46352	92183	51152	85878	30490	15974
53972	96642	24199	58080	35450	03482	66953	49521	63719	57615
14509	16594	78883	43222	23093	58645	60257	89250	63266	90858

Source: Reprinted from *A Million Random Digits with 100,000 Normal Deviates.* RAND (New York: The Free Press, 1955). Copyright © 1955 and 1983 by RAND. Used by permission.

37700	07688	65533	72126	23611	93993	01848	03910	38552	17472
85466	59392	72722	15473	73295	49759	56157	60477	83284	56367
52969	55863	42312	67842	05673	91878	82738	36563	79540	61935
42744	68315	17514	02878	97291	74851	42725	57894	81434	62014
26140	13336	67726	61876	29971	99294	96664	52817	90039	53211
95589	56319	14563	24071	06916	59555	18195	32280	79357	04224
39113	13217	59999	49952	83021	37709	53105	19295	88318	41626
41392	17622	18994	98283	07249	52289	24209	91139	30715	06604
54684	53645	79246	70183	87731	19185	08541	33519	07223	97413
89442	61001	36658	57444	95388	36682	38052	46719	09428	94012
36751	16778	54888	15357	68003	43564	90976	58904	40512	07725
98159	02564	21416	74944	53049	88749	02865	25772	89853	88714

Appendix F

Quality Management Resource Manuals

Carroll, J.G. 1992. *Monitoring with indicators: Evaluating the quality of patient care.* Gaithersburg, Md: Aspen Publishers, Inc.

This 8 1/2 x 11, three-ring hardcover notebook is a sequel to *Patient Care Adult Criteria,* first published in 1977. It is multidisciplinary in scope and includes indicators, thresholds, and criteria for monitoring such areas as administrative services, ambulatory care, diagnostic imaging, dietetic services (two indicators only: modified diets and uncontrolled diabetes), medicine, and surgery. The models included in the manual are not exhaustive. Rather, they are examples that can be used as guides for the monitoring and evaluation process.

Dakey, D.G. 1992. *Continuous quality improvement for dietetic services: A guide for implementation.* Available from Consulting Services/CMS, 1041 Wyoming Avenue, #2, Forty Fort, PA 18704. Phone (717) 288-6373.

This 125-page notebook gives step-by-step instructions on what is necessary to implement the ten-step process to make the transition from quality assurance to continuous quality improvement. The notebook is set up with dividers so that, as the process is implemented, materials are inserted to create the institution's quality notebook. Joint Commission standards for quality assurance are given, and readers are guided through the process of how to implement a plan. Documents include a sample policy with departmental responsibility for continuous quality improvement (CQI), sample scope of care, important aspect-of-care work sheet, ideas for indicators, practice guidelines, sample indicator status log, and sample data-collection form.

Ford, D.A., E. Liskov, and M.M. Fairchild. 1990. *The Yale–New Haven Hospital Nutritional Classification and Assessment Manual.* Write: Assistant Director, Clinical Nutrition, Department of Food and Nutritional Services, Yale–New Haven Hospital, 20 York Street, New Haven, CT 06504.

This manual arose out of a need for realistic and measurable standards of practice. The material in this 125-page, softcover, spiral-bound, 8 1/2 x 11 manual is used on a daily basis by dietitians at Yale–New Haven Hospital as the basis for patient-centered quality care. Chapters include nutritional classification and assessment of patients, initial assessment forms, metabolic/nutritional profile, Kardex cards, calculations, medications, interpretation of laboratory data, and nutrient information. Much of the material in this manual is summarized in the article, "Managing Inpatient Clinical Nutrition Services: A Comprehensive Program Assures Accountability and Success," by D.A. Ford and M.M. Fairchild, *Journal of the American Dietetic Association,* 1993;93:695–702.

Grant, A., and DeHoog, S. 1991. *Nutritional assessment and support.* **4th ed.** Authors: P.O. Box 75057, Northgate Station, Seattle, WA 98125.

Updated, revised, and expanded from earlier volumes, this 400-page, softbound manual includes guidelines for screening and assessment as well as standards for various clinical nutrition services, and some clinical monitor sheets. This manual also includes several forms such as nutrition screening and assessment, time management, productivity, discharge summary, and nutrition support flowsheet.

Jackson, R. 1992. *Continuous quality improvement for nutrition care.* Write: American Nutri-Tech, Inc., P.O. Box 1317, Amelia Island, FL 32034.

In 1988 Dr. Jackson published *Quality Assurance for Dietetic Services*. This book was to be a revision, but because of major changes in principles and practices of quality management she wrote a completely new book. This 240-page, softcover, spiral-bound, 8 1/2 x 11 book is designed to help students and practitioners understand, plan, and implement a continuous quality improvement program. Chapters include an introduction, multidisciplinary approach to nutrition care, "patient-centered" quality management, implementing a quality improvement program, performance-based indicators for monitoring quality, data-collection techniques, statistical analysis and team evaluation, and the effectiveness of quality improvement efforts. Several examples and forms are included.

Joint Commission on Accreditation of Healthcare Organizations. 1991. *An introduction to quality improvement in health care.*

This comprehensive manual offers step-by-step guidance for those who wish to use the ten-step process for monitoring and evaluation. The manual is not specific to nutritional services but contains many ideas that can be used by dietitians and food service managers.

Leebov, W., and C.J. Ersoz. 1991. *The health care manager's guide to continuous quality improvement.* Chicago: American Hospital Publishing, Inc.

This 215-page, 8 x 12 softbound book is an up-to-date guide to implementing a continuous quality improvement program in a health care setting. It is written primarily to assist middle managers in making the transition from quality assurance to "customer-driven management and continuous improvement." The book contains 13 chapters organized within four sections: revitalizing quality management, implementing quality management, the manager's tool kit, and meeting the quality management challenge. The information is practical, and the illustrations and examples are excellent.

Leebov, W. 1991. *The quality quest: A briefing for health care professionals.* Chicago: American Hospital Publishing, Inc.

This little 5 x 8, 40-page booklet is an abbreviated version of *The Health Care Manager's Guide to Continuous Quality Improvement*. It describes why quality is important and explains continuous quality improvement strategies. It gives suggestions on how to talk about quality in YOUR organization, assess quality attitudes, and analyze processes. Quality tools are summarized and illustrated. All concepts are applied to the health care setting.

Menashian, L., ed. 1991. *Standards of practice: Guidelines for the practice of clinical dietetics.* Fresno Community Hospital and Medical Center, Department of Nutrition and Food Service, P.O. Box 1232, Fresno, CA 93715. Copyright by Servicemaster.

This 6 x 8 softcover, spiral-bound, 160-page manual was developed by a cadre of registered dietitians to provide a single, convenient reference source of information useful in the daily practice of clinical dietetics. The manual includes standards for the nutritional care of patients, arranged in three columns: element, exception, and special instructions. Fourteen sections are given: cardiac, complications of pregnancy, diabetes mellitus, enteral and parenteral nutrition support, hemodialysis and CAPD, HIV positive, liver disease, pulmonary, rehabilitation, and renal. Standards included in this manual are very detailed.

Scholtes, P.R., and other Contributors. 1991. *The team handbook: How to use teams to improve quality.* Madison, WI: Joiner Associates Inc.

This softcover, spiral-bound book is filled with both practical information and useful forms. The book contains seven chapters, including the basics of quality improvement, setting the stage for a successful project, getting started, building an improvement plan, learning to work together, and team-building activities. The approach focuses on Deming's principles with emphasis on applications. Although the book is generic (not geared toward hospitals), it contains numerous ideas and exercises that can easily be used or adapted for use in nutrition services.

Smith, P., and A. Smith. 1988. *Superior nutritional care cuts hospital costs.* Write: Nutritional Care Management Institute, 6033 North Sheridan Road 15A, Chicago, IL 60660. Phone (312) 784-2826.

If your hospital has not instituted a protocol for routine screening and assessment, or if you want to show how identification and treatment of malnutrition saves money, this manual is for you. This 8 1/2 x 11, softcover, spiral-bound, 100-page manual gives a review of the literature on hospital malnutrition incidence, penalties, and costs. It describes the consequences of malnutrition, benefits of nutritional care, and how to initiate a screening program. A form is provided for calculating cost benefits of improved nutritional status.

Wooldridge, N.H., and N. Spinozzi, eds. 1990. *Quality assurance criteria for pediatric nutrition conditions: A model.* Chicago, IL: The American Dietetic Association.

Outcome and process criteria for nutrition assessment, support, and follow-up of selected pediatric conditions are described in this book prepared by the Quality Assurance Committee, Pediatric Nutrition Practice Group, a Practice Group of the American Dietetic Association. This 142-page manual is a comprehensive guide to quality care for pediatric patients. Most of the book deals with protocols and field-tested criteria with accepted levels of performance, exceptions, references, and audit sources for normal, healthy pediatric patients as well as problem situations such as neonatology, critical care, eating disorders, inborn errors of metabolism, developmental disabilities, oncology, and renal disease. Also included is a decision tree for both process and outcome criteria.

Appendix G

Glossary of Terms

Affective—The level of employee commitment based on affection, respect, like, or dislike.

Alienation—The lowest level of employee commitment; characterized by resentment and reluctance.

Aspect of care—Specific procedures or services provided by the department of nutrition and dietetics. Aspects of care include nutritional screening and assessment, diet instructions, nutrition counseling, nutritional intervention, periodic assessment of the effects of nutritional intervention, performance in compliance with standards of care, and the like.

Audit—A review of past performance. Quality assurance programs, now outmoded in favor of continuous quality improvement, were based on retrospective audits to ascertain whether performance was congruent with established standards.

Behavioral issues—Issues that relate to individual performance.

Benchmarks—The best processes and practices available in the industry.

Brainstorming—A group method used to generate ideas about a specific issue.

Brainstorming, free flow—The generation of ideas is unstructured, such that members speak spontaneously.

Brainstorming, structured—Ideas are presented in a structured, sequential manner.

Cause-and-effect diagram—A chart that graphically displays process dispersion.

Charter—A process that describes a quality improvement team's mission, boundaries, resources, members, and expectations for improvement.

Check sheet—A data-collection tool that measures how often, how long, or how many times an event takes place.

Clinical indicators—Patient outcomes that indirectly evaluate the quality of health care delivery.

Continuous quality improvement—Specific, structured, problem-solving methods that rely on data and group process tools.

Control chart—Statistical plots obtained from measuring the processes of work. The trends plotted on a control chart can identify problem areas before a crisis occurs.

Control limits—Upper (UCL) and lower (LCL) demarcations on a control chart that designate whether the variations in a process are due to natural or external causes. Control limits are usually set at two standard deviations from the mean.

Cross-functional processes—Work processes that require interdepartmental interactions.

Cross-functional team—A quality-improvement team, made up of members from two or more departments, that evaluates processes that span two or more departments.

Customer—Anyone who receives a service or product, whether it is inside or external to the organization.

Customer expectations—Preconceived ideas about what a product or service will be.

Delphi technique—A tool that organizes a group's ideas by the degree of importance and personal preference.

Departmental team—A quality-improvement team that evaluates problems that occur within a department; team members are selected from within the department.

Departmental champions—Individuals who practice the tenets of quality management without a directive from the organization's top management team.

Empowerment—The ability to fix problems as they occur.

Facilitator—A team member who functions as a disinterested third party; major functions are to influence how decisions are made, to promote effective group dynamics, and to attend to meeting ground rules.

Factoid—A piece of data that appears factual but is really a type of opinion that has come to be considered as fact.

FADE—A model action plan for effective problem solving: Focus, Analyze, Develop, Execute.

Fishbone analysis—A cause-and-effect diagram.

Flowchart—A visual representation of the series of steps that make up a work process.

FOCUS-PDCA—A model action plan that expands on the classic PDCA cycle by including preliminary steps: Find a process to improve, Organize a team, Clarify current knowledge, Understand process variation, Select the process improvement, Plan, Do data collection, data analysis and improvement, Check data, and Act to maintain the gain.

Focus group—A qualitative market research tool used to solicit feedback about a product or service.

Force field analysis—A graphic representation of the likely organizational impact that a proposed change will make.

Hassle factor—The degree of discomfort or unpleasantness created by a problem.

Health care delivery—The sum of the provision of support services, clinical products, and medical services.

Histogram—A bar graph that shows the frequency distribution of a measured event.

IMPROVE—A basic model action plan: Identify a problem, Measure, Prioritize, Research, Outline, Validate, and Execute.

Indicator—A measurable variable that relates to the structure, process, and outcomes of patient care.

Individual interviews—A qualitative market research tool used to solicit feedback about a product or service.

Instrumental—The level of employee commitment where work is viewed as the sum total of pay plus benefits.

Ishikawa diagram—A cause-and-effect or fishbone diagram.

Key process variable—The step most critical to achieving a quality result.

Key quality characteristic—The singular item that has the most impact on a desired quality outcome.

Knowledge issues—Issues related to the skill level of staff.

Market Research—The means by which customers' wants, needs, and expectations are determined.

Meeting ground rules—The preparations and regulations that go into coordinating and conducting effective quality-improvement meetings; includes details related to attendance, time management, participation, communication, decision making, and documentation.

Mind-set—Attitude.

Mission statement—A public statement that defines the organization's commitment to a quality initiative.

Model action plan—A framework for implementing quality-improvement activities; identifies and outlines the steps to take in chronological fashion.

Model of progress—A flowchart-style model action plan with the following steps: clarify goals, educate in team building, investigate, analyze data, seek solutions, take appropriate action, and close.

Monitoring—Process of using checklists or other forms for routine collection of data to determine whether performance is in line with established thresholds.

Morale—The level of employee commitment at which an individual identifies with an organization's mission; the relationship enhances the employee's self-respect.

Nominal group technique—A tool used to organize a group's ideas by degree of importance.

Normative—The average employee commitment level. These employees produce no more but no less than what is routinely expected.

Opportunity statement—A concise description of the process to be improved, its boundaries, and why it should be improved.

Aspen Publishers, Inc., 1994

Pareto chart—A bar graph that displays problem causes in relation to the degree of impact on a measured variable.

Patient-focused care—Multidisciplinary approach to patient care that maximizes patient convenience and decreases departmentalization. Services are brought to the bedside; personnel are expected to provide services previously considered the domain of other professionals. For example, respiratory therapists may draw blood and dietitians may take vital signs and dress wounds.

Peer review—A process in which colleagues evaluate each other's performance, usually in comparison to a fixed standard or set of criteria.

Problem statement—An opportunity statement.

Process dispersion—The relationships between events/causes and outcomes/effects.

Products, clinical—Diagnostic and treatment processes that are provided directly to patients.

Professional shopper—A qualitative market research tool that provides information about the psychological impact of a product or service.

Project requests plan—A formalized approach used to prioritize and select among complex problems, usually applied at the organizational level.

Project team—A work group that is involved in problem solving.

Quality, delivery—The customer's perception about the provision of a clinical product or medical service.

Quality, factual—The degree to which a product or service conforms to specifications.

Quality improvement council—A top management group that is charged with the successful implementation of a quality management program.

Quality, perceived—The degree to which a product or service meets the customer's expectations.

Quality steering team—Quality improvement council.

Resources—The collective assets or means needed to support quality-improvement activities, including funds, staff, managerial support, time, knowledge, and interest.

Run chart—A visual representation of how a measured event is changing over time so that trends can be assessed.

Service agreement—A written document between internal suppliers and customers that delineates details about the provision of a product or service.

Services, medical—Services that are provided to physicians by health care support personnel and assist in the delivery of patient care.

Services, support—Basic services, such as billing and cleaning, that all organizations must have.

Strategic plan—A document listing goals and/or actions that outlines how the department will position itself in the future and move forward to accomplish its mission.

Supplier—Anyone who makes a product or provides a service to an internal or external customer.

Survey—A quantitative market research tool used to gather information about a product or service.

Systems issues—A process that consists of a sequence of procedures that usually cross department lines.

Task force—A work group that uses structured problem-solving methods to examine an identified issue.

Team leader—A team member most familiar with the process being studied; major purpose is to influence what decisions are made and to guide the team to achieve a successful outcome.

Threshold—Target point for quality service. Threshold should be zero for undesirable sentinel events, such as food poisoning and unidentified malnutrition, and 100 percent for desirable events, such as all patients receiving total parenteral nutrition meet established criteria. Most thresholds for routine monitoring are set between 90 percent and 95 percent to allow for human error; goals may be set higher when thresholds are reached.

Total quality management—A philosophy that places quality at the center of an organization's values and operations.

Tracking—Monitoring a process to check results over time.

Traditional medical management—A model of health care delivery that emphasizes the practitioner's professional expertise, autonomy, and compliance with standards of care.

Transactional quality management—Programs that focus on the use of quantitative tools and structured problem-solving methods.

Transformational quality management—Programs that center on changing the values and culture of the organization.

Transition phase—The period of change between the shift from quality assurance to quality improvement.

Trend chart—Run charts.

Unsolicited feedback—A qualitative market research tool that comes from customer comments, suggestions, and/or complaints about a product or service.

Value added—The tangible benefit that is added to a product or service as the result of a supplier's participation in a work process.

Variability—The degree that a product or service differs from a measurable standard.

Work design—The sequence, content, and type of work processes that make up a job.

Work processes—A sequence of actions and interactions between customers and suppliers that results in a product or service.

Index

A

Accreditation standards, clinical manager, 200
Accreditation survey, evaluation, 239–240
Action plan, 151–153
   selection, 151–153
Affective commitment, defined, 367
Alienation, defined, 367
American Dietetic Association clinical indicator project, quality indicator, 319–326
Aspect of care, defined, 367
Assessment, 40
Audit, defined, 367

B

Bar graph, 130–133
Behavioral issues, defined, 367
Benchmark, 6, 92–94, 146
   defined, 367
Brainstorming, 70–72
   advantages, 72
   defined, 367
   facilitator, 71–72
   organizing ideas, 72
   problems, 72
   session procedures, 71
   success factors, 71–72

C

Care
   important aspects, 36–37
   scope, 36
Care component, conceptual model, 111
Case management, 219–220
Cause-and-effect diagram, 75–77
   advantages, 76–77
   defined, 367
   development, 75–76
   drawbacks, 76–77
Central tendency measure, data analysis, 143–144
Change, 206–216
   elements, 207–208
   framework, 206–207
   health care environment, 208, 209
   barriers, 209–210
   health care personnel, improving performance, 211–212
   nature of, 207–208
   patient compliance, 212
   refreezing, 207
   research utilization, 214–215
      individual vs. departmental approaches, 214
   resistance to, 210–211
   unfreezing, 207
Charter, defined, 367
Check sheet, 83–84
   application, 83
   defined, 367
   development, 84
Child care services
   quality indicators, 248
   standards, 248
Clinical indicator, defined, 367
Clinical manager, 191–201
   accreditation standards, 200
   communication, 193
   driving fear away, 192–193
   education, 200–201
   employee empowerment, 199
   employees' needs, 191–192
   expectation setting, 196–197
   function, 191–201
   program implementation, 199
   quality culture, 192
   quality rewards, 193–196
   resources, 193
   as role model, 197–199
   training, 200–201
Clinical nutrition patient acuity survey, 281, 282
Clinical pathway DRG 106: coronary artery bypass graft, 310–318
Clinical product, 9–10
   defined, 369
   nutrition, 20
   nutrition service, 11
   quality program, 19–20
Clinical staff
   continuous quality improvement, 191–204
   empowerment, 199–200
Collaboration, empowerment, 222–223
Communication, 179–189, 238–239
   clinical manager, 193
   flowchart, 186
   paper trail, 187
Compartmentalization, 217
Computer, data organization
   data management program, 136–137
   graphics programs, 137
   statistical programs, 137
Content quality, 11–12
Continuous quality improvement, 5, 33–43
   adult nutrition support service, 267
   bedside tables, 277
   calorie intake records, 272
   chart review work sheet, 262
   clinical staff, 191–204
   defined, 367
   departmental planning, 99–105
      cross-functional initiatives, 104–105
      important aspects of care, 101–103, 104
      patient satisfaction, 102–103
      quality management plan, 105
      responsibilities, 99–100
      role assignment, 99–100
      service scope, 100–101
   dialysis unit, chart review work sheet, 266
   enteral feedings, 264
   evaluating benefits of, 240
   height/weight documentation, 263
   Kardex cards, 265
   monitoring plan, 260
   monitoring schedule, 261
   nutrition counseling, 275
   patient education, 276
   patient satisfaction, 274
   patient snacks, 271
   RD assignments, 268
   timeliness of late trays, 273
Control chart, 91–92, 93
   case study, 92
   defined, 367
   development, 91
   interpretation, 91–92
Coronary artery bypass graft, 310–318

Cost-benefit ratio, FOCUS-PDCA cycle, 155, 156
Council on Practice Quality Management Task Force, current dietetic practice group projects/publications, 123–124
Criteria, 37
 thresholds, 37–38
Criteria-based performance standards, nutrition service, 64–65
Criteria form, 349
Cross-functional process, defined, 367
Cross-functional team
 decubitus ulcer, 226
 defined, 220, 367
 facilitator, 222
 flowchart, 221
 FOCUS-PDCA cycle, 161, 162
 height and weight measurement study, 223
 information flow, 223
 medical-surgical environment, 226
 neonatal intensive care, 226
 nutrition support team, 223
 problems, 227
 rehabilitation, 226
 start up, 220–222
 success factors, 226–227
Cultural transformation, 6
Customer
 defined, 368
 role definition, 24
Customer-driven products and services
 defining, 30
 information collection, 27
 provision, 27
Customer expectations, 29–32
 defined, 368
 external customer, 31–32
 internal customer, 32
Customer-supplier relationship, 22–24

D

Data, indicator, 38
Data analysis, 139–150, 144–150
 attributes, 139–140
 central tendency measure, 143–144
 comparing results with standards, 145
 dispersion measure, 144
 expected outcomes, 139
 frequency distribution, 143
 frequency table, 141, 142
 inferential statistics, 144
 interpreting change in results, 145–147
 interval data, 142–143
 measurement scale, 141–143
 methods, 140–144
 narrative information, 147
 nominal data, 141–142
 ordinal data, 142
 problem clarification method, 147–150
 purpose, 139
 ratio data, 143
 statistical analysis, 141–144
 summary data, 141
 tabulation, 140–141

Data collection, 125–130
 forms, 125–126, 127
  development, 126
  indicator links, 126
  number, 126
 monitoring frequency, 129–130, 131
Data collection summary sheet, 300, 301, 302
Data organization, 130–137
 computer
  data management program, 136–137
  graphics programs, 137
  statistical programs, 137
 graph, 130–133, 134, 135
 table, 130, 132–133, 134
Data retrieval form, 350
Day-care facility
 quality indicators, 248
 standards, 248
Decubitus ulcer, cross-functional team, 226
Delivery quality, 11–12
Delivery schedule, 278
Delphi technique, defined, 368
Demonstration project, 14–15
 disadvantages, 15
Departmental audit evaluation, patient-centered care, 218
Departmental champion
 defined, 368
 leadership, 48–49
Departmental communication
 quality steering committee, 181
 responsibilities, 181
Departmental planning, continuous quality improvement, 99–105
 cross-functional initiatives, 104–105
 important aspects of care, 101–103, 104
 patient satisfaction, 102–103
 quality management plan, 105
 responsibilities, 99–100
 role assignment, 99–100
 service scope, 100–101
Departmental team
 defined, 368
 FOCUS-PDCA cycle, 161–162
Departmentalization, 217
Dietetic technician, as star performer, 201–204
Dietitian
 as star performer, 201–204
 as trainer, 201
Dispersion measure
 data analysis, 144
 interval data, 144
Documentation, 40, 181–187
 form, 182–187
 nutrition service, 58, 60, 61, 62

E

Education
 clinical manager, 200–201
 quality program, 15–18
  comprehensive education advantages, 16–17
  concepts focus, 15–16
  how many trained, 16

 just in time training, 18
 skills-based training, 16
 targeted training, 17
 when trained, 17–18
 where trained, 17–18
 who trained, 16
Efficiency, quality-management structure, 233
Employee commitment, hierarchy, 192
Empowerment
 clinical staff, 199–200
 collaboration, 222–223
 defined, 368
Enteral nutrition, standards, 103
Evaluation
 accreditation survey, 239–240
 quality management program, 229–242
  communication, 238–239
  data processing, 234–235
  efficiency, 232–239
  meetings, 238
  monitoring issues, 233–234
  objectives, 230–232
  outcome, 230–232
  planning, 229–230
  professional qualifications, 233
  purpose, 229, 230
  quality functions, 233–239
  reports, 239
  staff feedback, 235–238
  strategic problems and opportunities, 238
 threshold, 38–40

F

Facilitator, 163–164
 brainstorming, 71–72
 cross-functional team, 222
 defined, 368
 multidisciplinary team, 222
Factoid, defined, 368
FADE, 151, 152
 defined, 368
Feedback, unsolicited, 28
Fishbone analysis, 75–77
 defined, 368
Flowchart, 77–81
 applications, 78, 79
 benefits, 80–81
 defined, 368
 development, 79
 formats, 78
 maximizing effectiveness, 79
Focus group, 28, 148–149
 defined, 368
FOCUS-PDCA cycle, 152, 153–177
 behavior issues, 172–173
 change expectations, 173
 cost-benefit ratio, 155, 156
 cross-functional team, 161, 162
 defined, 368
 departmental team, 161–162
 hassle factor, 155
 implementation, 173–177
 interest, 155
 knowledge, 155

knowledge clarification, 166–170
knowledge issues, 170–172
management support, 154
planning, 170–173
problem statement, 166–167
process selection, 153–154, 155–161
resources, 154–155
selection, 170–173
systems issues, 170
team, 161–166
time commitments, 155
Force field analysis, 81
  advantages, 81
  application, 81–82
  defined, 368
  graph development, 81
Form, documentation, 182–187
Frequency distribution, data analysis, 143
Frequency table, data analysis, 141, 142

G

Graph, data organization, 130–133, 134, 135
Graphic rating scale, 143
Group process, 67–82
Guidelines, quality management, 244–245

H

Hassle factor
  defined, 368
  FOCUS-PDCA cycle, 155
Health care, monitoring and evaluation, 288–290
Health care delivery, defined, 368
Health care environment, change, 208, 209
  barriers, 209–210
Health care personnel, improving performance, 211–212
Height and weight measurement study, cross-functional team, 223
Henry Ford Hospital Nutrition Services Continuous Monitors, 291–292
High-risk patient's evaluation, 283–285
Histogram, 86–89
  defined, 368
  interpreting, 88–89
  plotting, 87–88
Home care, 249–250
Hospice program, 250
  quality indicators, 251
Hospital, long-term care facility, differences, 246

I

IMPROVE model, 151
  defined, 368
Improvement, 40
Indicator, 37
  attributes, 107–109
  based on care delivery, 109–112
  based on seriousness, 109
  combination indicators, 112–113
  criteria, 119–120
  data, 38

data collection, 120–121
defined, 107, 368
desirable, 112
development, 113–122
direction of performance, 112–113
draft, 115–119
evaluation schedule, 121
indicator population description, 119
monitors, 120
terminology, 113–114
threshold, 37–38, 119
types, 109–113
undesirable, 112
who will develop, 113
Indicator development form, 114, 118
Individual interview, defined, 368
Inferential statistics, data analysis, 144
Information flow, cross-functional team, 223
Instrumental commitment, defined, 368
Interval data
  data analysis, 142–143
  dispersion measure, 144
  range, 144
  standard deviation, 144
Interview, 28
Ishikawa diagram, 75–77
  defined, 368

J

Joint Commission on Accreditation of Healthcare Organizations
  Agenda for Change, 33–34
  data focus, 34
  education, 34
  expectations, 34
  objectives, 33
  standards, 244–245
Just in time training, 18

K

Key process variable, defined, 368
Key quality characteristic, 167
  defined, 368
Knowledge deficiency, 211
Knowledge issues, defined, 368

L

Leadership, 47–50
  departmental champion, 48–49
  quality program, 14
  selection, 47–48
  self-assessment, 47, 48
Likert-type scale, 143
Line graph, 133
Listening, 25–26
Local agency standards, quality management, 245
Long-term care facility, 244–245
  hospital, differences, 246
  quality indicators, 247
  resident satisfaction, 246–247
Long Term Care Standards, 244

M

Management model, conflicts, 49–50
Managerial support, quality management program, 232–233
Market research, 25
  defined, 368
  tools, 27, 28
Measurement scale
  data analysis, 141–143
  statistical analysis, 141–143
Medical management model, quality management, transition, 50
Medical record, checklist for evaluating, 342
Medical service, 9–10
  defined, 369
  nutrition service, 11
Meeting
  agenda, 183
  ground rules, defined, 368
  minutes, 183–184
Menu review checklist, 270
Mind-set, defined, 368
Mission statement, defined, 368
Model action plan, defined, 368
Model of progress, 151–152, 153
  defined, 368
Modified Delphi technique, 73–75
  advantages, 74–75
  drawbacks, 75
  idea selection, 74
  success factors, 74
Monitoring, defined, 368
Monitoring and evaluation, health care, 288–290
Monitoring form, 341–356
Morale, defined, 368
Multidisciplinary audit, patient-centered care, 218
Multidisciplinary quality program, 252–253
Multidisciplinary team, facilitator, 222

N

Narrative information, data interpretation, 147
Neonatal intensive care, cross-functional team, 226
1994 Accreditation Manual for Hospitals, 217
Nominal data, data analysis, 141–142
Nominal group technique, 72–73
  advantages, 74–75
  defined, 368
  drawbacks, 75
  success factors, 74
Normative commitment, defined, 368
Nutrition manager
  medical management model, 49
    conflicts, 49–50
  traditional medical management model, 49
    conflicts, 49–50
  unique challenges, 49
Nutrition service
  clinical product, 11, 20
  criteria-based performance standards, 64–65
  Deming's 14 steps to quality, 10

documentation, 58, 60, 61, 62
environment, 55–65
institutional approach, 42–43
medical service, 11
mission, 55–56
monitoring and evaluation system, 65
nutritional risk screening method, 57–58, 59
patient acuity level, 58
policy and procedure manual, 57
productivity, 58–63, 64
  measurement, 59–62
  monitoring, 62–63
staffing, 63–64
standards, 56–57
statement of mission, 55–56
strategic plan, 56
support service, 11
supportive environment, 65
Nutrition support team, cross-functional team, 223
Nutritional risk screening method, nutrition service, 57–58, 59

O

Olin Teague Veterans' Center Quality Assurance Plan, 303–305
Omnibus Budget Reconciliation Act of 1987, quality management, 245
Opportunity statement, defined, 368
Ordinal data, data analysis, 142
Organization, quality indicators, 248–249
Organizational functioning, improving, 213–214
Outcome indicator, 110

P

Parallel audit, patient-centered care, 218
Parenteral nutrition, standards, 103
Pareto chart, 84–86
  applications, 85
  defined, 369
  graphing, 85
  success factors, 85–86
Participative problem solving, 6
Patient acuity level, nutrition service, 58
Patient-centered care
  components, 218–219
  defined, 369
  departmental audit evaluation, 218
  multidisciplinary audit, 218
  parallel audit, 218
Patient compliance
  change, 212
  improving, 212
Patient nutrition care monitor, 351–354
Patient satisfaction, 22–32
Patient satisfaction survey, 141
Peer review, defined, 369
Pie chart, 133
Policy and procedure manual, nutrition service, 57
Practice guidelines indicator, 107

Problem clarification method, data interpretation, 147–150
Problem-solving work group, 67–72
Problem statement
  defined, 369
  FOCUS-PDCA cycle, 166–167
Process dispersion, defined, 369
Process indicator, 110–112
Productivity, nutrition service, 58–63, 64
  measurement, 59–62
  monitoring, 62–63
Productivity code, 343
Professional association standards, quality management, 245
Professional shopper, 28
  defined, 369
Project requests plan, defined, 369
Project team, defined, 369

Q

Quality
  concepts, 5
  defined, 3, 35, 36
  employee commitment to, 191–204
  factual vs. perceived, 4–5
  measurement, 20–21
  quantitative vs. qualitative, 3–4
  structuring, 13
Quality assessment and improvement worksheet, 345–346
Quality assurance
  flowchart, 39
  patient's point of view, 41
  quality improvement, compared, 34–35
  transition, 40–41
Quality assurance annual appraisal, 355–356
Quality assurance improvement monitor development form, 347–348
Quality Assurance in Home Care Organizations and Hospice Programs, 250
Quality circle, 68
  action steps, 70
  advantages, 69
  characteristics, 68–69
  selection, 70
Quality cost management, 6
  steps, 7
Quality culture, clinical manager, 192
Quality improvement
  additional action steps, 41–42
  quality assurance, compared, 34–35
  ten-step process, 35–40
  transition, 40–41
Quality improvement council, 12
  defined, 369
Quality indicator, 107
  American Dietetic Association clinical indicator project, 319–326
  criteria, 119–120
  data collection, 120–121
  defined, 107
  development, 113–122
  draft, 115–119
  evaluation schedule, 121

indicator population description, 119
  monitors, 120
  terminology, 113–114
  threshold, 119
  University Hospital Consortium, 327–340
  who will develop, 113
Quality management
  guidelines, 244–245
  hospital-affiliated services, 244–254
  interdisciplinary approach, 252–253
  Joint Commission standards, 244–245
  local agency standards, 245
  medical management model, transition, 50
  overview, 3–8
  professional association standards, 245
  standards, 244–245
  state agency standards, 245
Quality management program
  evaluation, 229–242
    communication, 238–239
    data processing, 234–235
    efficiency, 232–239
    meetings, 238
    monitoring issues, 233–234
    objectives, 230–232
    outcome, 230–232
    planning, 229–230
    professional qualifications, 233
    purpose, 229, 230
    quality functions, 233–239
    reports, 239
    staff feedback, 235–238
    strategic problems and opportunities, 238
  future planning, 240–241, 242
  hospital-affiliated services, monitoring and evaluating, 253–254
  managerial support, 232–233
Quality-management structure, efficiency, 233
Quality management tool, quantitative, 83–95
Quality medical care
  components, 9
  definition, 9
Quality mission statement, 67
Quality patient care, multidimensional, 10–11
Quality program, 5–7
  areas of uncertainty, 12
  clinical product, 19–20
  education, 15–18
    comprehensive education advantages, 16–17
    concepts focus, 15–16
    how many trained, 16
    just in time training, 18
    participants, 16
    skills-based training, 16
    targeted training, 17
    when, 17–18
    where, 17–18
  expectations, 18–20
  health care implementation, 13–15
  leadership, 14
  monetary gains, 18–19
  organizational transformation, 19
  reducing costs, 19

structuring, 12–13
success, 7–8
top-down approach, 14
  criticisms, 14
  transaction, 13–14
  transformation, 13–14
work as process, 22–24
Quality Readiness Survey, 50–54
  interpretation, 53
  results utilization, 53
  scoring, 52
  work environment, 53–54
Quality steering committee, 12–13, 179
  department communication, 181
  department interactions with, 179–181
  responsibilities, 180
  roles, 180
Quality steering plan, defined, 369
Quality work design, 6
  steps, 7
Quantitative quality management tool, 83–95

R

Random digits table, 361–362
Random numbers table, 357–360
Range, interval data, 144
Rate-based indicator, 109
Ratio data, data analysis, 143
Record-keeping system, 181–182
Rehabilitation, cross-functional team, 226
Report, 40, 184–187
Research utilization, change, 214–215
  individual vs. departmental approaches, 214
Resource manual, 363–365
Resources, defined, 369
Responsibility, 36
Roles, definition, 220
Run chart, 89–91
  construction, 89–90
  defined, 369
  trend evaluation, 90–91

S

Sample
  defined, 127
  selection, 127–129
  size, 127–128
Scatter diagram, 94–95
Sentinel event, 109
Service agreement, 31, 32
  defined, 369
Skills-based training, 16
Staffing, nutrition service, 63–64
Standard, 22
Standard deviation, interval data, 144

Standards
  child care services, 248
  day-care facility, 248
  enteral nutrition, 103
  nutrition service, 56–57
  parenteral nutrition, 103
  quality management, 244–245
State agency standards, quality management, 245
Statistical analysis
  data analysis, 141–144
  measurement scale, 141–143
Statistical process control, 6
  steps, 7
Strategic plan
  defined, 369
  nutrition service, 56
Structure indicator, 112
Supervisory responsibility assignment, 269
Supplier
  defined, 369
  role definition, 24
Support service, 9–10
  defined, 369
  nutrition service, 11
Survey, 28
  defined, 369
  developing good questions, 29
Systems issues, defined, 369

T

Table, data organization, 130, 132–133, 134
Tabulation, data analysis, 140–141
Task force, 68
  action steps, 71
  advantages, 69–70
  characteristics, 68–69
  defined, 369
  selection, 70
Team leader, defined, 369
Threshold
  defined, 369
  evaluation, 38–40
Thurston-type scale, 143
Time study, 344
Total quality management
  defined, 5, 369
  success, 7–8
Tracking, defined, 369
Traditional medical management, defined, 369
Training, clinical manager, 200–201
Transaction, quality program, 13–14
Transactional quality management, defined, 369
Transformation, quality program, 13–14

Transformational quality management, defined, 369
Transition phase, defined, 369
Trend chart, 89–91
  construction, 89–90
  defined, 369
  trend evaluation, 90–91

U

University Hospital Consortium, quality indicator, 327–340
University of Chicago Hospitals Nutrition Services Quality Assurance Plan, 258–278
  forms, 259–278
  goal, 258
  important aspects of care, 258
  monitoring methods, 259
  objectives, 258
  overall responsibility, 258
  program scope, 258
  quality indicators, 259
University of Utah Ten-Step Monitoring Proram Nutrition Care Service, 286–290
Unsolicited feedback, defined, 369

V

Value added, 23, 25
  defined, 26, 370
Variability, defined, 370
Variation decision tree, 147, 148

W

Women, infants, and children program, 250–251
  quality indicators, 251
Work design, defined, 370
Work environment, Quality Readiness Survey, 53–54
Work group, readiness, 50
Work processes, defined, 370

Y

Yale-New Haven Hospital
  Clinical Quality Assurance Evaluation, 298–302
  Department of Food and Nutritional Services Quality Assurance Program, 279–285
  Quality Assurance Evaluation, assessment of your departmental quality assurance program, 306–309

About the Authors

M. ROSITA SCHILLER, RSM, PhD, RD, LD, is Professor and Director of the Medical Dietetics Division, School of Allied Medical Professions, College of Medicine, The Ohio State University, Columbus, Ohio. Previously, Dr. Schiller was director of the Career Mobility Program in Dietetics at Mercy College, Detroit, Michigan. She spent six years in dietetic practice and management at Manistee Community Hospital, Manistee, Michigan, and at St. Lawrence Hospital, Lansing, Michigan.

Dr. Schiller earned her PhD in Foods and Nutrition at Ohio State and her MS in Institutional Management from Michigan State University in East Lansing. She completed her undergraduate work at Mercy College in Detroit, and received her dietetic internship at Henry Ford Hospital, also in Detroit.

Dr. Schiller is known among dietetic circles for her work in role delineation, quality practice, leadership, and dietetic education. She has written numerous articles, and wrote *Handbook for Clinical Nutrition Services Management* with Judith Gilbride and Julie O'Sullivan Maillet. In addition to her teaching and writing, Dr. Schiller has facilitated several conferences and workshops in the United States, Germany, and South Africa. She is active in the American Society for Parenteral and Enteral Nutrition and The American Dietetic Association (ADA). Dr. Schiller has held several offices in the dietetic association; she currently serves on the ADA Board of Directors and is the 1993–1994 chairman of the ADA Council on Research.

KAREN MILLER-KOVACH, MBA, MS, RD, is a graduate of Bowling Green State University in Ohio. She completed the Coordinated Masters–Internship Program at University Hospitals of Cleveland and received her MS in Nutrition from Case Western Reserve University, Cleveland, Ohio. Ms. Miller-Kovach also holds an MBA in Executive Management from Baldwin-Wallace College. As Director of Nutrition Services for the Cleveland Clinic Foundation, Ms. Miller-Kovach was responsible for the department's quality management program. She has authored several publications and lectures frequently to nutrition professionals on the topic of managing a clinical nutrition business.

MARY ANGELA MILLER, MS, RD, LD, is a dietitian, lecturer, and author. She has 14 years of experience in the nutrition profession. She is currently Director of Nutrition and Dietetics at the Ohio State University Hospitals, where she administers all hospital nutrition care and food service operations. She was formerly Director of the Obesity and Risk Management Program at Dublin Medical Clinic, Dublin, Ohio.

Prior to practicing in the Columbus area, Ms. Miller was a Nutrition Care Coordinator for the Cleveland Clinic Foundation, where she managed the cardiac and critical care divisions. She is a featured speaker at local seminars and national conventions. She has conducted programs educating the public on cholesterol management and professional groups on nutrition therapy in intensive care. Among her accomplishments is an equation for calculating ideal body weight, which she developed and published, and which has become a reference standard in her field. Ms. Miller is the media spokesperson and past president of the Columbus Dietetic Association; she has been named in Who's Who of American Women.

Ms. Miller graduated with honors from both Youngstown State and Case Western Reserve Universities, where she earned her MS degree. She completed a professional internship at Indiana University Medical Center.

Printed in the United States
132803LV00002B/113-126/A